Membrane Processing for Dairy Ingredient Separation

IFT | Press

The *IFT Press* series reflects the mission of the Institute of Food Technologists – to advance the science of food contributing to healthier people everywhere. Developed in partnership with Wiley, *IFT Press* books serve as leading-edge handbooks for industrial application and reference and as essential texts for academic programs. Crafted through rigorous peer review and meticulous research, *IFT Press* publications represent the latest, most significant resources available to food scientists and related agriculture professionals worldwide. Founded in 1939, the Institute of Food Technologists is a nonprofit scientific society with 18,000 individual members working in food science, food technology, and related professions in industry, academia, and government. IFT serves as a conduit for multidisciplinary science thought leadership, championing the use of sound science across the food value chain through knowledge sharing, education, and advocacy.

IFT Press Advisory Group

WILEY Blackwell

Membrane Processing for Dairy Ingredient Separation

Edited by

Kang Hu
Global Dairy Ingredients, Land O'Lakes Inc., Arden Hills, Minnesota, USA

James M. Dickson
Department of Chemical Engineering, McMaster University, Hamilton, Ontario, Canada L8S 4L7

IFT Press WILEY Blackwell

Registered office: John Wiley & Sons, Ltd, The Atrium, Southern Gate, Chichester, West Sussex, PO19 8SQ, UK

Editorial offices: 9600 Garsington Road, Oxford, OX4 2DQ, UK
The Atrium, Southern Gate, Chichester, West Sussex, PO19 8SQ, UK
111 River Street, Hoboken, NJ 07030-5774, USA

For details of our global editorial offices, for customer services and for information about how to apply for permission to reuse the copyright material in this book please see our website at www.wiley.com/wiley-blackwell.

Library of Congress Cataloging-in-Publication Data applied for.

ISBN: 9781118590171

A catalogue record for this book is available from the British Library.

Wiley also publishes its books in a variety of electronic formats. Some content that appears in print may not be available in electronic books.

Cover image: White paint splash © portishead1/istockphoto.

Set in 10.5/12.5pt, TimesLTStd by Laserwords Private Limited, Chennai, India
Printed and bound in Malaysia by Vivar Printing Sdn Bhd

1 2015

Titles in the IFT Press series

- *Accelerating New Food Product Design and Development* (Jacqueline H. Beckley, Elizabeth J. Topp, M. Michele Foley, J.C. Huang, and Witoon Prinyawiwatkul)
- *Advances in Dairy Ingredients* (Geoffrey W. Smithers and Mary Ann Augustin)
- *Anti-Ageing Nutrients: Evidence-based Prevention of Age-Related Diseases* (Deliminda Neves)
- *Bioactive Compounds from Marine Foods: Plant and Animal Sources* (Blanca Hernandez-Ledesma and Miguel Herrero)
- *Bioactive Proteins and Peptides as Functional Foods and Nutraceuticals* (Yoshinori Mine, Eunice Li-Chan, and Bo Jiang)
- *Biofilms in the Food Environment* (Hans P. Blaschek, Hua H. Wang, and Meredith E. Agle)
- *Calorimetry in Food Processing: Analysis and Design of Food Systems* (Gönül Kaletunç)
- *Coffee: Emerging Health Effects and Disease Prevention* (YiFang Chu)
- *Food Carbohydrate Chemistry* (Ronald E. Wrolstad)
- *Food Industry Design, Technology and Innovation* (Helmut Traitler, Birgit Coleman, and Karen Hofmann)
- *Food Ingredients for the Global Market* (Yao-Wen Huang and Claire L. Kruger)
- *Food Irradiation Research and Technology, second edition* (Christoper H. Sommers and Xuetong Fan)
- *Foodborne Pathogens in the Food Processing Environment: Sources, Detection and Control* (Sadhana Ravishankar, Vijay K. Juneja, and Divya Jaroni)
- *Food Oligosaccharides: Production, Analysis and Bioactivity* (F. Javier Moreno and Maria Luz Sanz
- *Food Texture Design and Optimization* (Yadunandan Lal Dar and Joseph M. Light)
- *High Pressure Processing of Foods* (Christopher J. Doona and Florence E. Feeherry)
- *Hydrocolloids in Food Processing* (Thomas R. Laaman)
- *Improving Import Food Safety* (Wayne C. Ellefson, Lorna Zach, and Darryl Sullivan)
- *Innovative Food Processing Technologies: Advances in Multiphysics Simulation* (Kai Knoerzer, Pablo Juliano, Peter Roupas, and Cornelis Versteeg)
- *Mathematical and Statistical Methods in Food Science and Technology* (Daniel Granato and Gastón Ares)
- *Membrane Processes for Dairy Ingredient Separation* (Kang Hu and James M. Dickson)
- *Microbial Safety of Fresh Produce* (Xuetong Fan, Brendan A. Niemira, Christopher J. Doona, Florence E. Feeherry, and Robert B. Gravani)
- *Microbiology and Technology of Fermented Foods* (Robert W. Hutkins)
- *Multiphysics Simulation of Emerging Food Processing Technologies* (Kai Knoerzer, Pablo Juliano, Peter Roupas and Cornelis Versteeg)
- *Multivariate and Probabilistic Analyses of Sensory Science Problems* (Jean-François Meullenet, Rui Xiong, and Christopher J. Findlay)

- *Nanoscience and Nanotechnology in Food Systems* (Hongda Chen)
- *Nanotechnology and Functional Foods: Effective Delivery of Bioactive Ingredients* (Cristina Sabliov, Hongda Chen, and Rickey Yada)
- *Natural Food Flavors and Colorants* (Mathew Attokaran)
- *Nondestructive Testing of Food Quality* (Joseph Irudayaraj and Christoph Reh)
- *Nondigestible Carbohydrates and Digestive Health* (Teresa M. Paeschke and William R. Aimutis)
- *Nonthermal Processing Technologies for Food* (Howard Q. Zhang, Gustavo V. Barbosa-Cánovas, V.M. Balasubramaniam, C. Patrick Dunne, Daniel F. Farkas, and James T.C. Yuan)
- *Nutraceuticals, Glycemic Health and Type 2 Diabetes* (Vijai K. Pasupuleti and James W. Anderson)
- *Organic Meat Production and Processing* (Steven C. Ricke, Ellen J. Van Loo, Michael G. Johnson, and Corliss A. O'Bryan)
- *Packaging for Nonthermal Processing of Food* (Jung H. Han)
- *Practical Ethics for the Food Professional: Ethics in Research, Education and the Workplace* (J. Peter Clark and Christopher Ritson)
- *Preharvest and Postharvest Food Safety: Contemporary Issues and Future Directions* (Ross C. Beier, Suresh D. Pillai, and Timothy D. Phillips, Editors; Richard L. Ziprin, Associate Editor)
- *Processing and Nutrition of Fats and Oils* (Ernesto M. Hernandez and Afaf Kamal-Eldin)
- *Processing Organic Foods for the Global Market* (Gwendolyn V. Wyard, Anne Plotto, Jessica Walden, and Kathryn Schuett)
- *Regulation of Functional Foods and Nutraceuticals: A Global Perspective* (Clare M. Hasler)
- *Resistant Starch: Sources, Applications and Health Benefits* (Yong-Cheng Shi and Clodualdo Maningat)
- *Sensory and Consumer Research in Food Product Design and Development* (Howard R. Moskowitz, Jacqueline H. Beckley, and Anna V.A. Resurreccion)
- *Sustainability in the Food Industry* (Cheryl J. Baldwin)
- *Thermal Processing of Foods: Control and Automation* (K.P. Sandeep)
- *Trait-Modified Oils in Foods* (Frank T. Orthoefer and Gary R. List)
- *Water Activity in Foods: Fundamentals and Applications* (Gustavo V. Barbosa-Cánovas, Anthony J. Fontana Jr., Shelly J. Schmidt, and Theodore P. Labuza)
- *Whey Processing, Functionality and Health Benefits* (Charles I. Onwulata and Peter J. Huth)

WILEY Blackwell

Contents

Acknowledgment

This book would not have been possible without the work from all the contributors. It is their unique background and experience on membrane technology that bring the depth and color to this book. We express our appreciation to the publisher, John Wiley & Sons, particularly to David McDade, Fiona Seymour, Audrie Tan, and Lea Abot, for their patience and assistance during the past three years.

As well the book editors, Kang Hu and Jim Dickson would like to thank the continued support of McMaster University. Kang Hu acknowledges the support of Land O'Lakes. Jim Dickson wishes to thank the hard work of all the students that have worked with him over the years many of whom continue to work in membrane science and engineering.

Kang Hu and James M. Dickson

Preface

According to the Food and Agriculture Organization of the United Nations, 2011, about 730 million tonnes of milk were produced annually from around 260 million dairy cows.[1] The vast majority of this production goes to feeding humans in various forms including raw milk, various processed milk products, and cheese. However, many valuable components exist within milk that are or can be produced industrially, such as lipids, proteins, minerals, and vitamins, by using various separation processes on milk and/or milk by-products. Membrane-based separation processes have proven to be effective in recovering these products and this book is concerned with such processes.

Dairy ingredients, such as lipids, proteins, peptides, lactose, and dairy minerals, provide a wide variety of potential for product development in the food, chemical, and pharmaceutical industries, due to the functional, nutritional, and biological characteristics, which are strongly influenced by the processes used for manufacturing. Among all dairy ingredient processes, separation processes are quite essential for fractionating ingredients from fluid milk. These processes could be as simple as just reducing pH to coagulate casein proteins or as complicated as isolating immunoglobulin G (IgG) by chromatography. Membrane separation processes have been applied in the dairy industry for many decades. One of the earliest applications was an ultrafiltration process to make soft cheese by Maubois, Macquot, and Vassal (1969) due to several advantages of ultrafiltration including increased yield and low energy consumption.[2] Compared to traditional membrane applications, such as water treatment and desalination, ingredient separation from fluid milk has particular challenges due to the high solid content and complicated composition of fluid milk, which increases the complexity of separation processes. With the continued development on membrane materials and membrane element configurations, almost all types of membrane processes have now been successfully applied in the dairy industry for solid concentration, ingredient separation, and waste recovery.

[1] Food and Agriculture Organization of the United Nations (May 2012) *Food Outlook – Global Market Analysis*, pp. 8, 51–54.

[2] Cheryan, M. (1998) *Ultrafiltration and Microfiltration Handbook*, Technomic Publishing Company, Inc.

This book provides detailed information on the development of a variety of membrane separation technologies in dairy ingredient separation as they have evolved over the past decades. The approach in this book is to view membrane separation processes for dairy ingredient separation from a chemical engineer's point of view; that is, rather than just viewing a membrane as a tool to obtain the ingredients, the book also addresses questions of mathematical modeling, design, and optimization of the treatment systems. The text also presents in-depth knowledge of the mechanisms of each membrane separation process, as well as the membrane and module types applied in the dairy industry. Model equations are given to help the audience understand the processes and to help predict results (rejection, fractionation, and flux), and what factors are important for process control for these systems.

Microfiltration membranes, with relatively larger pore sizes (in the range of 0.1 to 10 μm) than other membrane processes, are typically used for dairy protein fractionation. In Chapter 1, the use of microfiltration processes for serum protein (or native whey protein) fractionation from skim milk is illustrated. Comparing this to traditional cheese-whey processes to obtain this protein ingredient, there are no changes in pH and ionic strength during the microfiltration process, resulting in less denaturation of the protein and more purity of the product streams, which then provides better functionalities of the ingredients.

Ultrafiltration is probably the most widely used membrane process in dairy ingredient manufacturing. Chapter 2 provides a comprehensive review of the current state of knowledge in using polymeric ultrafiltration membranes for separating dairy ingredients (with an emphasis on lactose recovery). The factors affecting lactose (also called milk sugar) separation/fractionation and the effects on yield are discussed in detail. The concept and principles of integrated membrane bioreactor to attain the functional dairy products are also presented. Finally, the future and challenges in separating milk sugar for a production of low-lactose milk and other value-added products are discussed. Chapter 3 focuses on the fouling occurring during ultrafiltration. In this chapter, the mechanism of ultrafiltration fouling and the factors affecting fouling are described, and certain fouling-control strategies are suggested. Chapter 4 proposes that placing a charge on the ultrafiltration membrane fundamentally changes separation mechanism, allowing fractionation of proteins with a similar size but different net charges. This chapter spells out the utility of charged ultrafiltration membranes for dairy protein fractionation, and also for traditional protein concentration applications where a significant flux increase is observed compared to uncharged ultrafiltration membranes.

Nanofiltration is used in the dairy industry to remove sodium chloride from solutions. This can include the de-salting of milk protein concentrates, whey, or lactose solutions. The approach can also be used to recover important minerals, notably calcium and magnesium. Chapter 5 outlines the key features of such membrane systems, including the type of membranes usually employed and the mineral species of interest during these filtration processes.

In dairy ingredient manufacturing processes a large amount of water is consumed and this water ends up in waste streams with pollutants that must be further treated. In Chapter 6, reverse osmosis membrane processes are reviewed and discussed for the application to purifying dairy waste streams to produce a more concentrated waste stream and a clean permeate water stream for reuse. Such preconcentration before evaporating and drying saves energy consumption and reduces the residence time in heat environment, resulting in lower operation costs, and reduces denaturation of the milk proteins. The permeate water from reverse osmosis is almost "pure" water that can be reused for process diafiltration and system clean-in-place (CIP) cleaning; in CIP the permeate water is used to generate cleaning solutions for membrane systems and other processing equipment. Thus considerable advantages on cost saving and environmental sustainability are realized.

Besides the above widely used membrane processes, several unique and emerging membrane separation processes have also been applied for ingredient production. Chapter 7 deals with the current status of pervaporation membranes for dairy aroma concentration. The fundamentals of the pervaporation process for aroma extraction are presented and the membranes suitable for dairy aroma enrichment are documented. The various aspects related to the aroma recovery (e.g., the nature of aroma compounds, permselectivity of the membrane, and the interactions between the aroma compounds and the membrane materials) are discussed. Chapter 8 examines protein purification and analysis methods using membrane chromatography. This chapter discusses some of the more conventional applications such as purification and polishing of proteins before moving on to newer and lesser known applications such as analysis of protein aggregates and the use of membrane stacks as bioreactors for enzymatic and synthetic modification of proteins. Chapter 9 reviews electrodialysis that has recently been applied to separate efficiently certain dairy ingredients. The principles of conventional electrodialysis as well as mass transport phenomenon through ion-exchange membranes are first described. Then hybrid electrodialysis processes using bipolar and filtration membranes are highlighted. After that, recent results on conventional and advanced electrodialysis processes for the separation, purification, and fractionation of dairy ingredients from milk, whey, or milk by-products are presented.

This book should prove to be of value to anyone working in the development and design of membrane-based systems for recovery of products in the dairy industry. Overall, the contents of this book cover the majority of membrane separation processes used for dairy ingredient separation and product stream production. It includes membrane-based separation technologies that have been recently commercialized, which is of substantial interest to the dairy industry. As well the book includes some cutting-edge technologies that have been thoroughly researched and have a great potential to be commercialized in the near future. The target audience are those professionals, such as process and system design engineers and ingredient product developers. This contemporary information and experience-based knowledge will be of great use to those professionals in membrane and membrane processing industries, leading to greater potential for the commercialization of the membranes and the processes in dairy related industries described in this book.

Kang Hu
James M. Dickson

List of contributors

Abhiram Arunkumar, PhD
Department of Chemical and Biological Engineering
1605 Linden Drive
Madison
WI 53706, USA

Laurent Bazinet, PhD
Institute of Nutrition and Functional Foods (INAF)
Dairy Research Center (STELA)
Department of Food Sciences and Nutrition,
Université Laval,
Pavillon Paul Comtois
Québec, Canada G1V 0A6

Sheng Dai, PhD
School of Chemical Engineering
The University of Adelaide
Adelaide
South Australia 5005,
Australia

James M. Dickson, PhD
Department of Chemical Engineering
McMaster University
Hamilton
Ontario, Canada L8S 4L7

Mark R. Etzel, PhD
Department of Chemical and Biological Engineering
1605 Linden Drive
Madison
WI 53706, USA

Xianshe Feng, PhD
Department of Chemical Engineering
University of Waterloo
Waterloo
Ontario, Canada N2L 3G1

Raja Ghosh, PhD
Department of Chemical Engineering
McMaster University
1280 Main Street West
Hamilton
Ontario, Canada L8S 4L7

Kang Hu, PhD
Department of Chemical Engineering
McMaster University
Hamilton
Ontario, Canada L8S 4L7
and
R&D
Land O'Lakes Inc.
Arden Hills
MN 55126, USA

Bo Jin, PhD
School of Chemical Engineering
The University of Adelaide
Adelaide
South Australia 5005, Australia

Dharmesh Kanani, PhD
Principal Scientist
Process Sciences
Natrix Separations Inc.
5295 John Lucas Drive, Unit 6
Burlington
Ontario, Canada L7L 6A8

Sandra E. Kentish, PhD
Department of Chemical and Biomolecular Engineering
The University of Melbourne
Melbourne
Victoria 3010, Australia

Suwattana Pruksasri, PhD
Department of Biotechnology
Faculty of Engineering and Industrial Technology
Silpakorn University
Nakhon Pathom 73000, Thailand

Gwynneth Rice, PhD
Beca Pty Ltd
L4, 5 Queens Rd,
Melbourne
Victoria 3004, Australia

Panida Sampranpiboon, PhD
Department of Chemical and Material Engineering
Rangsit University
Pathumthani 12000, Thailand

Masoumeh Zargar, PhD Candidate
School of Chemical Engineering
The University of Adelaide
Adelaide
South Australia 5005, Australia

Boya Zhang, PhD Candidate
Department of Chemical Engineering
University of Waterloo
Waterloo
Ontario, Canada N2L 3G1

1 Microfiltration for casein and serum protein separation

Kang Hu[1], James M. Dickson[1], and Sandra E. Kentish[2]

[1] *Department of Chemical Engineering, McMaster University, Hamilton, Ontario, Canada, L8S 4L7*
[2] *ARC Dairy Innovation Hub, Department of Chemical and Biomolecular Engineering, The University of Melbourne, Parkville, Victoria 3010, Australia*

1.1 INTRODUCTION OF MICROFILTRATION

Microfiltration(MF), probably the oldest membrane separation technology, was developed between the First and the Second World Wars in Germany for the purpose of bacteria removal (Zsigmondy and Bachmann, 1922). Generally, MF membranes have a pore size ranging from 0.1 to 10 μm. This size range encompasses a wide variety of natural and industrial particles, such as colloids, bacteria, and red blood cells.

MF is a pressure-driven separation process, which is similar to other widely used membrane processes such as ultrafiltration, nanofiltration, and reverse osmosis. Compared to these processes, MF is typically operated at a relatively lower pressure and is mainly applied for larger particle separation and fractionation.

In this section, the principle of MF is introduced. This includes the introduction of MF membranes and processes, the mechanism of cross-flow MF, and membrane fouling.

1.1.1 Microfiltration membranes and processes

MF membranes can be synthesized from a wide variety of materials, normally categorized as either organic, such as polymers, or inorganic, such as ceramic materials. Polymeric materials can be hydrophobic, including polytetrafluoroethylene (PTFE), poly(vinylidene fluoride) (PVDF) and polypropylene (PP), or hydrophilic, including polycarbonate (PC) and poly(ethersulfone) (PES) (Mulder, 1996a). Ceramic membranes are typically made from alumina (Al_2O_3), zirconia (ZrO_2), and titania (TiO_2). During membrane fabrication, some hydrophobic materials can

Membrane Processing for Dairy Ingredient Separation, First Edition.
Edited by Kang Hu and James M. Dickson.

be surface-modified to become hydrophilic, as required by specific applications. Gekas and Hallstrom (1990) reviewed these materials and summarized both the advantages and disadvantages of them. They suggested that comparing the two materials, polymer membranes are generally less expensive and have a higher area/volume ratio than ceramic membranes. On the other hand, polymer membranes bind protein more readily and have a wider pore size distribution. Ceramic membranes have exceptional thermal and chemical resistance and a much longer membrane life.

Various techniques have been employed to fabricate microporous membranes for MF. For example, ceramic membranes could be prepared by sintering layers on supporting materials. Polymeric membranes, benefiting from current polymer processing technology, could be prepared by various methods including: melt stretching, track etching, phase inversion, and casting (Glimenius, 1985; Mulder, 1996b).

Depending upon the materials and techniques used to prepare the membranes, MF membrane pore structure varies significantly. Figure 1.1 illustrates some typical examples of membrane porous surface structures obtained with different fabrication methods and materials. From the images, membrane pores created by stretching (a) are not circular, but the manufacturing process is relatively simple. Pores created by track etching (b) are cylindrically shaped with uniform dimensions but with lower porosity, while pores created by phase inversion (c) have a much higher porosity (or pore density). For ceramic membranes, sintering results typically in a nodular structure (d).

The MF process is pressure-driven and generally is carried out in two modes: dead-end filtration or cross-flow filtration. As illustrated in Figure 1.2a, in the dead-end MF, an applied pressure pushes the feed stream toward the membrane perpendicularly. The solvent and some smaller particles permeate through the membrane due to the pressure difference across the membrane, but larger particles are retained and then accumulate and deposit on top of the membrane. As a consequence, a fouling layer is formed, resulting in a decrease in filtration permeate flux. As the fouling increases with time, the permeate flux decreases drastically. When the fouling is sufficiently large, no more permeate can be collected and the membrane needs to be replaced. Due to the nature of the process, dead-end MF is economically feasible only when applied to rather dilute and/or valuable solutions, since the fouled membranes are difficult to clean and are normally just discarded. Dead-end MF is typically conducted in batch mode and mostly applied on a laboratory scale, using cylindrical vessels such as syringes as the filtration cell.

In cross-flow MF (Figure 1.2b), the feed stream flows tangentially to the membrane surface on the feed side. Similar to the dead-end mode, the solvent and some small particles transport through the membrane and

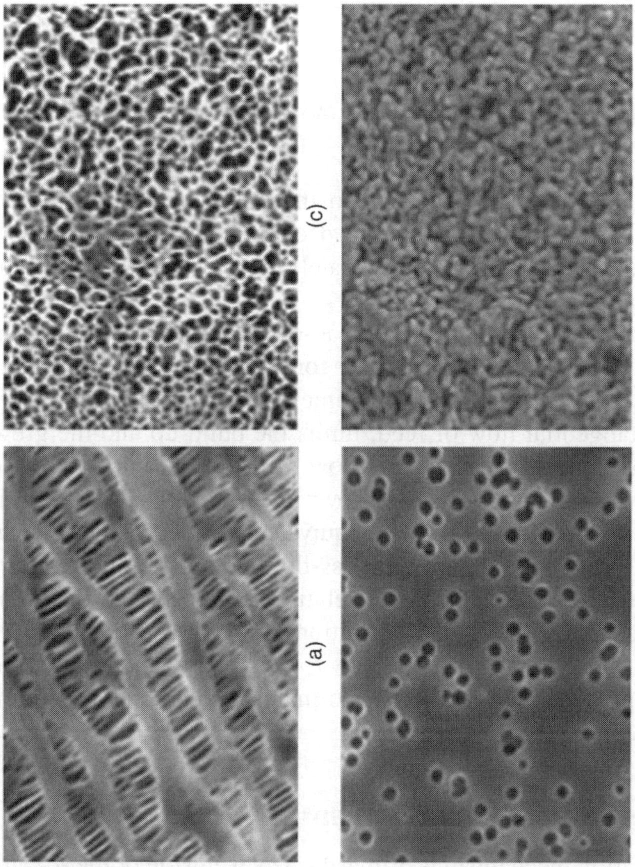

Figure 1.1 Microfiltration membrane surface images. (a) Polymeric membranes fabricated by melt-stretch (from Barbe, Hogan, and Johnson, 2000. Reproduced with permission of Elsevier). (b) Polymeric membranes fabricated by track-etching (Millipore Product Catalogue, 2013). (c): Polymeric membranes fabricated by phase inversion (Ying, Kang, and Neoh, 2002. Reproduced with permission of Elsevier). (d) Ceramic membranes fabricated by sintering (Zhang, Zhong, and Xing, 2013. Reproduced with permission of Elsevier).

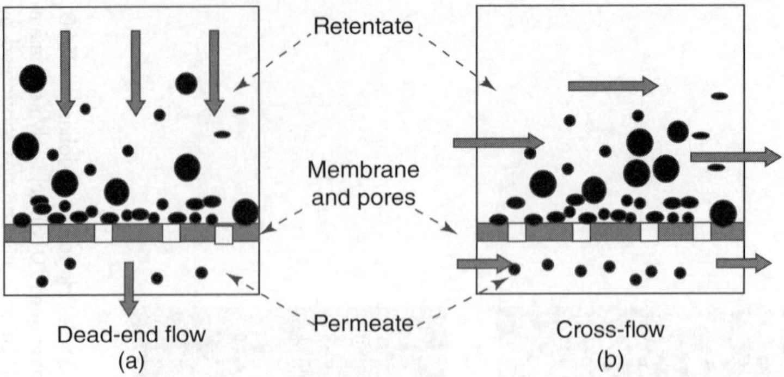

Figure 1.2 (a) Dead-end flow and (b) cross-flow MF.

are collected in the permeate side due to the pressure difference across the membrane. The remainder of the feed continuously flows along the membrane surface and is removed separately as a concentrated solution called the retentate.

On the retentate side, similar to the dead-end mode, the retained particles accumulate at the membrane surface forming a fouling layer, resulting in a decrease of permeate flux over time. However, a shearing effect, induced by the tangential flow of feed, limits the build-up and the growth of the fouling layer. Thus, during cross-flow MF, after an initial rapid formation, the growth of the layer slows down or even stops. At this point, the accumulation of particles becomes equivalent to the amount of particles lifted back to the bulk stream by cross-flow. The hydraulic resistance generated by the fouling layer becomes relatively constant and the permeate flux becomes stable. Under such conditions, cross-flow MF can operate continuously at a steady-state filtration flow for a long operating time, which substantially improves the separation performance and broadens the industrial application.

1.1.2 Cross-flow MF mechanisms

In MF, the separation mechanism is primarily size exclusion or the so-called "sieve effect." Particles smaller than the pore size may pass through the membrane but particles larger than the pore size are rejected. A great number of parameters influence this separation performance, including the cross-flow velocity, applied pressure, physical and chemical properties of the membranes and particles, interactions between particles and particles, and interactions between particles and membrane materials. Considering all these parameters, many mathematical models have been developed to describe the transport mechanisms of cross-flow MF. All these models have their limitations owing to the fact that the phenomena occurring on the membrane surface and interface are far from clear, especially when various types of particles are present in the feed solution, adding to the complexity. However, utilizing these models can improve understanding of the process and provide guidance on the process design of such operations.

Earlier attempts to modeling cross-flow MF started with the concentration polarization model, since concentration polarization affects membrane performance for almost all types of membrane separations. As illustrated in Figure 1.3, during MF the particles are rejected by the membrane and then accumulate near the membrane surface. Thus, the concentration of particle i at the membrane surface ($C^B_{i,m}$) is always higher than that in the bulk solution (C^B_i). This phenomenon, just outside the membrane surface, is called "concentration polarization". The formed polarized layer (boundary layer), with thickness z, on the membrane feed side has some undesirable effects on the membrane separation. The higher concentration of particles at the membrane surface can cause increased fouling (binding of particles to the membrane surface). In addition, the concentration polarization invariably leads to reduced flux since the layer has a relatively large resistance to permeation (Zeman and Zydney, 1996) and can cause either increased rejection or decreased rejection as the fouled layer can be more restricting to solute permeation (hence increasing rejection) and the higher concentration in the boundary layer translates into a higher concentration in the permeate layer (hence decreasing rejection).

On the retentate side, the concentration build-up on the membrane surface generates a diffusive flow back of the particles from the membrane to the

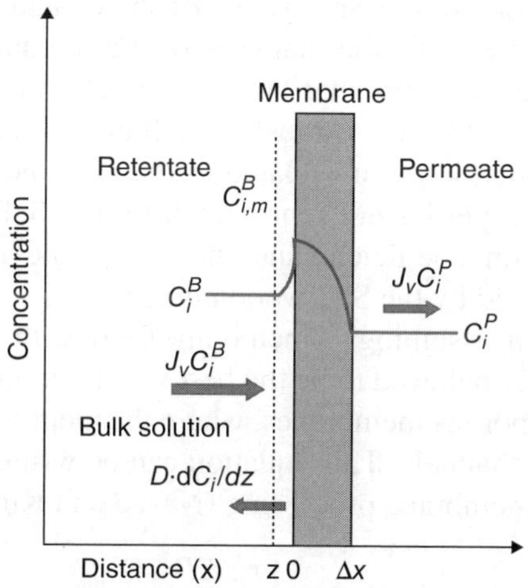

Figure 1.3 Concentration polarization effects of a microfiltration membrane with particle concentration profiles.

bulk of the feed solution. After a period of time, a steady state is achieved and the flow of particles i to the membrane surface ($J_V C_i^B$) is balanced by the particle flow through the membrane ($J_V C_i^P$) plus the diffusive flow from the membrane surface back to the bulk ($D\,\mathrm{d}C_i/\mathrm{d}z$). The mass balance is given by

$$J_V\, C_i^P = J_V\, C_i^B + D\frac{\mathrm{d}C_i}{\mathrm{d}z} \tag{1.1}$$

where J_V is the volumetric permeate flux, D is the diffusion coefficient of particles in the solution, C_i^P is the permeate concentration, and $\mathrm{d}C_i/\mathrm{d}z$ is the concentration gradient through the boundary layer. The apparent membrane rejection of species i is defined as

$$R = \left(1 - \frac{C_i^P}{C_i^B}\right) \tag{1.2}$$

Integration of Equation (1.1) gives (Kimura and Sourirajan, 1967)

$$J_V = k \ln\left(\frac{C_{i,m}^B - C_i^P}{C_i^B - C_i^P}\right) \tag{1.3}$$

where k is the mass transfer coefficient in the concentration polarization layer. For turbulent flow, a semi-empirical relationship for the mass transfer coefficient in pipes and channels has been described as (for example, from Mulder, 1996c)

$$Sh = \frac{k\, d_h}{D} = a\, Re^b Sc^c \tag{1.4}$$

where Sh, Re, and Sc are the Sherwood, Reynolds and Schmidt numbers, respectively, d_h is the hydraulic diameter of the membrane flow channel, and a, b, and c are constants that have to be determined by experiment. From Equation (1.4), the mass transfer coefficient k is a function of the feed flow velocity that is in turn related to the Reynolds number, the feed solution physical properties are represented by the Schmidt number, and the particle diffusion coefficient and the membrane module shape and dimensions are related by the Sherwood number.

In an ideal system, assuming no membrane fouling, the Hagen–Poiseuille equation is generally believed to be the best way to describe the fluid transport through microporous membranes, where the membrane pores are modeled as cylindrical channels. This equation can be written as one form that is broadly used in membrane processing (Nakao and Kimura, 1981):

$$J_V = \frac{r_P^2\, \Delta P}{8\mu\left(\Delta x / A_K\right)} \tag{1.5}$$

where r_p is the capillary pore radius of the membrane, μ is the viscosity of the solution at the operating temperature, ΔP is the transmembrane pressure (TMP), which is the pressure difference across the membrane, and $\Delta x/A_K$ is the ratio of the membrane effective thickness to the membrane porosity. Using the Hagen–Poiseuille equation, the permeate flux can be simply related to the membrane structure and the operating conditions, such as pressure and temperature. From Equation (1.5), permeate flux is proportional to TMP, pore size, porosity, and indirectly to temperature, as a temperature increase reduces the viscosity. Conversely, the permeate flux is inversely proportional to the membrane effective thickness that generates a hydraulic resistance to the fluid flux. In the real world, however, flux does not conform exactly to the model, owing to other factors such as the nature of the membrane–particle interactions and particularly due to fouling.

In this concentration polarization model, the particle convective flux toward the membrane is balanced by Brownian back-diffusion, as illustrated in Equation (1.1) and Figure (1.3). In addition, for cross-flow MF, shear stress, induced by high-speed cross-flow, has an effect on lifting particles away from the layer, increasing particle back-diffusion to the bulk solution. Thus, several modifications integrate these effects by using different approaches to obtain the diffusion coefficient (Ripperger and Altmann, 2002). For example, Zydney and Colton (1986) assumed that the shear-induced particle diffusion was the predominant transport mechanism at cross-flow MF and the diffusion coefficient could be expressed as $D = 0.03\ r^2\tau$, where r is the particle radius and τ is the wall shear rate. Agashichev (2006) proposed a model that could analyze the effect of shear stress on the degree of concentration polarization and enabled calculation of the concentration profile where shear-dependent behavior took place.

Overall, it is difficult to develop a model sufficient enough for all types of cross-flow MF applications, since the ideal condition can never be achieved and the understanding on what occurs on the membrane surface or the fouling mechanism is far from complete (Gekas and Hallstrom, 1990). Thus, some assumptions have to be made, such as the adsorption of particles on the membrane surface. These assumptions simplify calculations but increase the discrepancy between models and experimental results.

1.1.3 Fouling in cross-flow microfiltration membrane operations

The major technical hurdle that limits the applications of MF is membrane fouling. Once substantial fouling occurs, the permeate flux declines drastically and the particle rejection increases, resulting in undesired separation results. Then the operation needs to stop for cleaning, which typically consumes time and generates waste streams. Thus, membrane fouling increases the complexity of the process operation, changes the product quality, and adds production downtime and cost.

Membrane fouling is influenced by a variety of factors, as presented in Figure 1.4. These factors include membrane internal pore-blocking, membrane surface cake layer formation and growth, and concentration polarization. To illustrate the fouling phenomenon, a typical permeate flux versus time curve of cross-flow MF is presented in Figure 1.4a. On a commercial scale, cross-flow MF operation is generally started by processing with water and then replacing the water with the feed stream. At time zero, the permeate flux is basically at the highest value. As the feed stream is introduced to the system, a drastic decrease in permeate flux is observed. This usually takes place within minutes. Following this, the permeate flux decreases continuously but more slowly than before. Finally, the permeate flux stabilizes at a steady-state value.

The drastic decline of permeate flux when the feed solution is first replaced with water is due to the rapid blocking of membrane pores (Song, 1998). As illustrated in Figure 1.4b, once the particles are trapped inside

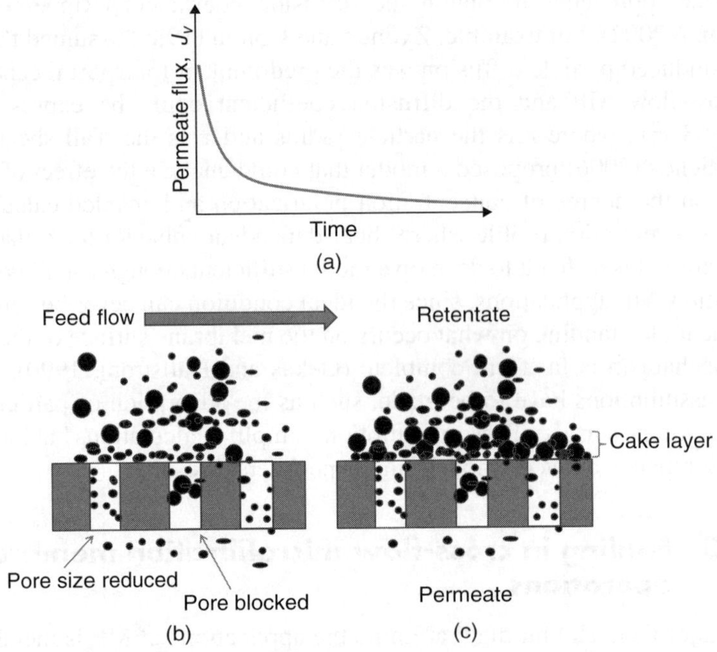

Figure 1.4 (a) Cross-flow microfiltration permeate flux decreases with time due to membrane fouling. Membrane fouling caused by pore plugging and cake layer formation: (b) initial deposition of particles and concentration polarization followed by (c) cake layer formation.

the membrane, the effective number of membrane pores and the pore size are reduced. The permeate flux decreases immediately due to the reduction of r_P and A_K, as indicated by Equation (1.5). Pore-blocking is a typical phenomenon for porous membranes, especially for microporous membranes used for MF. This phenomenon is influenced by both membrane–particle interactions and by the shape and size of the particles compared to the pore size. Thus fouling is much more significant when the particles and the pores are similar in both shape and size (Belfort, Davis, and Zydney, 1994).

Following the rapid decline, the permeate flux decreases exponentially, due to the formation and the growth of the cake layer. As presented in Figure 1.4c, particles retained by the membrane due to the sieve effect and the particle–membrane adsorption accumulate and deposit on the membrane surface and create the cake layer. This layer creates a hydraulic resistance to permeate flux and particle transport. However, the cake layer resistance can be minimized in a commercial process by employing a higher flow velocity and/or lower feed concentration. In other words, mass transfer can be improved by a higher Reynolds number, as indicated in Equation (1.4).

After the exponential decrease, the permeate flux keeps decreasing but the rate starts to level off as the cake layer approaches a constant thickness. When the cake layer "thickness" stabilizes a steady-state flux is obtained.

Development of an overall theory of cake layer formation is difficult because of the many factors that impact on the layer formation and growth. These factors include operation conditions and the interactions between membrane and particles and between particles and particles. For example, if the membrane material is hydrophobic, then hydrophobic particles are readily adsorbed. At a certain pH and ionic strength, some particles may also coagulate on the membrane surface and form a gel layer.

Overall, it is the particles in the feed that cause the fouling. Thus, the permeate flux, reflecting the convective flow of the solution (including particles) from bulk toward the membrane, is the critical factor that controls the fouling. Field et al. (1995) proposed a critical-flux hypothesis; if the initial permeate flux is less than a certain value, a constant filtration rate is achieved since the slow growth of the fouling could be diminished by exerting a cross-flow velocity that lifts the particles back to the bulk. Brans et al. (2004) proposed that in membrane filtrations, such as milk MF or ultrafiltration, the permeate flux did not always increase linearly with TMP, as given by Equation (1.5), due to the fouling. Three regimes could be distinguished with a different flux dependency on TMP (Figure 1.5).

In the filtration zone I, the permeate flux is low with only slight fouling. The flux increases linearly with the TMP, as expected from Equation (1.5). As the permeate flux increases, more particles are carried toward the membrane, but the growth of the cake layer is limited by the cross-flow shear

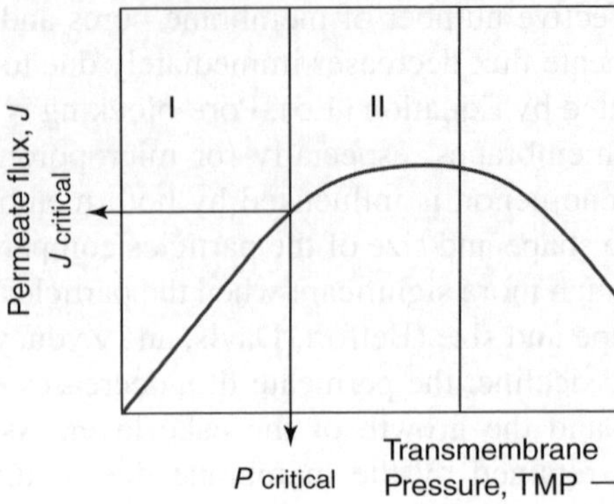

Figure 1.5 Critical flux regimes: flux dependency on transmembrane pressure (from Brans et al., 2004. Reproduced with permission of Elsevier).

effect. In this zone I, a constant filtration flux is obtained but the low permeate flux may require a larger membrane area. As the TMP increases, the permeate flux exceeds the critical flux and reaches zone II, membrane fouling becomes considerable and the rate of permeate flux declines. In this zone II, the positive effect of TMP on the permeate flux is still greater than the negative effect of cake layer growth on the permeate flux. However, the increase in fouling and the possible compaction of the cake layer produces a permeate flux that does not increase linearly with the TMP.

When the permeate flux exceeds the limitation by further increasing the TMP (zone III), the permeate flux starts to decline, instead of increase, with the driving force TMP. In real operations in zone III, when the pressure is increased, a short-term initial increase in flux is observed followed by a fast decline. Then the permeate flux becomes even lower than it was before the pressure was increased. This is because the initial higher permeate flux carries more particles toward the membrane and then adds to and compresses the cake layer, resulting in an increase in the hydraulic resistance. Compared to zone II, the cake layer fouling is much more severe and becomes irreversible. When the processing reaches zone III, the cake layer is so dense that even the permeate composition can change as the fouling layer causes increased rejection of particles.

It is widely recognized that the MF process is more effective when conducted at a lower permeate flux. The initial permeate flux should be controlled below the "critical flux" value to avoid rapid fouling; of course, for different feed streams and membranes, the critical flux value varies. The configurations of membrane element and system also affect the critical flux value as the mass transfer coefficient can be altered. The start-up procedure is also important for the purpose of avoiding irreversible fouling. Once the operation has reached zone II or III, it is difficult to return to zone I even though the TMP is reduced, since irreversible fouling has been created.

1.1.4 Application of cross-flow microfiltration

Cross-flow MF has been applied in a wide variety of industrial processes. For example, in the food and beverage industry, cross-flow MF is used to sterilize fluids. By removing microorganisms in a low-temperature membrane process, the product quality can be improved, compared to a high heat treatment such as conventional ultrahigh heat treatment or pasteurization. Cross-flow MF is also used to clarify fruit juice, soybean sauce, wine and beer, etc., as fluid turbidity can be significantly reduced by removing undesired particles, without retention of valuable smaller size ingredients. In the water/wastewater treatment industry, cross-flow MF is used to prepare clean potable water as it removes bacteria that may trigger a health hazard, or is used to remove particles in the polishing process of ultrapure water production. In biotechnology, cross-flow MF is applied to cell harvesting and recovery or acts as a part of a membrane bioreactor. A detailed review and case studies of these applications can be found in Cheryan (1998).

In the dairy industry, cross-flow MF is mainly used to remove bacteria and spores from milk or to polish processing water. Cross-flow MF can reduce the amount of bacteria and spores without affecting the taste of milk, resulting in both a longer shelf life and better quality. Cross-flow MF is also used in the dairy culture process for removal of microorganisms. Compared to these applications, protein separation/fractionation by MF in the dairy industry is relatively new. In the following sections, the application of cross-flow MF to skim milk casein and serum protein separation is reviewed and discussed.

1.2 CASEIN MICELLES AND SERUM PROTEINS IN SKIM MILK

Bovine milk, as one of the main protein sources for human beings, contains roughly 3.2% (w/w) true protein, 3.4% fat, 4.9% lactose, and 0.7% ash in water, and these values vary slightly with the milk source during a year. Milk proteins are broadly defined as two categories based on the chemical composition and physical properties. The first category is casein proteins, which contain phosphorus and coagulate at pH 4.6, the isoelectric point of the proteins. The second category is commonly called whey proteins, which do not contain phosphorus and remain soluble at pH 4.6. Among the 3.2% total protein in milk, casein proteins are around 2.6% and whey proteins are around 0.6% (Fox and McSweeney, 2003).

In the dairy industry, casein proteins are used to produce cheese. When rennet or acid is added to milk, the casein proteins coagulate and form a solid curd, while the solution left contains mainly whey proteins and lactose. Over the past 20 years, whey, which used to be a waste by-product of cheese manufacture, has become a valuable dairy-based ingredient due to its functionality and nutritional properties. Whey is used in a wide range of products, including infant formulas, food supplements, energy bars, and sport beverages. Commercial whey protein products are available in a variety of forms, such as whey protein concentrates (WPCs), whey protein isolate (WPI), dried sweet dairy whey (DSW), etc.

Compared to the traditional cheese whey process, the MF membrane separation process is an alternative way to produce whey proteins. In the membrane process, casein proteins are rejected on the retentate side of membrane, while whey proteins (including mainly β-lactoglobulin, α-lactalbumin, and bovine serum albumin (BSA)) pass through the membrane and can be recovered on the permeate side. In this case, the whey proteins are also called "native whey proteins" or "serum proteins" as they are obtained in a different process and the composition is different from traditional whey proteins. For MF membranes, this is a physical separation process so that there is no physical or chemical change to the proteins. Thus, the soluble serum proteins, recovered in the permeate side, have the same pH as milk, unlike traditional cheese whey, which is relatively acidic. The serum protein solution does not contain casein peptides, enzymes, fat, and denaturized whey proteins. Therefore, the technological and economic value is higher than that for traditional cheese whey. For example, Maubois (2002) suggested that the gelation, foaming and solubility of serum proteins were significantly improved compared to conventional WPI and WPC. On the other hand, the high purity of the serum protein solution is a better starting material for further isolation of individual functional proteins. The serum proteins also have a higher nutritional value than that of cheese whey, due to the different amino acid composition. Boirie et al. (1997) demonstrated that the serum proteins were oxidized more rapidly in human beings, suggesting a potential solution for the needs of athletes or people wishing to follow a slimming diet.

Meanwhile, concentrating casein proteins using an MF membrane process can improve the cheese-making process. Maubois (2002) suggested that the rennet coagulation was improved since the coagulation time could be significantly reduced, curd firmness kinetics accelerated, and final firmness increased. Consequently, casein and fat retention in the cheese curd was significantly improved and the cheese yield increased 2 to 4%. Concentrated casein proteins could be recombined with cream to make cheese or could be dried as a native casein ingredient for food applications. From native casein it was possible to manufacture pure β-casein or other proteins by further processing (Maubois and Ollivier, 1997).

Separation of casein and serum proteins using MF is based on the physical and chemical properties of the two types of proteins. The composition and molecular weight of major casein proteins and serum proteins in milk are given in Table 1.1 (Fox and McSweeney, 2003). Although the casein proteins themselves have a relatively low molecular weight, about 95% of the casein in milk exists as suspended colloidal particles, called casein micelles. This makes the separation feasible based on the particle size difference. Casein micelles are generally spherical in shape with a diameter ranging from 0.05 to 0.5 μm, with the majority falling between 0.13 and 0.16 μm (Fox and McSweeney, 2003). At this size, the micelles are much larger than the whey proteins present (listed in Table 1.1) and are easily rejected, while the remaining proteins pass through the membrane. Previous works on casein and serum protein separation used membranes ranging from larger pore size UF membranes to 0.5 μm pore size MF membranes, depending upon the objective of the separation. Skim milk, which is when cream (or fat) is removed from whole milk, is a better starting material for casein and serum protein separation, since fat droplets foul membranes rapidly.

A typical process for casein and serum protein separation from skim milk using MF is illustrated in Figure 1.6. Skim milk is introduced to an MF system. At stage one, the skim milk is generally concentrated to a certain volume reduction. Some serum proteins, with lactose, are removed to permeate. The casein proteins, as well as the rest of the serum proteins and lactose, are retained in the retentate. If it is a continuous process mode, there will be two outlets: permeate flow and retentate flow, at a certain flow rate ratio. A second stage (or more stages) is normally added to further remove serum proteins from the retentate. Diafiltration water is added to the feed side stream to prevent concentration increasing during the filtration. All the permeates are combined together and sent for further treatment.

Table 1.1 The composition and molecular weight of the major milk proteins (Fox and McSweeney, 2003).

Proteins	Concentration (g/L)	Molecular weight, *D*
α_{s1}-Casein	10.0	24 000
α_{s2}-Casein	2.6	25 000
β-Casein	9.3	24 000
κ-Casein	3.3	19 000
ϒ-Casein	0.8	12 000–20 000
β-Lactoglobulin	3.2	18 000*
α-Lactalbumin	1.2	14 000
BSA	0.4	66 000
Immunoglobulins	0.7	150 000–900 000

*Normally present as a dimer in milk solution with MW of 36 000.

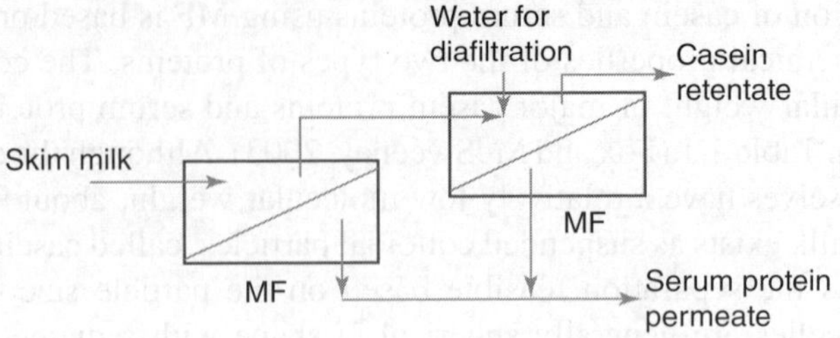

Figure 1.6 Casein and serum protein separation from skim milk using MF.

1.3 EFFECTS OF PERMEATE FLUX AND SHEAR STRESS ON SEPARATION PERFORMANCE

The application of cross-flow MF on separation of serum proteins from casein micelle dates back to the late 1980s. It was found that the major constraint of this application is the cake layer fouling. As suggested by Le Berre and Daufin (1996), after the formation of a cake layer, the membrane pores no longer govern the membrane separation. The formed cake layer, with its own characteristics of thickness and porosity, dominates the separation performance.

The permeate flux and the surface shear stress (or surface flow velocity) are the most important factors in determining the properties of the cake layer (see Section 1.1). In this section, we illustrate how the two parameters influence the cake layer growth. This understanding will provide guidance for new membrane development, system and process design, and operation.

The permeate flux governs the transport of proteins from the bulk stream to the membrane surface by convection. The higher the permeate flux, the more the proteins move to and accumulate on the membrane surface, which then potentially results in a thicker cake layer and higher resistance. Thus, the permeate flux is often operated at a relatively low value, such as lower than a "limiting permeate flux," which is similar to the concept of "critical flux" introduced by Field et al. (1995). Correctly selecting the initial TMP so that the critical flux value was not exceeded, the rate of fouling could be greatly reduced. Therefore, a cross-flow MF system should be designed to be able to provide a higher limiting permeate flux value, and during the operation the permeate flux should be constrained to be no higher than the limiting value.

The limiting permeate flux is influenced by several factors, including the cross-flow velocity, protein properties, and membrane module geometry. Samuelsson et al. (1997a) reviewed several models for skim milk cross-flow MF and concluded that the shear-induced diffusion model was able to predict the limiting flux value very well, given as

$$J_{lim} = 0.072\,\frac{\tau}{\mu}\left(\frac{\Phi_w \alpha^4}{\Phi_b L}\right)^{1/3} \tag{1.6}$$

where J_{lim} is the limiting permeate flux, Φ_w is the particle volume fraction in the boundary layer immediately above the fouling layer on the membrane surface, α is the particle radius, Φ_b is the volume fraction in the bulk, and L is the membrane element length. Clearly from Equation (1.6), a higher limiting permeate flux value can be achieved by using a shorter membrane element or housing, operating at a higher temperature to reduce viscosity (μ) and operating at a higher shear stress (τ) that can reduce the thickness of the cake layer by "lifting off" the proteins from the cake layer to promote the diffusion back to the bulk stream.

In order to raise the limiting permeate flux value by exerting a high shear stress, a high feed flow rate (or high surface flow velocity) is needed, which produces a high-pressure drop from the inlet to the outlet on the retentate side of the membrane. The Darcy–Weisbach equation can relate the pressure drop at a given length (ΔP_L), for skim milk flowing in a membrane element at a certain velocity, to the frictional force that is a function of the shear stress, as shown:

$$\Delta P_L = P_{inlet} - P_{outlet} = f\frac{L}{D}\rho v^2/2$$

and

$$f = 8\frac{\tau}{\rho v^2} \tag{1.7}$$

where P_{inlet} and P_{outlet} are the pressures at the inlet and outlet of the membrane element on the retentate side, respectively, ρ is the density of the feed stream, v is the velocity in the flow channel, D is the hydraulic diameter (tube diameter or height of a channel), and f is the Darcy friction factor for laminar or turbulent flow. Then the wall shear stress on the membrane surface can be obtained by reorganizing Equation (1.7):

$$\tau = \frac{D}{4L}(P_{inlet} - P_{outlet}) \tag{1.8}$$

As illustrated by Equation (1.8), the shear stress can be increased by simply increasing the pressure drop from the membrane element inlet to the outlet, or by a greater D/L.

There are several methods that can be applied to restrain the permeate flux below the limiting flux value. Based on Equation (1.5), firstly, the permeate flux can be reduced by modifying the membrane properties, such as increasing the effective membrane thickness or reducing the membrane porosity. This approach is the focus of membrane manufacturers and is discussed in

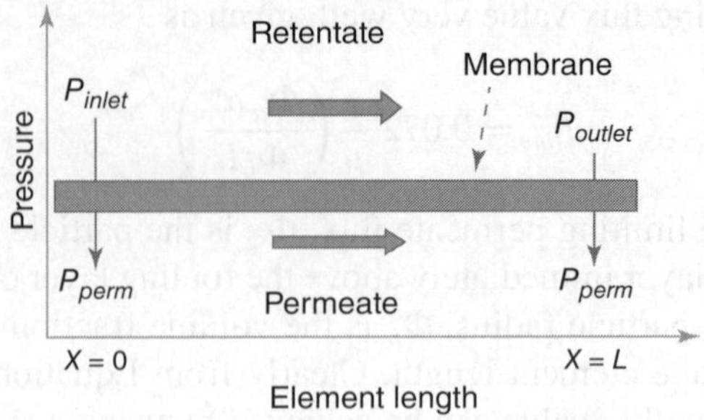

Figure 1.7 Illustration of the TMP along a cross-flow membrane.

the next two sections. Secondly, once the membrane is manufactured, the permeate flux can be reduced by conducting the filtration at a lower TMP driving force. This approach is the focus of membrane process designers.

A depiction of a membrane element is illustrated in Figure 1.7. Distribution of the pressure along the element length is given below. The TMP at the inlet and outlet are given as

$$\text{at } x = 0,\ \Delta P = P_{inlet} - P_{perm}$$
$$\text{at } x = \text{L},\ \Delta P = P_{outlet} - P_{perm}$$

Assuming the permeate pressure is approximately constant then

$$\Delta P = \frac{P_{inlet} + P_{outlet}}{2} - P_{perm} \tag{1.9}$$

where P_{perm} is the average pressure at the permeate side and L is the length of the element. Typically the P_{perm} in cross-flow MF is set to be just low enough to maintain permeate flow and to reduce energy consumption. The ΔP is high at the element inlet and progressively decreases from the inlet to the outlet, so that the permeate flux is also higher at the inlet than at the outlet. Thus, to maintain a low permeate flux (or low TMP), one needs to keep not only a low outlet pressure but also a low inlet pressure along a membrane element.

Since the growth of the cake layer is dependent upon the competition between the particle transfer to the membrane surface by convection (J_V) and the "lift" away from the membrane surface by shear (τ), the ratio of J_V/τ is an operating factor typically used to evaluate the membrane separation performance. Gesan, Daufin, and Merin (1995) proposed that for skim milk MF, a low J_V/τ could increase the overall effective operating time and more serum proteins could be recovered. However, the attempts to prevent cake layer growth, by applying both a low permeate flux and a high shear stress simultaneously, were not fully successful. This is because high shear stress requires a high pressure drop along the membrane element, so that the TMP at the inlet end of the module increases. As a consequence, the permeate flux at the inlet end may become higher than the critical flux and the separation moves to zone II or III as in Figure 1.6, resulting in failure of the separation.

In summary, casein and serum protein separation performance can be improved generally by operating at a low permeate flux and high shear stress. To achieve this goal, one can select membranes with a specific design (see the next section) and/or operate the system at optimized conditions. Note that both decreasing the permeate flux and increasing the shear stress have disadvantages in practical work: a lower permeate flux increases the membrane area required, implying an increase in capital investment, while a higher shear stress requires higher flow velocity, implying greater energy consumption. Adjusting the operating temperature to improve the separation performance should be carefully considered, because of the impact on reducing the permeate flux (Equation (1.5)) and increasing the limiting permeate flux value (Equation (1.6)), by altering the viscosity, have opposing effects. In addition, temperature can affect the structure of the protein molecules, with higher temperatures leading to denaturation. Therefore, in a practical dairy process, one should select an operation temperature to not only maximize the permeate flux below the limiting permeate flux value but also to minimize the growth of microorganisms and to avoid protein denaturation.

1.4 SEPARATION OF CASEIN AND SERUM PROTEINS USING CERAMIC MEMBRANE MF

Ceramic membrane processes have been applied to the dairy industry for more than three decades. Many dairy processors, particularly in Europe, believe that ceramic membranes are the only ones that satisfy all the requirements of dairy applications, such as strong mechanical resistance that allows high recirculation velocities and a wide tolerance to pH (0.5 to 13.5) that allows cleaning in place using strong caustic and acidic solutions. A few typical ceramic membrane elements and a module are shown in Figure 1.8. The ceramic membrane element has a tubular configuration and the type of channel can be altered as necessary (Figure 1.8a). This configuration allows a high flow velocity and thus high shear. However, the membrane area/volume ratio is low, which is typically a problem with tubular membranes. To improve this, more elements are installed in a membrane module. For example, a Pall Membralox module contains up to 60 elements, which gives about 21 m^2 membrane area and 240 m^2/m^3 packing density (Figure 1.8b).

For the casein and serum protein separation from skim milk, three types of ceramic membrane processes have been applied commercially. The most

Figure 1.8 Commercial ceramic membrane elements (a) and module (b). Adapted from Pall Corporation product data sheet (Pall Corporation Catalogue, 2007).

widely used is the uniform TMP process (UTP). This process is being replaced more recently by the grade permeability process (also called gradient porosity, GP, and the Isoflux process. In the following section, the three processes are reviewed, after an initial discussion of conventional MF processing.

1.4.1 Conventional microfiltration processing

Studies on the use of conventional cross-flow ceramic MF for skim milk protein fractionation are rather limited, since it is commonly believed that the process is not practical. As illustrated by Figure 1.9a, in a conventional process the feed stream is recycled on the retentate side while the permeate is collected on the other side of the membrane (alternative process arrangements, shown in Figure 1.9, are discussed below). The TMP is much higher at the inlet and then progressively decreases to the outlet, resulting in a higher permeate flux at the inlet and lower at the outlet. Thus, the high TMP at the inlet, caused by the high pressure drop required to maintain high shear, may result in rapid fouling at the inlet end.

Saboya and Maubois (2000) reviewed conventional ceramic membrane processes and suggested that, even though the chemical and heat stability of these membranes met the requirements of the dairy industry, the rapid permeate flux decline and the selectivity change during the process were considered unsuitable for industrial scale-up.

Pouliot, Pouliot, and Britten (1996) investigated the separation performance of a conventional ceramic membrane for casein and serum protein separation from skim milk. The tests were conducted at a bench-top scale with a 0.22 μm pore size ceramic membrane (Ceraflo, Millipore) and

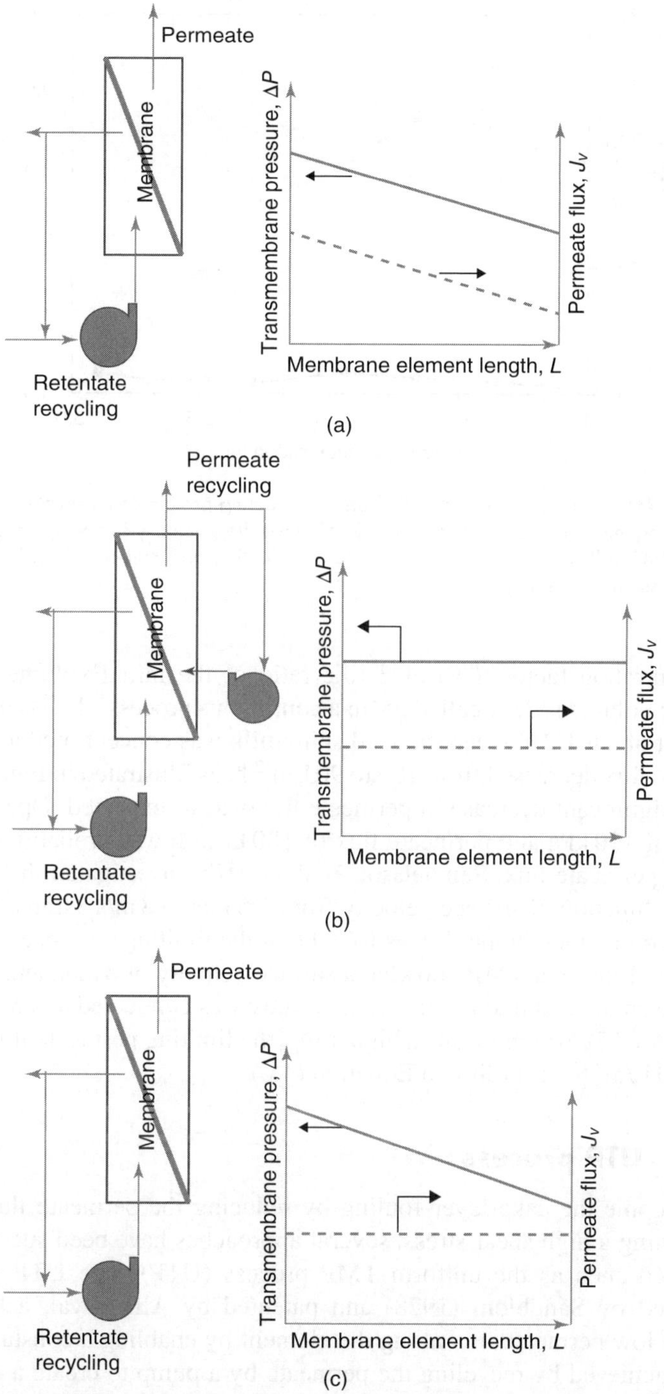

Figure 1.9 Process illustration and the changes of TMP (solid line) and J_V (dotted line) with L for (a) conventional, (b) UTP, and (c) GP processes.

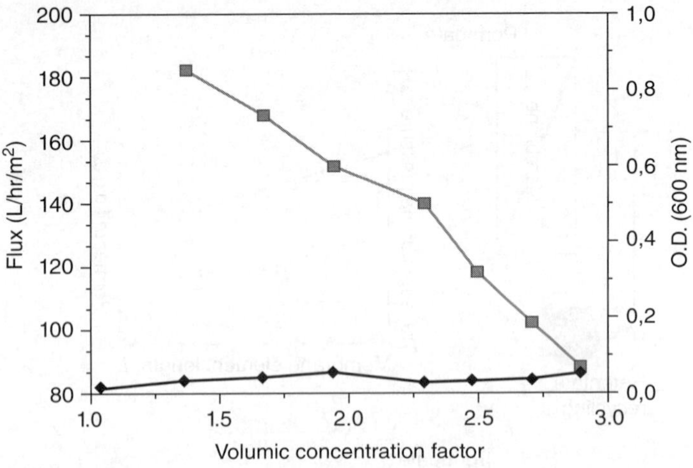

Figure 1.10 Permeate flux and absorbance (indicating protein concentration) of the permeate during microfiltration of skim milk at 50 °C, cross-flow velocity of 6.9 m/s and 190 kPa average TMP. ■ flux; ◆ absorbance (from Pouliot, Pouliot, and Britten, 1996. Reproduced with permission of Elsevier).

a concentration factor (CF) of 3 (the ratio of the initial volume to final retentate volume is 3, or called 3× in a continuous process). It was observed that after about 1.2 h, when the feed skim milk was concentrated to 3×, the permeate flux decreased from 180 to 90 L/m^2 h, as illustrated in Figure 1.10.

The significant decrease in permeate flux was as expected. Operating at a TMP of 190 kPa and permeate flux of 180 L/m^2 h was probably over the limiting permeate flux. Samuelsson et al. (1997b) investigated the limiting flux as a function of surface velocity from 1.5 m/s to 8 m/s using a conventional ceramic membrane. It was found that the limiting permeate flux was about 140 L/m^2 h at TMP 100 kPa, a surface velocity of 8 m/s, and a CF of 1.15. Given the fact that Pouliot et al.'s study was conducted at a higher CF (3 versus 1.15), which means a higher Φ_b, the limiting permeate flux is less than 140 L/m^2 h, according to Equation (1.6).

1.4.2 UTP process

To overcome the cake layer fouling by reducing the permeate flux while maintaining a high shear stress, several approaches have been successfully developed such as the uniform TMP process (UTP). The UTP process, developed by Sandblom (1978) and patented by Alfa-Laval, achieves a uniform low permeate flux along the element by enabling a constant TMP. This is achieved by recycling the permeate by a pump to create a pressure drop along the membrane on the permeate side (Figure 1.9b). As a result, the TMP is identical from the inlet to the outlet and so is the permeate flux. This design brings a substantial advantage to the operation, as the permeate flux can be kept low along the entire length of the element, while still maintaining a high surface flow velocity (or shear stress) and hence a high pressure drop along the element on the retentate side.

Vadi and Rizvi (2001) compared the performance of the UTP versus conventional processes for skim milk separation up to a CF 10 using a 0.2 μm pore size Membralox ceramic membrane; the membrane area was 0.184 m^2 at 50 °C. It was found that the conventional process resulted in a rapid membrane fouling and could not operate beyond CF 6 due to the drastic permeate flux decline. They suggested that the cake layer formed during the conventional process mode was more compact and durable (harder to erode) than in the UTP mode. Compared to the conventional mode, the UTP mode exhibited a lower initial permeate flux, but the flux decline with time was significantly slower, enabling the concentration of suspensions to higher concentration factors.

Le Berre and Daufin (1996) investigated the effect of operation conditions, such as shear stress and permeate flux, on casein and serum protein separation performance by a UTP process, using a 0.1 μm pore size Membralox membrane at 50 °C and CF 2. In this study, the experiments were conducted at various constant permeate fluxes (30–110 L/m^2 h) and wall shear stresses (40–110 Pa), which resulted in cross-flow velocities of 3.0–7.2 m/s. They found that up to a permeate flux of around 90 L/m^2 h and wall shear stress of 110 Pa, the separation showed a slow increase in fouling and a long membrane operating time (>6 h). The transmission (defined as 100% − R) of serum protein was high (70–80%) and almost all of the caseins were retained. Thus, they concluded that the performance was satisfactory only when the J_V/τ_W ratio was below a critical value of about 1.0 L/m^2 h Pa. They also suggested that the operating permeate flux could still be increased beyond 90 L/m^2 h without worsening the cake layer fouling by increasing the wall shear stress. However, the cake layer porosity would decrease since larger micelles were removed and replaced by small micelles, which resulted in a decrease of serum protein transmission. A further study was conducted by Gesan-Guiziou, Boyaval, and Daufin (1999) and a J_V/τ_W ratio of 0.95 L/m^2 h Pa was obtained. They concluded that the critical ratio characterized a transition from reversible protein deposition to irreversible deposition (Figure 1.11). This has important implications in determining the optimal operating conditions for a stable filtration process with maximum productivity and reduced membrane cleaning frequency.

Nelson and Barbano (2005) investigated the MF performance of serum protein recovery from skim milk using a UTP process at a CF 3 and larger membrane area than that of Le Berre and Daufin (1996). In this study, MF was conducted by a Membralox system with a 0.1 μm pore size and total

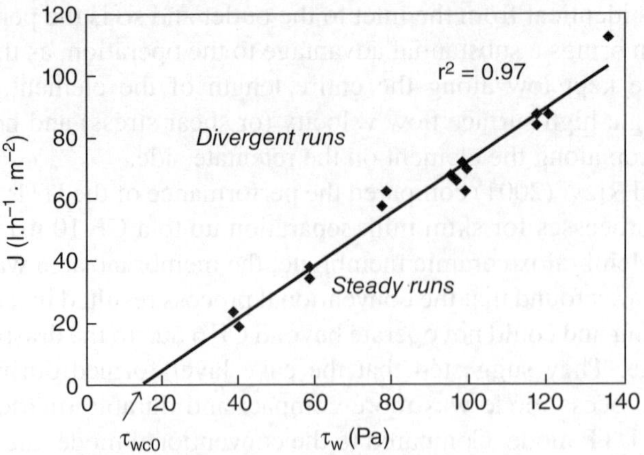

Figure 1.11 Evaluation of the critical operating ratio J/τ during MF of skimmed milk: permeation flux, J, versus wall shear stress, τ (from Gesan-Guiziou, Boyaval, and Daufin, 1999. Reproduced with permission of Elsevier).

membrane area of 1.7 m^2 with one step of concentration followed by two steps of diafiltration. The permeate flux could be maintained at 53 L/m^2 h during a 6 h processing time due to the low and uniform TMP. This permeate flux was less than that in Le Berre and Daufin (1996) as the retentate concentration was higher. About 56% of the total serum protein was removed to the permeate in the concentration step and overall 95% of the total serum protein was collected after the two stages of diafiltration.

The UTP process has been successfully commercialized and used for casein and serum protein separation in dairy processing since the late 1980s. However, there are several disadvantages associated with this approach. Firstly, since a permeate recycling pump is applied, a capital cost for the permeate loop is added to the system, and an energy consumption of the additional pump adds to the operation costs. Secondly, in the permeate side of the element, plastic beads are placed to fill the void volume. This is to increase the resistance and increase process stability by reducing pressure fluctuations. However, the beads may cause contamination to the permeate serum protein stream. Thirdly, the UTP operation is relatively complicated as the pressure drops along the element on both the permeate side and the retentate side, which should always be adjusted to match each other over the processing time.

1.4.3 GP process

Another approach to achieve low and uniform permeate flux is to modify the structure of the membranes. From Equation (1.5), it can be seen that,

Figure 1.12 A commercial GP system with three modules installed in a parallel manner. The image is provided by Pall Corporation.

for a given feed stream, the permeate flux (J_V) can be altered by changing the membrane structure, such as the porosity (A_K) and the effective thickness (Δx).

Based on this principle, Garcera and Toujas (2002) developed a graded permeability ceramic membrane called Membralox GP, which was able to obtain a uniform permeate flux along the membrane length without recycling the permeate. As illustrated by Garcera and Toujas (2002), the element incorporated a variation in the porosity of the membrane support matrix along the length of the element. For instance, along the membrane the mean porosity was progressively decreased from 30.5% at the outlet to 10.5% at the inlet, where 30% is the typical value of ceramic membranes. This porosity modification made it possible for a uniform low permeate flux along the membrane length because the high TMP at the inlet end was compensated by the low porosity, as illustrated in Equation (1.5) and presented in Figure 1.9c. Compared to the UTP mode, GP operation is much simpler and the energy consumption is lower. A commercial GP system is presented in Figure 1.12.

Zulewska, Newbold, and Barbano (2009) compared the efficiency of the GP mode versus the UTP mode on serum protein removal from skim milk. In the UTP mode, they used Membralox conventional membranes with a pore size of 0.1 μm and surface area of 1.7 m^2. Inside the permeate section of the module, polymeric beads were filled to reduce the dead volume and to act as a buffer for pressure changes. The pressure drop from the inlet to the outlet was 191 kPa but the TMP was able to be maintained at 25 kPa along the membrane length due to the UTP mode. The experiments were conducted at 50 °C with CF 3 and the permeate flux was set at 50 L/m^2 h. In the GP process, Membralox GP membranes, having the same pore size and membrane area as the conventional one, were used but a higher permeate flux (70 L/m^2 h) was achieved. The pressure drop from the outlet to the inlet was around 198 kPa and the TMP was around 225 kPa at the inlet and 13 kPa at the outlet.

During the 6 h run time, it was found that the permeate flux and TMP of the UTP process were very stable. This was because the permeate flux was controlled to be low and the shear stress was high. As a result, the process was running at a point lower than the limiting permeate flux or in the steady region, as illustrated in Figure 1.11, where the fouling is low and stable operation is possible. For the GP process, the TMP at both the inlet and outlet increased slightly over time in order to keep a constant permeate flux. The increase in TMP for the GP membranes during the run probably indicated that fouling was occurring and growing during the process, which implied that the permeate flux set point was close or even slightly higher than the limiting permeate flux. The protein transmission was also compared and it was found that serum protein transmission for the two processes were about the same (minor casein proteins were found in both permeates). After the 3× concentration, more than 60% of the total whey proteins were recovered in the permeate side for both cases.

1.4.4 Isoflux process

The third approach that is able to achieve a uniform permeate flux was developed by Grangeon et al. (2002) and was patented by Tami Industries. The so-called Isoflux ceramic membrane adopts a similar concept to the GP process by altering the membrane structure to compensate for the TMP change. For the Isoflux membrane, the effective thickness of the membrane selective layer (rather than the porosity as in the GP process) is progressively increased from the outlet to the inlet to compensate for the high TMP at the inlet end. As given in Equation (1.5), when the ratio of $\Delta P/\Delta x$ is maintained constant along the element, a uniform permeate flux is expected.

Grangeon et al. (2002) compared the water permeate flux along an element of a conventional ceramic membrane and an Isoflux membrane. In the experiment, the elements were divided into four sections. For the conventional ceramic membrane with a uniform thickness around 50 μm, as the average TMP reduced from 174 to 111 kPa from the inlet to the outlet, the permeate flux reduced from 4.5×10^3 L/m^2 h at the first section to 3.1×10^3 L/m^2 h at the last section. For the Isoflux membrane, the surface layer thickness progressively decreased from 89 to 45 μm from the inlet to the outlet. When the TMP was reduced from 174 to 111 kPa, the permeate flux was approximately constant at 3.4×10^3 L/m^2 h for all four sections.

The practical application of the Isoflux ceramic membrane in the dairy industry was reviewed by Skrzypek and Burger (2010). The most common applications included: (a) bacteria and spore reduction in milk, (b) bacteria and spore reduction in cheese brine, and (c) micellar casein separation from milk. For example, Isoflux MF membranes with a pore size 0.14 μm were used in a plant for separating casein and serum protein to manufacture micellar casein concentrates. The system contained four filtration modules with a total membrane area of 198 m^2 and was operated at CF of 2 to 6. Unfortunately, detailed operation parameters were not given in their review.

1.5 SEPARATION OF CASEIN AND SERUM PROTEINS USING POLYMERIC MEMBRANE MF

Polymers are the most popular material for the synthesis of separation membranes. Firstly, there is an abundance of polymer materials available for membrane development; secondly, polymer properties can be modified readily to meet various requirements of the separation; and, thirdly, polymer materials are generally less expensive, which suggests a lower cost of investment.

Spiral wound is the most prevalent configuration for commercial polymeric membrane MF elements in the dairy industry due to a high capability to handle a wide variety of feed streams and the high flexibility to scale-up by changing the element diameter and length. This configuration provides a greater membrane area/volume ratio than for ceramic systems, up to 1000 m^2/m^3 (Mulder, 1996d). The structure of a spiral wound element is illustrated in Figure 1.13. Basically, the element is constructed by rolling up several membrane leaves, separated by feed spacers, around a central permeate collection pipe. For each leaf (also called a membrane envelope), there are two piece of flat sheet membrane, which are separated by a permeate spacer, glued together on three sides and open to the central pipe for permeate collection.

In operation, the feed stream flows along the membrane element (parallel to the central pipe) in the feed spacers with a pressure drop from the element inlet to the outlet. Driven by the pressure difference across the membrane thickness and based on size exclusion, small particles and solvent transport through the membrane, flow in the permeate spacers, and are collected in the central pipe. In a commercial industrial system, several elements can be installed serially in one pressure vessel (housing) and a number of vessels can be mounted in series and parallel, sharing a feed pump or pumps.

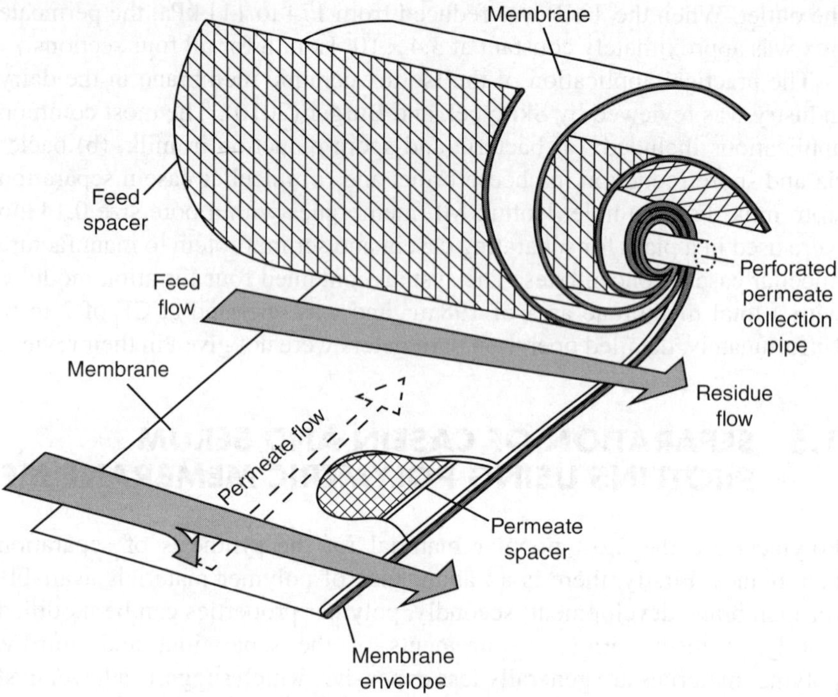

Figure 1.13 The structure of a polymeric spiral wound membrane element. (Adapted from Lin et al., 2013. Reproduced with permission of Elsevier).

A spiral wound membrane system used for dairy processing is presented in Figure 1.14.

In the dairy industry, spiral wound MF membranes are mainly used for treatment of low fouling streams, such as process water. Application on dairy streams is rather limited owing to the fact that the microporous membrane structure is easily fouled and cleaning is difficult. Jost and Jelen (1997) and Saboya and Maubois (2000) reviewed the dairy stream clarification and purification by MF membranes and suggested that the available polymeric membranes at the time, fabricated from polysulfone and polycarbonate, were not satisfactory in terms of the membrane separation performance and the chemical and mechanical stability. Recently, however, a new type of MF membrane, fabricated from PVDF, has promoted the commercialization of larger-scale polymeric membrane systems for dairy processing, by taking advantage of the high stability and low protein adsorption of this material.

The principles that are used to guide the operation of a ceramic membrane process can also be applied to a polymeric membrane process. Thus, the spiral wound polymeric membrane process for casein and serum protein

Figure 1.14 A spiral wound system used for dairy processing. © GEA Process Engineering Inc.

separation should also be operated at a high cross-flow velocity and a low and uniform permeate flux along the membrane length.

The cross-flow velocity in spiral wound polymeric membrane elements is much lower (around only one-tenth) than that in tubular ceramic membrane elements due to the configuration of the elements. The turbulence, introduced by the feed spacers, could limit the growth of the cake layer (Schwinge et al., 2004). A low permeate flux can be obtained by using a low TMP. However, the spiral wound element cannot achieve a uniform permeate flux along the membrane length, since the TMP is always higher at the inlet than that at the outlet. Thus, when installing three elements in one housing, which is a typical commercial membrane filtration system design, the relatively high TMP at the inlet may cause rapid fouling. This limits the scale-up of the spiral wound membrane system for these applications.

Lawrence et al. (2008) evaluated the performance of polymeric membranes for casein and serum protein separation, using 0.3 μm and 0.5 μm pore size PVDF membranes (Parker Hannifin Corporation) with a 1.1 mm (0.043 in) spacer. The process was operated at 50 °C and 10 °C and surface velocities from 0.31 m/s to 0.52 m/s. Similar to the conventional ceramic membrane operation, the TMP at the inlet was much higher than that at the outlet and the TMP value was the overall average ΔP along the membrane element length. It was found that at low temperatures and TMP of 150 kPa, the permeate flux decreased from 17 L/m^2 h to 6 L/m^2 h when the retentate reached CF 3 after about 7 h. The rejection of casein proteins and the transmission of serum proteins as a function of TMP were also investigated. It was found that the casein rejection increased with the TMP, from 96% to ~100%, as the TMP increased from 50 kPa to 258 kPa. This is because the cake layer density increases with the TMP. For the same reason, the serum protein transmission decreased as the TMP increased, from 22% to 1%, as the TMP increased from 50 kPa to 258 kPa. After a 1:1 volume diafiltration, about 50% of the total serum proteins was removed to the permeate side. The authors concluded that the use of polymeric membranes for the production of a casein protein concentrate and enrichment of serum protein was feasible. However, both the permeate flux and the selectivity were lower than those of ceramic membrane systems.

Beckman et al. (2010) evaluated the casein and serum protein separation performance using the same 0.3 μm pore size membrane as above, but with a larger membrane area. The experiments were conducted at 50 °C and TMP of 100 kPa. The system was operated at a bleed-in mode to a CF 3, which means that the concentration in the membrane retentate side was relatively stable, unlike the above study. It was found that the permeate flux was maintained at 15 L/m^2 h over the 2.5 h run time and 70% of all serum proteins were removed to the permeate side after two diafiltrations. The majority of casein proteins were again able to be retained.

1.6 COMPARISON OF CERAMIC MEMBRANE SYSTEM AND POLYMERIC MEMBRANE SYSTEM

The application of ceramic membranes and polymeric membranes on the casein and serum protein separation has been thoroughly investigated over the last two decades. Generally, to achieve a long and stable operating time with an acceptable separation performance, a low permeate flux and a high cross-flow surface velocity are desired. Ceramic membrane systems, such as UTP, GP, and Isoflux, are able to achieve these requirements and have been successfully applied on an industrial scale. Polymeric membrane systems appeared feasible for this application based on laboratory and pilot plant scale studies, but scale-up to an industrial scale might still have some technical limitations (Karasu et al., 2010). Industrial-scale spiral wound polymeric membrane systems normally use housings containing several elements, which creates a substantial pressure drop from the inlet to the outlet of the housing. As a consequence, to keep a low permeate flux with a low TMP at the inlet, having a sufficient area for reasonable recoveries then becomes impossible (see the discussion in Section 1.5).

The protein separation performance using the ceramic and the polymeric membrane systems was quite different. Beckman et al. (2010) found the recovery of serum protein, the protein collected in the permeate, at a 3× concentration was 38% using a 0.3 μm pore size PVDF membrane and 68% using a 0.1 μm pore size ceramic membrane with a UTP mode. This difference is because the rejection of serum protein is higher by the PVDF membrane than by the ceramic membrane, although the PVDF membrane has a much larger pore size. Similar phenomena were also observed by Zulewska, Newbold, and Barbano (2009). As discussed previously, in MF skim milk separation, it is the cake layer that dominates the separation, rather than the membrane itself, and this is controlled by the ratio of J_V/τ_W. Karasu et al. (2010) showed that the two systems provide comparable results if they are run under comparable levels of shear stress and TMP. However, as there is no equivalent approach to the UTP concept available for polymeric systems, an effective balance between high hydraulic pressure drops and low TMPs is difficult to achieve.

Despite the drawbacks in separation performance, the capital investment of polymeric membrane systems is much lower than that of ceramic membrane systems. Thus, dependent upon the requirement on the quality and cost of the final products, the polymeric membrane systems may have advantages over ceramic systems in certain applications when the high protein selectivity and high serum protein recovery are not critical.

In this section, an economic analysis of a ceramic membrane system and a polymeric membrane system for casein and serum protein separation is conducted in terms of capital investment and operation cost. Typically, the capital investment for a membrane filtration system mainly includes themembrane module, pump and control system and the CIP (clean-in-place) system (Singh and Cheryan, 1998). For the purpose of simplification, in this section we combine these three parts as the "membrane system". The operation cost is also simplified by only including a membrane replacement cost and cleaning cost. The labor cost is not included since the objective of this estimation is to compare the two processes, rather than make an estimation of a cost for protein production.

In this analysis, costs are based on a continuous multistage membrane system fractionating 20 m^3/h of skim milk in a 22 h/day and 350 day/yr operation. This assumption leaves a 2 h/day membrane cleaning cycle and 15 day/yr downtime for system maintenance. The multistage membrane system includes concentration stages, which concentrates skim milk to CF 3, and diafiltration stages, which continue the filtration until a 90% serum protein recovery is reached. The permeate flux is 71 L/m^2 h for the ceramic membrane system and 16 L/m^2 h for the polymeric membrane system, adopted from Zulewska, Newbold and Barbano (2009). Assuming a 100% retention of casein proteins, the mass balance is calculated based on serum protein

Table 1.2 Estimation of membrane area required to recover 90% serum protein from a feed stream 20 m^3/h, using a multistage MF system including concentration and diafiltration stages.

System	Ceramic membrane system	Polymeric membrane system
Concentration stages, 3× continuous process		
Permeate flow rate (m^3/h)	13.3	13.3
Membrane area (m^2)	188	834
Retentate flow rate (m^3/h)	6.7	6.7
Serum protein concentration in retentate (%)	0.91	1.11
Diafiltration stages		
Permeate flow rate (m^3/h)	27	34
Retentate flow rate (m^3/h)	6.7	6.7
Serum protein concentration in retentate (%)	0.18	0.18
Membrane area (m^2)	381	2153
Sum of membrane area (m^2)	569	2987

recovered in the permeate. The serum protein rejection is assumed to be 52% after the 3× concentration for the ceramic membrane system and is 69% for the polymeric membrane system. The detailed calculation results are given in Table 1.2. For the purpose of simplification, the density of milk is assumed to be 1 kg/L.

Based on the obtained membrane area for the two systems, the capital costs are estimated based on the assumptions below and is presented in Table 1.3.

- \$250/m^2 for a spiral wound polymeric membrane system and \$2200/m^2 for a ceramic membrane system, which are at the upper end of the typical range. A factor of 1.2 is given to include CIP and other unknown costs.
- Membrane replacement cost is \$40/m^2 for a polymeric membrane and \$1500/m^2 for a ceramic membrane.
- Life time is 1.5 yr for a polymeric membrane and 7 yr for a ceramic membrane, as given by membrane manufacturers.
- Cleaning cost is \$4/m^2yr.

In summary, the ceramic membrane system has a much higher permeate flux and serum protein transmission than the polymeric membrane system, resulting in less membrane area required. However, the capital investment for the ceramic membrane system is higher mainly due to the high cost of the membrane. Overall, for the ceramic membrane system versus the

Table 1.3 Capital investment and operation cost for the ceramic membrane system compared to the polymeric membrane system for the recovery of casein from skim milk using microfiltration systems.

	Ceramic membrane system	Polymeric membrane system
Membrane system (MM$)	1.5	1.1
Membrane replacement ($k/yr)	122	79.6
Cleaning cost ($k/yr)	2.3	11.9
Energy consumption	Medium	Low

polymeric membrane system, capital investment is about 27% higher, the membrane replacement cost is about 35% higher, and energy consumption is also higher. The ceramic system has a much lower cleaning cost due to the smaller membrane area. The serum protein obtained from the polymeric membrane system has a lower purity due to the wider pore size distribution, which may result in lower quality of the product or higher cost for potential further individual protein isolation. The polymeric membrane system requires more water for diafiltration to remove sufficient serum protein from skim milk, resulting in a more dilute permeate stream. A more dilute permeate stream increases the work load for the downstream concentration steps, either a membrane process or evaporation, when serum protein powder product is to be manufactured, suggesting an addition to the overall cost.

Nomenclature

a = constant, dimensionless
A_K = membrane porosity
b = constant, dimensionless
c = constant, dimensionless
C_i^B = bulk solution of particle i, mol/m^3
$C_{i,m}^B$ = concentration of particle i at the membrane surface, mol/m^3
C_i^P = permeate concentration of particle i, mol/m^3
dC_i/dz = concentration gradient through the boundary layer
$\mathrm{d_h}$ = hydraulic diameter of the membrane flow channel, m
D = diffusion coefficient, m^2/s, or the hydraulic diameter (tube diameter or height of a channel), m
f = Darcy friction factor for laminar or turbulent flow, dimensionless
J_{lim} = limiting permeate flux, m^3/m^2 s

J_V = volumetric permeate flux, m^3/m^2 s
k = mass transfer coefficient, m/s
L = membrane element length, m
ΔP = pressure difference across the membrane, kPa
P_{inlet} = pressures at the inlet of the membrane element on the retentate side, kPa
P_{outlet} = pressures at the inlet and outlet of the membrane element on the retentate side, kPa
P_{perm} = average pressure at the permeate side, kPa
r = particle radius, m
r_p = capillary pore radius of the membrane, m
R = apparent membrane rejection
Re = Reynolds number, dimensionless
Sc = Schmidt number, dimensionless
Sh = Sherwood number, dimensionless
TMP = transmembrane pressure, kPa
v = velocity in the flow channel, m/s
Δx = membrane effective thickness, m
z = boundary layer thickness, m

Greek letters

α = particle radius in Equation (1.6), m
μ = viscosity of the solution, kPa s
ρ = density of the feed stream, kg/m^3
Φ_b = volume fraction in the bulk
Φ_w = particle volume fraction in the boundary layer immediately above the fouling layer on the membrane surface
τ_W = membrane wall shear stress, Pa

References

Agashichev, S.P. (2006) Concentration polarization in cross-flow microfiltration under the conditions of shear-induced diffusion. *Desalination*, **200**, 346–348.

Barbe, A.M., Hogan, P.A., and Johnson, R.A. (2000) Surface morphology changes during initial usage of hydrophobic microporous polypropylene membranes. *Journal of Membrane Science*, **172**, 149–156.

Beckman, S.L., Zulewska, J., Newbold, M., and Barbano, D.M. (2010) Production efficiency of micellar casein concentrate using polymeric spiral-wound microfiltration membranes. *Journal of Dairy Science*, **93**, 4506–4517.

Belfort, G., Davis, R.H., and Zydney, A.L. (1994) The behavior of suspensions and macromolecular solutions in crossflow microfiltration. *Journal of Membrane Science*, **96**, 1–58.

Boirie, Y., Dangin, M., Gachon, P., Vasson, M.P., et al. (1997) Slow and fast dietary proteins differently modulate postprandial protein accretion. *Proceedings of the National Academy of Sciences of the United States of America*, **94**, 14930–14935.

Brans, G., Schroen, C.G.P.H., van der Sman, R.G.M., and Boom, R.M. (2004) Membrane fractionation of milk: state of the art and challenges. *Journal of Membrane Science*, **243**, 263–272.

Cheryan, M. (1998) *Ultrafiltration and Microfiltration Handbook*, Technomic Publishing Company, Inc., pp. 345–484.

Field, R.W., Wu, D., Howell, J.A., and Gupta, B.B. (1995) Critical flux concept for microfiltration fouling. *Journal of Membrane Science*, **100**, 259–272.

Fox, P.F. and McSweeney, P.L.H. (eds.) (2003) *Advanced Dairy Chemistry*, Vol. 1, *Proteins*, third edition, Kluwer Academic/Plenum Publishers, New York.

Garcera, D. and Toujas, E. (2002) Graded permeability macroporous support for crossflow filtration. US patent, US6375014.

Gesan-Guiziou, G., Boyaval, E., and Daufin, G. (1999) Critical stability conditions in crossflow microfiltration of skim milk: transition to irreversible deposition. *Journal of Membrane Science*, **158**, 211–222.

Gekas, V. and Hallstrom, B. (1990) Microfiltration membranes, cross-flow transport mechanisms and fouling studies. *Desalination*, **77**, 195–218.

Gesan, G., Daufin, G., and Merin, U. (1995) Performance of whey crossflow microfiltration during transient and stationary operating conditions. *Journal of Membrane Science*, **104**, 271–281.

Glimenius, R. (1985) Microfiltration-state of the art. *Desalination*, **53**, 363–372.

Grangeon, A., Lescoche, P., Fleischmann, T., and Ruschel, B. (2002) Cross-flow filter membrane and method of manufacturing it. US patent US6499606.

Jost, R. and Jelen, P. (1997) Cross-flow microfiltration – an extension of membrane processing of milk and whey. *Bulletin of the International Dairy Federation*, **320**, 9–15.

Karasu, K., Glennon, N., Lawrence, N.D., Stevens, et al. (2010) A comparison between ceramic and polymeric membrane systems for casein concentrate manufacture. *International Journal of Dairy Technology*, **63**, 284–289.

Kimura, S. and Sourirajan, S. (1967) Analysis of data in reverse osmosis with porous cellulose acetate membranes used. *American Institute of Chemical Engineers Journal*, **13** (3), 497–503.

Lawrence, N.D., Kentish, S.E., O'Connor, A.J., et al. (2008) Microfiltration of skim milk using polymeric membranes for casein concentrate manufacture. *Separation and Purification Technology*, **60**, 237–244.

Le Berre, O. and Daufin, G. (1996) Skim milk crossflow microfiltration performance versus permeation flux to wall shear stress ratio. *Journal of Membrane Science*, **117**, 261–270.

Lin, H., Thompson, S.M., Serbanescu-Martin, A., et al. (2013) Dehydration of natural gas using membranes. Part II: Sweep/countercurrent design and field test. *Journal of Membrane Science*, **432**, 106–114.

Maubois, J.L. (2002) Membrane microfiltration: a tool for a new approach in dairy technology. *Australian Journal of Dairy Technology*, **57**, 92–96.

Maubois, J.L. and Ollivier, G. (1997) Extraction of milk proteins, in *Food Proteins and Their Application*, (eds. S. Damodaran and A. Paraf), Marcel Dekker Inc., New York, pp. 579–595.

Millipore product catalogue, www.millipore.com/catalogue/module/c153# (12 December 2013).

Mulder, M. (1996) *Basic Principles of Membrane Technology*, second edition, Kluwer Academic Publishers, pp. 288–293 (a), pp. 72–77 (b), pp. 421 (c), and pp. 468–469 (d).

Nakao, S. and Kimura, S. (1981) Analysis of solutes rejection in ultrafiltration. *Journal of Chemical Engineering of Japan*, **14** (1), 32–37.

Nelson, B.K. and Barbano, D.M. (2005) A microfiltration process to maximize removal of serum proteins from skim milk before cheese making. *Journal of Dairy Science*, **88**, 1891–1900.

Pall Corporation Catalogue, www.pall.com/pdfs/Fuels-and-Chemicals/PIMEMBRAEN.pdf (12 December 2007).

Pouliot, M., Pouliot, Y., and Britten, M. (1996) On the conventional cross-flow microfiltration of skim milk for the production of native phosphocaseinate. *International Dairy Journal*, **6**, 105–111.

Ripperger, S. and Altmann, J. (2002) Crossflow microfiltration – state of the art. *Separation and Purification Technology*, **26**, 19–31.

Saboya, L.V. and Maubois, J.L. (2000) Current developments of microfiltration technology in the dairy industry. *Lait*, **80**, 541–553.

Samuelsson, G., Huisman, I.H., Tragardh, G., and Paulsson, M. (1997a) Predicting limiting flux of skim milk in crossflow microfiltration. *Journal of Membrane Science*, **129**, 277–281.

Samuelsson, G., Dejmek, P., Tragardh, G., and Paulsson, M. (1997b) Minimizing whey protein retention in cross-flow microfiltration of skim milk. *International Dairy Journal*, **7**, 237–242.

Sandblom, R.M. (1978) Filtering process. US patent US4105547.

Schwinge, J., Neal, P.R., Wiley, D.F., et al. (2004) Spiral wound modules and spacers. Review and analysis. *Journal of Membrane Science*, **242**, 129–153.

Singh, N. and Cheryan, M. (1998) Process design and economic analysis of a ceramic membrane system for microfiltration of corn starch hydrolysate. *Journal of Food Engineering*, **38**, 57–67.

Skrzypek, M. and Burger, M. (2010) Isoflux ceramic membranes – Practical experiences in dairy industry. *Desalination*, **250**, 1095–1100.

Song, L. (1998) Flux decline in crossflow microfiltration and ultrafiltration: mechanisms and modeling of membrane fouling. *Journal of Membrane Science*, **139**, 183–200.

Vadi, P.K. and Rizvi, S.S.H. (2001) Experimental evaluation of uniform transmembrane pressure crossflow microfiltration unit for the concentration of micellar casein from skim milk. *Journal of Membrane Science*, **189**, 69–82.

Ying, L., Kang, E.T., and Neoh, K.G. (2002) Covalent immobilization of glucose oxidase on microporous membranes prepared from poly(vinylidene fluoride) with grafted poly(acrylic acid) side chains. *Journal of Membrane Science*, **208**, 361–374.

Zeman, L.J. and Zydney, A.L. (1996) *Microfiltration and Ultrafiltration, Principles and Applications*, Marcel Dekker, Inc., p. 350.

Zhang, H., Zhong, Z., and Xing, W. (2013) Application of ceramic membranes in the treatment of oilfield-produced water: Effects of polyacrylamide and inorganic salts. *Desalination*, **309**, 84–90.

Zulewska, J., Newbold, M., and Barbano, D.M. (2009) Efficiency of serum protein removal from skim milk with ceramic and polymeric membranes at 50 °C. *Journal of Dairy Science*, **92** (4), 1361–1377.

Zsigmondy, S. and Bachmann, R. (1922) Filter and method of producing same. US patent 1,421,341.

Zydney, A.L. and Colton, C.K. (1986) Concentration polarization model for filtrate flux in cross flow microfiltration of particulate suspensions. *Chemical Engineering Communications*, **47**, 1–21.

2 Dairy stream lactose fractionation/concentration using polymeric ultrafiltration membrane

Suwattana Pruksasri

Department of Biotechnology, Faculty of Engineering and Industrial Technology, Silpakorn University, Nakhon Pathom 73000, Thailand

2.1 INTRODUCTION

Milk consists of various functional components that are beneficial to health. The fractionation of milk components would enable the utilization of milk in the manufacture of many food and functional food products such as cheese and bioactive peptides (Brans et al., 2004). Membrane separation processes have gained popularity in the dairy industry at approximately 40%, in which 10% are in milk protein standardization (Daufin et al., 2001). The worldwide market for membrane technology increased from 250 million in 2001 to 800–850 million in 2010 (Daufin et al., 2001; Lipnizki, 2010). The advanced characteristics, such as low energy consumption, easy operation, possibility to operate at room temperature, and less product deterioration, are the main advantages of the membrane separations. Membrane technology is currently well integrated in the dairy industry. The dairy applications of membrane technology include: isolation of serum proteins from whey, separation of milk fat from whole milk, reduction of bacteria and spores from skim milk (cold pasteurization), concentration of casein micelles from skim milk (pre-treatment in cheese manufacturing), and much more, as shown in Table 2.1 (Brans et al., 2004).

2.2 ULTRAFILTRATION MEMBRANE

Ultrafiltration is basically a size-exclusion pressure-driven process. It not only fractionates but also concentrates the macromolecules in the feed stream, such as colloids, microorganisms, proteins, and lipids. These large molecules are generally retained in the retentate while the small molecules, such as sugars, minerals, and vitamins, pass through the membrane into the

Membrane Processing for Dairy Ingredient Separation, First Edition.
Edited by Kang Hu and James M. Dickson.

Table 2.1 Applications of membrane processes in the dairy industry (modified from Cheryan and Alvarez, 1995; Rosenberg, 1995; Brans et al., 2004; Pouliot, 2008).

1. Alternatives to unit operations
• Centrifugation: separation and fractionation of milk fat
• Cold pasteurization: removal of bacteria and spores
• Evaporation: concentration of casein micelles from skim milk
• Electrodialysis: demineralization of whey
2. Fractionation of milk/ whey
• Isolate individual serum proteins: α-lactalbumin, β-lactoglobulin, bovine serum albumin, immunoglobulins, lactoferrin, transferrin, and peptides
3. Formulating new products
• UF cheeses, MF cheeses
• Whey-based beverages
4. Standardization of milk products
5. On-farm ultrafiltration of milk: to reduce transportation and refrigeration costs

permeate. Ultrafiltration membranes reject macromolecules with different sizes depending on the membrane pore size, which can range between 0.001 and 0.2 μm or between the molecular weight values of 1000–1 000000 daltons (Garcia et al., 1999). The driving force for transport across the membrane is the pressure differential across the membrane. Since ultrafiltration membranes allow small molecules like sugars and salts to pass through the membrane, osmotic pressures of the ultrafiltration system are small compared to the reverse osmosis. Therefore, the operating pressures of ultrafiltration processes are in the range of 2 to 10 bars; in some cases, high operating pressures up to 25 to 30 bars have been operated as well (Kulkarni, Funk, and Li, 2001). Ultrafiltration processes are widely used membrane processes for the following purposes: (1) to clarify the feed solution, (2) to concentrate the rejected solutes, and (3) to fractionate solutes.

Typically, ultrafiltration membrane materials can be classified into two groups: polymeric and inorganic membranes. The polymeric membranes are based on polysulfone, polyethersulfone, cellulose acetate, regenerated cellulose, polyamides, polyvinylidenefluoride, and polyacrylonitrile. While the inorganic membranes are based on γ-alumina/α-alumina, borosilicate glass, pyrolyzed carbon, and zirconia/stainless steel or zirconia/carbon (Kulkarni, Funk, and Li, 2001).

2.2.1 Ultrafiltration module and configuration

The selection of membrane modules in each process depends on the nature of solutes to foul or plug the membrane pores and the characteristics of the retentate stream. Membrane configurations can be divided into two main groups: (1) a flat sheet membrane, which is used in spiral-wound or plate-and-frame modules, and (2) a membrane in cylindrical geometry, which is used in a tubular or hollow-fiber membrane. However, tubular and hollow-fiber membranes are different in tube diameters. Hollow-fiber membranes have much smaller diameters than the tubular membranes. The filtration mode also has significant effects on the membrane performance. There are two general filtration modes: dead-end and cross-flow filtration. In dead-end filtration, the feed flow is perpendicular to the membrane surface while in cross-flow filtration, the feed flows tangentially across the membrane by the applied pressure. The dead-end filtration is simple but causes the deposition of particles on the membrane surface, leading to a low permeate flux. Therefore, cross-flow filtration is used to minimize this effect because the tangential flow carries the particles along the flow.

2.2.2 Ultrafiltration membrane properties

2.2.2.1 Flux

The volumetric permeate flux, J, defined as the permeate volume per filtration area per unit time, is calculated by

$$J = \frac{V_p}{At} \tag{2.1}$$

where J is the permeate flux, V_p is a collected permeate volume, A is a filtration area, and t is the filtration time.

According to Darcy's law, the convective flux through the membrane without the osmotic pressure effects is presented as the following equation:

$$J = \frac{\Delta P}{\mu_p R_m} \tag{2.2}$$

where ΔP is the pressure drop across the membrane (Pa), μ_p is the permeate viscosity (Pa s), and R_m is the resistance of flow through the membrane (m^{-1}).

If there are no concentration polarization and membrane fouling, the only resistance toward the mass transfer is the membrane resistance. However, the effect of concentration polarization and fouling is usually accounted for by the extra membrane resistance because of the deposited milk components on the membrane surface and in the pores (Makardij, Chen, and Farid, 1999). Equation (2.2), therefore, can be modified to include the effect of fouling in the resistance of flow through the membrane term

$$J = \frac{\Delta P}{\mu_p (R_m + R_f)} \tag{2.3}$$

where R_f is the resistance due to the fouling which varies with pressure, feed concentration, and cross-flow velocity at a lower transmembrane pressure. The value of R_f can be much higher than R_m so the permeate flux becomes independent of membrane permeability (Kulkarni, Funk, and Li, 2001). Increasing the applied transmembrane pressure results in an increment of fouled layer deposited on the membrane surface or within the pores. The pores may then be completely blocked so the permeate flux through the membrane is reduced.

For small solutes molecules, the osmotic pressure across the membrane should be considered. Thus, the permeate flux can be modified to

$$J = \frac{\Delta P - \Delta \pi}{\mu_p (R_m + R_f)} \tag{2.4}$$

where $\Delta \pi$ is the osmotic pressure across the membrane.

2.2.2.2 Concentration polarization

Concentration polarization is a common phenomena in all membrane processes. During the filtration process, the retained solutes accumulate near the membrane surface to a certain concentration in which the solute concentration at the surface is much higher than that of the feed solution. The accumulated concentration can increase to its saturation limit, resulting in the formation of a gel layer. According to Fick's law, the concentration build-up at the membrane surface causes back-diffusion of the solute to the bulk solution. The boundary layer of concentration polarization also causes resistance towards the mass transfer of solute from the bulk solution to the membrane surface, resulting in reduced permeate flux and reduced apparent rejection (Kulkarni, Funk, and Li, 2001). The deteriorating effect of concentration polarization can be minimized by increasing the mixing of the bulk feed solution either by increasing the feed flow rate to attain a turbulent flow or including the stirrer in the feed channel.

By applying a mass balance to the solute, mathematical terms of concentration polarization can be defined as:

Solute carried toward membrane

= solute moving away from wall by diffusion

+ solute permeating the membrane

or

$$JC = -D\frac{\mathrm{d}C}{\mathrm{d}x} + JC_p \tag{2.5}$$

where D is the solute diffusion coefficient, x is the coordinate that is used to measure boundary layer thickness, C is the solute concentration in the bulk solution, and C_p is the solute concentration in the permeate.

When the above equation is integrated over the concentration boundary layer and assuming that the solute diffusion coefficient is constant and the amount of the solute passes through the membrane is negligible, then

$$JC = -D\frac{dC}{dx} \tag{2.6}$$

Finally,

$$J(x) = \frac{D}{\delta(x)} \ln \frac{C_w}{C_b} \tag{2.7}$$

where C_w is the concentration at the membrane wall, C_b is the concentration in the bulk solution, and $\delta(x)$ is the boundary-layer thickness. When the concentration polarization is very low, the solute concentration at the membrane wall is nearly the same as the solute concentration in the bulk solution ($C_w = C_b$). This is due to the term

$$\frac{D}{\delta(x)J} \gg 1.$$

2.2.2.3 *Fouling*

Membrane fouling generally refers to the deposition of some feed components on the membrane surface or within membrane pores. In general, the feed components that cause the major fouling effects are present in low amounts (Makardij, Chen, and Farid, 1999). Fouling can not only decrease the permeate flux but can also change the chemical properties of the membrane, leading to the changes in membrane selectivity. Examples of membrane fouling are adsorption, pore blockage, precipitation, and cake formation (Koh, Ashokkumar, and Kentish, 2013). Both concentration polarization and membrane fouling have negative effects on permeate flux. These two phenomenon, however, are quite different. As mentioned earlier, concentration polarization depends on operating conditions such as transmembrane pressure, temperature, feed concentration, and feed flow velocity, but does not depend on time. In contrast, membrane fouling is partially dependent on these parameters but is more time dependent (Kulkarni, Funk, and Li, 2001).

2.2.2.4 *Rejection*

The observed rejection of components during membrane filtration is calculated for each component by

$$R_i = 1 - \frac{C_{p,i}}{C_{r,i}} \tag{2.8}$$

where $C_{p,i}$ is the concentration of component i in the permeate whereas $C_{r,i}$ is the corresponding concentration in the retentate.

In a case of a batch process, the rejection of components can be calculated from either permeate or retentate concentrations and the volume concentration factor. The rejection is then calculated by (Kulkarni, Funk, and Li, 2001)

$$R_i = \frac{\ln(C_r/C_f)}{\ln(VCF)} \tag{2.9}$$

or as

$$R_i = \frac{\ln[VCF - (C_p/C_f)(VCF - 1)]}{\ln(VCF)} \tag{2.10}$$

$$VCF = \frac{V_0}{V_r} \tag{2.11}$$

where C_r is the retentate concentration, C_f is the initial feed concentration, C_p is the permeate concentration, V_0 is the initial feed volume, and V_r is the remaining retentate volume. The volume concentration factor (VCF) is calculated according to Equation (2.11).

The rejection of components is mainly affected by many factors, such as pore diameter or the molecular weight of the membrane, chemical compositions of the membrane, the interactions between membrane surfaces, and feed materials. The operating parameters, for example, transmembrane pressure, temperature, and feed flow velocity, also affect the membrane rejection (Smith, 2013).

2.2.2.5 Selectivity

Selectivity ($S_{i,j}$) or the separation factor is used to compare the separation capacity of two components, calculated from the rejections of solutes i (R_i) and j (R_j):

$$S_{i,j} = \frac{R_i}{R_j} \tag{2.12}$$

2.2.2.6 Diafiltration

Diafiltration is an ultrafiltration membrane technique used for clarifying, removing, or lowering concentration of solutes or salts from the feed or retentate solution. It involves the addition of a solvent (usually water) to the retentate or feed in order to minimize the problem of low fluxes at high concentrations and to get a better removal of small solutes that are permeable across the membrane and through the permeate. During the diafiltration process, the concentration of solutes in the feed stream is diluted such that small molecules are easily removed through the permeate. However, the addition of water causes increased volume of the process. The energy requirement is, thus, higher in order to achieve the same degree of concentration. The addition of water can be done either by sequential concentration followed by dilution in several stages or by continuously adding water to balance the permeate being lost. When diafiltration is performed in the sequential concentration/dilution mode, the retentate concentration can be expressed as (Kulkarni, Funk, and Li, 2001)

$$C_r = C_0(VCF)^{1+n(R_i-1)} \tag{2.13}$$

where C_r is the retentate concentration, C_0 is the original concentration, VCF is the volume concentration factor calculated from Equation (2.11), n is the number of sequential diafiltration stages, and R_i is the observed solute rejection.

If diafiltration is performed by adding water at the same rate as the permeate fluxes, the retentate concentration can be expressed in terms of a dilution factor (DF) with respect to the original feed volume (Kulkarni, Funk and Li, 2001):

$$DF = \frac{\text{volume of liquid added as diluent}}{\text{initial feed volume}} \tag{2.14}$$

and

$$C_r = C_o \exp[-DF(1 - R_i)] \tag{2.15}$$

where C_r is the retentate concentration, C_0 is the original concentration, and R_i is the observed solute rejection.

2.3 ULTRAFILTRATION ON LACTOSE FRACTIONATION/CONCENTRATION

2.3.1 Lactose fractionation/lactose-free milk

Milk is a nutritionally ideal food containing high-quality proteins and calcium. The composition of cow's milk includes 87.3% water, 3.3% protein comprising of 2.6% casein and 0.7% whey protein, 4.1% milk fat, 0.7% minerals, and 4.6% lactose (de Wit, 2001). The main carbohydrate in milk is lactose, which presents at a concentration of 4.5–5%. In recent years there has been growing interest in lactose-free milk because a large number of people, approximately 70% of the world population, who suffer from lactose intolerance because of the lack of or low intestinal lactase activity (Adhikari et al., 2010). It has been found that most of Asians, Native Americans (more than 90%), Africans (80–100%), and Southern Europeans (more than 80%) are lactose intolerant (Mlichova and Rosenberg, 2006). The lactase deficiency is dependent on both ethnic and racial groups of the population. Maldigested lactose is subsequently fermented by the intestinal microflora, resulting in the production of short chain fatty acids and gas. Flatulence, bloating, abdominal pain, cramps, and diarrhea are common symptoms occurring after milk consumption (Hertzler, Huynh, and Savaiano, 1996). It has been reported that the market of lactose-free dairy products in both the United States and Europe was $900m in 2011 (Prescott, 2012). The booming market is currently the European market, with an estimated growth of 75% between 2012 and 2016, followed by the US market (Valio, 2013). The US market of lactose-free dairy products is also anticipated to increase from $500m to $650m by 2016. However, the lactose-free dairy markets in Asia and South America are still underdeveloped. Table 2.2 lists the current available lactose-free products on the market worldwide. Today, there are three major lactose-free companies, which are Arla Foods' Lactofree, Valio's Zero Lactose, and McNeil's Lactaid (Prescott, 2012). The lactose content in commercial lactose-free milk products is usually less than 0.25 g of lactose per 100 gram of milk (Adhikari et al., 2010). In 2001, Valio Ltd Launched a lactose-free milk drink containing less than 0.01% lactose but tasted like normal milk (Jelen and Tossavainen, 2003; Harju, Kallioinen, and Tossavainen, 2012). According to the EU Directive, this product is called a "milk drink" because lactose can only be removed from milk by enzymatic hydrolysis (Jelen and Tossavainen, 2003). Even though there is no common international limit for the claim "lactose-free", the consumption of 2 grams of lactose is proven to have no intolerance symptoms compared with the consumption of a lactose-free product (Hertzler, Huynh, and Savaiano, 1996; Harju, Kallioinen, and Tossavainen, 2012). It has also been reported that the consumption of lactose up to 6 grams did not affect intolerance symptoms. The lactose maldigestion and intolerance symptoms occurred after the consumption of lactose of 12 grams or higher (Hertzler, Huynh, and Savaiano, 1996). The authorities of the Scandinavian (Nordic) countries have set the limit of lactose at less than 10 mg of lactose per 100 g of milk for the claim "lactose-free" (Harju, Kallioinen, and Tossavainen, 2012).

2.3.2 Methods of lactose removal

2.3.2.1 Hydrolysis of lactose with enzymes

The hydrolysis of lactose has been studied since the mid-1970s when the first lactase enzymes were commercially available. Lactase (β-D-galactosidase, β-D-galactosidegalactohydrolase, EC 3.2.1.23) is an enzyme that hydrolyzes D-galactosyl molecules and can be isolated from natural sources such as fungi, bacteria, yeasts, plants, and animal cells (Husain, 2010; Harju, Kallioinen, and Tossavainen, 2012). The optimum pH of the enzymes is in the pH range of 2.5–7.0 and depends on enzyme

Table 2.2 Current available lactose-free products on the market worldwide.

Brand	Company, Country	Method of lactose removal	Lactose-free products	References
Liddells	Murray Goulburn Co-Operative Co. Ltd, Australia	Enzymatic	Milk, cheese, cream, yoghurt, dairy spread	http://www.liddells.com.au/
Prairie Farms	Prairie Farms Dairy, Inc., USA	n/a	Milk	http://www.prairiefarms.com/
LACTAID®	McNeil Nutritionals, USA	Enzymatic	Milk, cottage cheese, ice cream, eggnog	http://www.lactaid.com/
LAND O LAKES® Dairy Ease®	Land O' Lakes, Inc., USA	Enzymatic	Milk	http://www.landolakes.com/
H-E-B Mootopia	H-E-B Co., USA	Membrane + enzymatic	Milk	http://www.heb.com/page/healthy-primo-picks/heb-mooptopia
Smart Balance®	Boulder Brands, Inc., USA	Enzymatic	Milk	http://www.landolakes.com/
GreenValley Organics	Redwood Hill Farm, USA	n/a	Yoghurt, kefir, sour cream	http://greenvalleylactosefree.com/
Hiland	Hiland Dairy, USA	Enzymatic	Milk	http://hilanddairy.com/
Organic Meadow®	Organic Meadow Inc., Canada	n/a	Milk	http://organicmeadow.com/
ArlaLactofree	Arla Foods Ltd., Denmark	n/a	Milk, cream, cheese	http://www.arla.com/
HYLA, Eila®	Valio, Finland	Chromatographic/membrane	Milk, milk powder, cheese, butter, yoghurt	http://www.valio.com/
Meggle	RAJO, A.S., slovakia	n/a	Milk	http://www.rajo.sk
GAZi	Garmo AG, Germany	n/a	White cheese, Kashkaval semi-hard cheese	http://www.gazi.de

sources (pH 2.5–5.4 for fungal lactases and pH 6.0–7.0 for yeast and bacterial lactases) (Mlichova and Rosenberg, 2006; Husain, 2010). The hydrolysis of lactose by an enzymatic reaction is not a complicated process and does not require special equipment (Zadow, 1986). However, there are several factors affecting lactose hydrolysis, pH, reaction temperature, substrate or lactose concentration, enzyme activity, enzyme dosage, and enzyme cost, that should be considered. Enzymatic hydrolysis of lactose either from whey or milk has been studied in various strategies. The less enzyme dosage, the higher incubation time is desired to obtain the complete conversion. This is not applicable for milk since the long contact time at 35–45 °C may cause extensive microbial growth. The enzymatic hydrolysis of lactose at refrigeration temperature for an overnight reaction has also been studied by Zadow (1986) to reduce the enzyme cost (Harju, Kallioinen, and Tossavainen, 2012). Several studies have performed the enzymatic hydrolysis of lactose either before or after heat treatment (pasteurization, UHT) (Dahlqvist et al., 1977; Chen, Hsu, and Chiang, 2002; Harju, 2004; Mendoza, Olano, and Villamiel, 2005). Different procedures of lactose hydrolysis can drastically affect milk properties and sensory characteristics. The hydrolysis of lactose into glucose and galactose increases the sweetness of the milk product because glucose is much sweeter than lactose. It has also been stated that the degree of lactose hydrolysis at 70% increases the sweetness to the same degree as an addition of about 2% sucrose (Harju, Kallioinen, and Tossavainen, 2012). Furthermore, increasing or reducing monosaccharides (glucose and galactsose) causes more undesirable by-products (lactulose and furosine) from the Maillard reaction, which occurs not only in the heating process but also during the storage time (Ruiz-Matute et al., 2012). The amount of furosine in UHT lactose-hydrolyzed milk samples increases during storage at room temperature as a result of the Maillard reaction (Messia, Candigliota, and Marconi, 2007). The isomerization of glucose to fructose and of galactose to tagatose could also occur after UHT treatment (Mendoza, Olano, and Villamiel, 2005; Messia, Candigliota, and Marconi, 2007). Also, the Maillard reaction causes a considerable loss of available lysine in milk, which is an essential amino acid (Mendoza, Olano, and Villamiel, 2005). It is recommended that the UHT lactose-hydrolyzed milk or milk with high reducing sugars should be stored at 4 °C (Messia, Candigliota, and Marconi, 2007). The hydrolysis of lactose by an enzymatic reaction has some limitations in terms of organoleptic characteristics and product quality. There are several suggestions on how to avoid these limitations. For example, numerous studies have also recommended heat treatment prior to the hydrolysis of lactose. In addition, the degree of lactose hydrolysis should be between 80 and 90% in order to avoid excessive sweetness (Mendoza, Olano, and Villamiel, 2005). The sweetness of lactose-hydrolyzed milk can be reduced by an addition of potassium chloride (Flynn, Bakal, and Snyder, 1994) or potassium salt of an organic acid such as citrate, malate, gluconate, or lactate at 15-45 mmol/L (Harju, Kallioinen, and Tossavainen, 2012). The removal of lactose by β-galactosidase increases the cost of the milk product by approximately 0.06–0.08 dollars per liter (Shakeel-Ur-Rehman, 2009).

The commercial hydrolyzed lactose (HYLA®) skim milk powder was introduced in 1979 with the application of soluble β-galactosidase to hydrolyze lactose in milk in which about 80% of the milk lactose is hydrolyzed after 1 week of storage at 18–20 °C in the UHT plant. In order to achieve a complete hydrolysis of lactose, it is therefore recommended that the hydrolyzed milk should be stored at ambient temperature during its 3 months of storage (Tuure and Korpela, 2003).

2.3.2.2 Chromatographic separation

Chromatographic separation of lactose from milk has been patented since 1989 (Harju, 1989). Lactose is fractionated from milk while the salts remain in the protein fraction or in the protein/fat fraction. Harju (1989) patented a specific chromatographic separation of lactose from milk using a strong cation exchange resin having a polystyrene divinyl benzene backbone and containing sulfonic acid groups. The proteins and minerals can bind to oppositely charged resins whereas lactose cannot bind and pass through the system. The column is filled with resin in the sodium salt form. Then the resin is balanced in order to obtain a balanced ionic form that corresponds to that of milk. The operating temperature of the chromatographic separation is in the range from 50 to 80 °C to avoid bacterial growth. A temperature exceeding 80 °C is still possible in term of resin characteristics. However, whey proteins start to denature at high temperatures. This method should therefore operate at temperatures from 55 to 70 °C. Concentrated milk or skim milk is then pumped through the column. The lactose fraction is collected at the bottom of the bed while proteins and minerals are eluted with demineralized water. In this way, lactose is fractionated as a pure fraction whereas salts remain in the protein fraction. The remaining salts are further removed by ultrafiltration. However, the chromatographic process has some disadvantages in terms of the time consumed and the costly process (Shakeel-Ur-Rehman, 2009). The chromatographic separation has been combined with enzymatic hydrolysis of lactose and reduction of lactose by membrane technology in an industrial process for production of reduced lactose and lactose-free dairy products (Shakeel-Ur-Rehman, 2009; Harju, Kallioinen, and Tossavainen, 2012).

2.3.2.3 Membrane separation

Membrane filtration systems such as microfiltration, ultrafiltration, nanofiltration, and reverse osmosis are widely used to reduce and/or modify the proportion of lactose in milk and milk products. The ultrafiltration membrane is an attractive process for reducing lactose in milk. Fats and proteins are retained in the retentate while lactose, minerals, and salts can easily pass through the membrane. However, salts attached to the proteins such as calcium, magnesium, and phosphate are still retained in the retentate. It has been presented that the retention of lactose during ultrafiltration may be up to 10% (Shakeel-Ur-Rehman, 2009). There have been many researches produced on lactose removal using ultrafiltration alone or with a combination of β-galactosidase and/or nanofiltration.

In 2002, Chen, Hsu, and Chiang compared two processing conditions for manufacturing low-lactose milk containing oligosaccharides. The first method used *Aspergillus oryzae* β-galactosidase to catalyze lactose in concentrated milk into oligosaccharides directly whereas the second method was performed by separating lactose in milk from milk proteins by a hollow fiber cartridge ultrafiltration system (with a molecular weight cut-off of 50 kDa with an effective membrane area of 5 ft^2). The inlet and outlet pressures were maintained at 138 and 103 kPa, respectively, with a flow rate of 25.28 kg/min. Approximately 30% of the lactose in skim milk was still retained in the retentate. The permeate was then concentrated and the lactose in the permeate was further catalyzed by *Aspergillus oryzae* β-galactosidase into oligosaccharides. The maximum yield of total oligosaccharides at 31.1% was obtained from the reaction of 25.3% lactose in permeate with the ratio enzyme to substrate (E/S) of 6.7% at 50 °C for 3.5 hours. Finally, the hydrolysate was mixed with the retentate to obtain the low-lactose milk containing oligosaccharides. However, the final yield of total oligosaccharides in the final reconstituted milk was 22.0%, which does not differ from that in the whole milk from the first method (without the ultrafiltration (UF) system). This UF technique, however, provided a better product quality since the milk proteins and other flavor substances were not in direct contact with the enzyme itself. It has also been presented that the quantity of milk albumins and globulins decreased with the addition of β-galactosidase (Parlapanova, 1990).

2.3.3 The patented processes of lactose-free milk product

Lange (2005) patented a process for producing a lactose-free milk that does not add the sweet taste to the milk. Figure 2.1 presents an overview process for the production of a lactose-free milk invented by Lange with The Agropur Cooperative, Canada. Cow milk with the lactose content of 4.6–4.9% was standardized prior to the thermal treatment at 72.8 °C for 16 seconds (in the case of milk containing 3.25% fatty matter). The purposes of thermal treatment are to eliminate the contaminated pathogenic

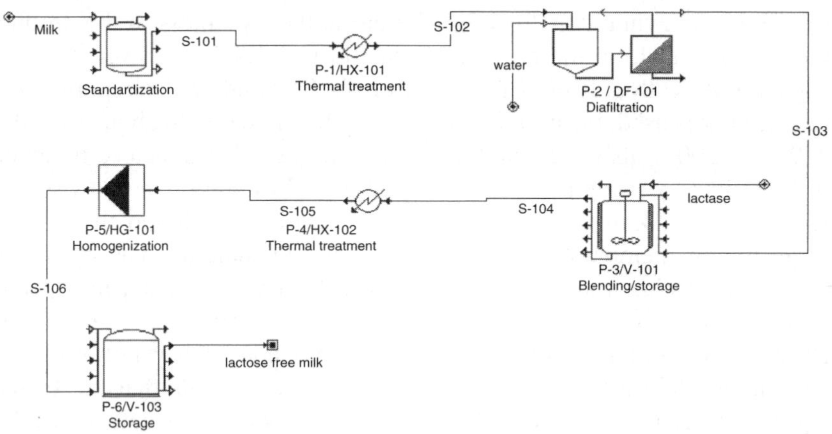

Figure 2.1 Schematic flow diagram of the process for a lactose-free milk product patented by Lange (modified from Lange, 2005).

bacteria in milk and to increase the permeation speed or the permeate flux in the next ultrafiltration step by reducing the deposited layer on the membrane surface. Next, the ultrafiltration was carried out with a Romicon™ PM50 filtration membrane with transmembrane pressure, cross-flow velocity, and operating temperature at 1.0 bar, ≥7 m/s and ≥50 °C, respectively. The permeate flux was 30 L/m^2 h. The diafiltration by adding pure water into the retentate at the same speed as the permeate flux was also included in this ultrafiltration step. During this step, lactose was reduced from 4.6 to 4.9% to approximately 3%, which results in the ratio of lactose to protein of 1:1. The author claimed that this ratio of lactose to protein with the fatty matter content at about 3.25% is the key factor for the production of lactose-free milk without any changes in sweetness. After the UF process, the UF permeates were subjected to an enzymatic hydrolysis and thermal treatment. These two processes are interchangeable. If the thermal treatment is performed prior to the enzymatic hydrolysis, the incubation temperature was <5°C. In contrast, if the enzymatic hydrolysis is performed prior to the heat treatment, the incubation temperature was <10°C, while the incubation time of these two processes were controlled at 24–48 h. The incubation temperature for the heat treatment followed by the enzymatic reaction process is decreased because the enzyme is not deactivated so it still continually hydrolyzes lactose in the milk product. To avoid the phase separation between the lipid and aqueous phase, the lactose-free milk product is homogenized to disperse all fat molecules. The recommended diameter of fat vesicles should be less than 1 μm. Finally, the final lactose-free milk may be further processed, such as by the addition of flavor. This process included the lactose removal by ultrafiltration to about

3% prior to enzymatic hydrolysis. Therefore, the sweetness of the product obtained from ultrafiltration and enzymatic hydrolysis is lower than the original milk with only enzymatic hydrolysis. The author also reported that if skim milk is used, the protein content may be increased to about 3.8–4%.

Wang (2005) also presented a patented process for a lactose-removed milk product by using a sequential membrane filtration process, as shown in Figure 2.2. Skim milk containing 5% lactose was filtered through an ultrafiltration membrane with an MWCO of 900 daltons. The operating parameters, such as transmembrane pressure, feed flow velocity, and operating temperature, were controlled at 45 psi, 22 gpm, and 10 °C, respectively. The molecules of large molecular weight compounds, such as fats, proteins, some nonprotein nitrogen (NPN) materials, and minerals that are bound to large molecules, are retained in the retentate whereas small molecular weight compounds, such as lactose, minerals, some NPN materials, and water, can pass through the membrane. The diafiltration technique was also applied. The UF permeate was then subjected to a nanofiltration system with an MWCO between 150–300 daltons. The transmembrane pressure, feed flow velocity, and operating temperature were operated at 220 psi, 4 gpm, and 10 °C, respectively. The nanofiltration permeate, which contains some minerals and water, was pumped back to the ultrafiltration retentate tank to serve as a diafiltration media. The process was continued until the lactose percentage in the ultrafiltration retentate was approximately 0.07%. This process can remove 99% of lactose when compared to the initial skim milk. However, the amount of protein in the final product at 4.44% is higher than that in the initial skim milk (3.40%), accounting for 30% more protein than the milk feedstock. Additionally, the calcium content in the lactose-removed milk product (136 mg/100 g) is higher than that of the skim milk feedstock (117 mg/100 g) at approximately 16%. This product has no sweetness while containing a natural milk flavor. As can be seen, this process requires the nanofiltration membrane to recover minerals and salts in the ultrafiltration permeate. The final compositions of this lactose-removed milk are 4.44% total nitrogen protein, 0.07% lactose, 0.31% fat, 136 mg/100 g Ca, 45 mg/100 g Na, 147 mg/100 g K, 16 mg/100 g Mg, and 110 mg/100 g P.

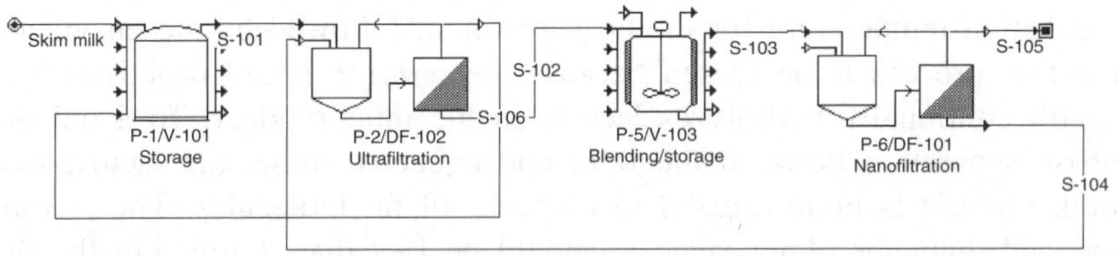

Figure 2.2 Schematic flow diagram of the process for a lactose-free milk product patented by Wang (Wang, 2005).

Whereas skim milk contains 3.4% total nitrogen protein, 5% lactose, 0.25% fat, 117 mg/100 g Ca, 49 mg/100 g Na, 151 mg/100 g K, 13 mg/100 g Mg, and 90 mg/100 g P. The lactose-removed milk still contains all minerals found in milk with a slight increase in minerals.

Tossavainen and Sahlstein (2013) patented processes for producing lactose-free milk products using ultrafiltration, nanofiltration, and reverse osmosis for concentration of milk without any effects on the organoleptic characteristics of the milk product, as presented in Figure 2.3. Milk was standardized to the desired fat and protein contents and further pasteurized at 72 °C for 15 seconds and ultrafiltered at 50 °C using GR61PP membranes with an MWCO of 20 000 daltons. Then the ultrafiltration permeate, which mainly consisted of lactose and minerals, was filtered through Nanomax-50 NF membrane (Millipore) (NaCl retention <65%) at room temperature until the retentate volume was one-fourth of the original feed volume. Lactose (about 90% of the dry matter) was retained in the NF retentate and can be further used as a low-mineral lactose fraction. The NF permeate, which consisted mainly of univalent minerals and urea at 0.1–0.3% dry matter, was further concentrated using a reverse osmosis membrane (Nanomax-95 with NaCl retention >94%, Millipore) until the reverse osmosis retentate volume was one-tenth of the nanofiltration permeate feed volume. The minerals were then concentrated in the reverse osmosis retentate. Typically, reverse osmosis retentate consists of 0.5–2% ash with 1–3% dry matter. The previous ultrafiltration retentate was mixed with reverse osmosis retentate, water, and β-D-galactosidase (HA-lactase, Denmark). The hydrolysis reaction was carried out at 10 °C for 24 hours, which depends on types of commercial enzymes. Finally, the obtained milk product contained less than 0.01% of lactose. The final composition of this lactose-free milk is 3.35% total protein, <0.01% lactose, 3% glucose and galactose, 1.5% fat, 0.79% ash, with 8.84% dry matter, while the

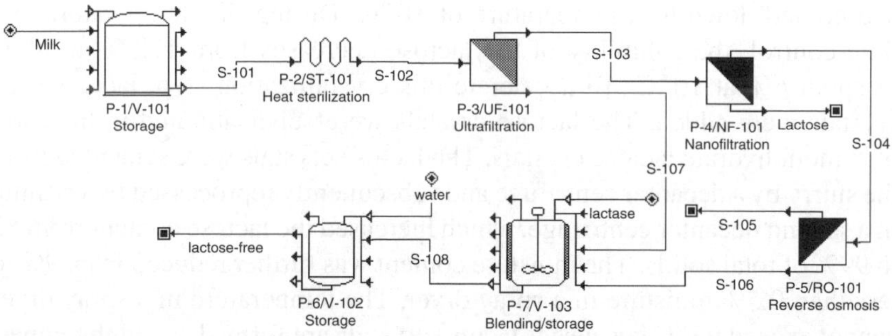

Figure 2.3 Schematic flow diagram of the process for a lactose-free milk product patented by Tossavainen and Sahlstein (modified from Tossavainen and Sahlstein, 2013).

semi-skimmed milk contains 3.3% total protein, 4.64% lactose, 1.5% fat, 0.79% ash, with 10.39% dry matter. As can be seen, this lactose-free product has a composition similar to that of ordinary semi-skimmed milk. The tastes of this lactose-free product are not different from those of semi-skimmed milk. It includes all ultrafiltration and nanofiltration systems and enzymatic hydrolysis.

2.3.4 Lactose fractionation from whey

Whey is a nutritious liquid from the cheese manufacturing process, containing whey proteins, lactose, vitamins, minerals, and growth factors. Typically, cheese whey contains 5–6% lactose, 0.8–1% protein, and trace amounts of fat (Souza et al., 2010). Generally, 1 kg of cheese is made from 10 L of milk and the remaining of 8–9 L is a whey by-product. Whey ingredients are then recovered using different techniques, such as membrane filtration, ion exchange, and the electrodialysis process. Cloidt and Lehman (2007) patented a process for the preparation of lactose from whey. Whey proteins were separated from the whey by ultrafiltration. Then the permeate containing lactose was concentrated by reverse osmosis in order to facilitate the evaporation of water and the crystallization of lactose (Souza et al., 2010). In general, there are two basic methods used to recover lactose from whey, depending on the whey composition: (1) crystallization in concentrated whey and (2) crystallization in a deproteinated whey obtained from ultrafiltration permeate (de Wit, 2001).

de Wit (2001) reported a process for lactose recovery from sweet cheese whey using the crystallization technique. The sweet whey coming from the cheese processes still contains 0.2–0.5% milk fat, depending on the cheese type, and curd fines. These can be removed by centrifugal separation, obtaining the skimmed whey with a clear yellowish color. Then the skimmed whey was pasteurized and concentrated by evaporation to obtain final solids at 60–65%. The concentrated whey was transferred to a crystallization tank and cooled down to a temperature of 10 °C. During this temperature and time control, the solubility of the lactose decreases from 17.5% at 50 °C to about 6% at 10 °C. To accelerate this crystallization step, lactose seed crystals were added. The lactose crystals were supersaturated in the form of α-monohydrate lactose crystals. The lactose crystals were separated from the slurry by a decanter centrifuge and subsequently reprocessed by washing in a second decanter centrifuge, which increased the lactose content from 80 to 99% of total solids. The moisture content was further reduced from 9% to less than 0.5% moisture in a spray dryer. The temperature in a spray dryer cannot exceed 95 °C because β-lactose crystals are formed at a high temperature, which is undesirable. Finally, the dried lactose was grounded, sieved to the desired size, and packed; this product is known as edible lactose for use in the food industry. If a high degree of purity is desired, more refining processes such as adsorption by activated carbon and crystallization are necessary to achieve high product purity for pharmaceutical grade.

Souza et al. (2010) evaluated four integrated processes including microfiltration, ultrafiltration, ion exchange, reverse osmosis, and spray drying for the recovery and purification of lactose from cheese whey. The recovery and purity of the final product were evaluated. In process 1-3, cheese whey was first centrifuged to remove suspended solids and then subjected to the first ultrafiltration system, which is hollow fiber polyethersulfone membranes with a nominal MWCO of 500 kDa. The operating pressure and temperature were controlled at 1.5 bar and 25 °C, respectively. The protein and lactose rejections of this process were 0.805 and 0.145, respectively. The lactose purity was 88.5%, which is relatively low. Then the ultrafiltration permeate was subjected to the second ultrafiltration system followed by an additional diafiltration step. The authors compared two types of ultrafiltration: (1) a spiral wound polyether sulfone membrane with an MWCO of 5–8 kDa and (2) a cassette polyethersulhone membrane with an MWCO of 5 kDa. The results indicated that the ultrafiltration membrane with the smaller MWCO (the cassette membrane at 5 kDa) tends to show a better retention for protein and a lower lactose rejection. The ultrafiltration permeate from the second ultrafiltration system was further purified by ion exchange, reverse osmosis, and spray drying. The lactose purities of these two processes were improved. Process 4, which consisted of microfiltration (a titanium oxide tubular membrane with a nominal pore diameter of 0.2 μm), ultrafiltration (MWCO of 5 kDa), ion exchange, and reverse osmosis, was shown to be the most effective process among the four processes investigated. The lactose purity of 99.8% and overall lactose recovery of 74% were obtained from this process. The authors suggested that the lost lactose in the retentate, membrane modules, and ion exchange could be further recovered to improve overall efficiency. The process for lactose removal from whey is rather simple since the solid matter in whey is lower; thus the membrane fouling of whey is much lower than that of milk.

2.3.5 Factors affecting lactose fractionation/separation

Factors affecting lactose fractionation include the operating pressure and temperature, feed flow velocity, and feed concentration.

2.3.5.1 Operating pressure

Operating pressure is an important factor in membrane fractionation and separation because it is the main driving force for the transport of molecules across the membrane. Pressure significantly affects the permeate flux. The permeate flux increases with increasing operating pressure until it reaches a certain critical transmembrane pressure. The further increase in operating pressure may cause no change in the permeate flux. At this critical transmembrane pressure, the permeate flux is independent of the operating pressure. The critical pressure may vary depending on the membrane morphology, membrane material, and type of filtering fluid (Makardij, Chen and Farid, 1999). Atra et al. (2005) studied the preconcentration of light milk by ultrafiltration using an FS10 polyvinyldifluoride membrane with an MW cut-off of 6–8 kDa. The permeate flux depended on the transmembrane pressure until the transmembrane pressure reached 2–4 bar. At that point, the protein molecules deposited on the membrane surface causing concentration polarization. It has also been reported that the type of membrane also influences the permeate flux. Membranes with larger open pore sizes have higher permeate fluxes and tend to have concentration polarization problems. As a result, the pressure independent region develops at lower transmembrane pressures (Kessler, 2002; Atra et al., 2005). The protein rejection also decreased with an increase in the transmembrane pressure (Atra et al., 2005). Limsawat and Pruksasri (2010) studied the effects of transmembrane pressure on permeate flux and rejections of lactose and protein. It has been reported that there is a linear correlation between the transmembrane pressure and permeate flux up to the transmembrane pressure at 4.5 psig for the ultrafiltration of commercial low-fat UHT milk in a cross-flow hollow fiber unit (Quixstand-benchtop system) with a molecular weight cut-off of 5000 daltons.

2.3.5.2 Operating temperature

The operating temperature mainly affects the properties of the feed stream, especially on the milk feed. An increase in the permeate flux often occurs as the temperature increased because of the decrease in milk viscosity. As the temperature increases, the mass transfer coefficient also increases. This improves the transfer of milk components from the membrane surface back into the bulk solution. However, the operating temperature should not exceed the specified temperature range of the membrane and the denaturation temperature of milk proteins. This may damage both the membrane and milk proteins permanently. The maximum operating temperature for concentration of whey by ultrafiltration is 50 °C (Atra et al., 2005). The operating temperature could be controlled to minimize the microbial growth.

2.3.5.3 Feed flow velocity

The flow rate or feed flow velocity can also affect the permeate flux and the filtration time to reach a certain protein concentration in the retentate. An increase in feed flow velocity usually results in a decrease in the concentration polarization, the deposits of protein molecules on the membrane surface. Therefore, the pressure independence region occurs at a higher transmembrane pressure (Kessler, 2002; Atra et al., 2005). At a high feed flow velocity, the deposited protein molecules are removed from the membrane surface continuously (Atra et al., 2005). Additionally, the hydraulic resistance of the fouling layer is low, resulting in the high mass transfer of solutes through the boundary layer.

Clarke and Heath (1997) compared the permeate fluxes in the flat-plate and spiral-wound membrane systems for the ultrafiltration of skim milk. Permeate fluxes increased with the increase in both transmembrane pressure and cross-flow velocity in both modules. However, the spiral-wound membrane system presented a higher permeate flux than the flat-plate system in every cross-flow velocity. The high cross-flow velocity or the increase in Reynolds number leads to a significant increase in the mass transfer coefficient or a decrease in the boundary-layer resistance, which attributed to an increase in the permeate flux.

The lactose and protein rejections at different feed flow rates of ultrafiltration of commercial low-fat UHT milk are studied in a cross-flow hollow fiber unit (Limsawat and Pruksasri, 2010). The permeate flux increases with increasing feed flow rate. The maximum lactose recovery is also obtained at the high feed flow rate of 1.72 L/min.

2.3.6 Problems occurring in the ultrafiltration of milk

Since milk consists of many components both small and large molecules, the changes in operating conditions, such as temperature, pressure and feed flow rate, will affect the final milk products. During filtration of milk, the initial permeate flux decline may be attributed to protein adsorption (James, Ying, and Chen, 2003). It was reported that hydrophobic membranes (polysulfone, polypropylene, polyvinyldienefluoride, and polytetrafluoraethylene) adsorb more proteins than hydrophilic membranes (cellulose acetate and polyacrylonitrile membranes) (Makardij, Chen, and Farid, 1999). Fouling, in some cases, may be reversible by membrane cleaning but some foulings are irreversible in which a membrane replacement is necessary. Two major foulants in the dairy industry are proteins and minerals, especially calcium phosphate (Rice et al., 2009). Proteins in milk can deposit and adsorb on to the membrane surface or within the pores. A gelatinous cake layer is formed, thus creating a flux resistance toward the flow. The protein adsorption is influenced by the hydrophobicity of the membrane and the electrostatic effects between the charged membrane and proteins. Additionally, calcium plays a major role in the protein cake formation by the protein–protein and protein–membrane bridges. The precipitation of salts also occurs when their concentration in the feed stream is much higher than their solubility, forming a scale within the pores or on the membrane surface. Calcium phosphate salt is the major salt foulant in the dairy industry because of its supersaturation in the milk aqueous phase. Its solubility decreases with an increase in temperature (Rice et al., 2009). Also, the precipitation of calcium phosphate salt increases with increasing pH. Therefore, the operations at low pH and low temperature should be designed to avoid the likelihood of calcium phosphate scaling in the dairy filtration.

Generally, caseins, the major group of proteins in milk, consist of submicelles linked together via colloidal calcium phosphate bridges (Hausmann et al., 2013). They are relatively hydrophobic, making them poorly soluble in water. Caseins are found in milk as a suspension of particles called casein micelles. Besides caseins, whey, the remaining milk proteins, consist of α-lactalbumin, β-lactoglobulin, and bovine serum albumin. It is expected that these proteins are attributed to the adsorption on the membrane surface via hydrophobic interactions. It has been reported that the typical adsorbance of the individual proteins on nonporous hydrophobic surfaces would be between 1 and 4 mg/m^2 depending on whether there are other components on the membrane surfaces (Waninge et al., 2005; Hausmann et al., 2013). It has been presented that lactose alone does not cause membrane fouling (Hausmann et al., 2013; Koh, Ashokkumar, and Kentish, 2013). Between two major proteins in milk, caseins have a tendency to deposit on the membrane surface more extensively than whey proteins. Also, the membrane fouling develops faster when salts are present in the solution.

Microbial fouling or biofouling is caused by the formation of biofilms, which are extracellular polymeric substances (EPS) on membrane surfaces. It reduces the process performance and increases the operation and maintenance costs and also the potential of product contamination by causing the flux decline (Tang et al., 2009). The produced EPS consists of polysaccharides, resulting in a high negative charge density. This may protect the bacterial cells from hydraulic shear and from chemical cleaning. The hydrophilic membranes have been reported to resist fouling better because of the reduced hydrophobic interactions between the biofoulants and the hydrophobic surface (Asatekin et al., 2007; Tang et al., 2009; Chiag et al., 2012). The availability of membrane materials with biofouling resistance is desirable. To enhance the biofouling resistance, the modification of membranes can be achieved through polymer blending of methyl cellulose, glycerine, and polyethylene glycol, etc., during membrane formation (Kim and Lee, 1998), the grafting technique, by employing hydrophilic polymers to the membrane casting solution creating hydrophilic surfaces (Asatekin et al., 2007) or the adding of antimicrobial additives such as the nanoparticles TiO_2, SiO_2, Al_2O_3, and carbon nanotubes (Chen et al., 2013). It has been reported that incorporating silver nanoparticles into polysulfone ultrafiltration membranes enhanced antimicrobial properties towards *Escherichia coli* K12, *Pseudomonas mendocina* KR1, and the MS2 bacteriophage (Zodrow et al., 2009). Tang et al. (2009) investigated the types of microorganisms involved in biofilm formations on dairy ultrafiltration membranes. It has been found that the predominant isolates recovered from the membrane surfaces in dairy plants were Gram-negative bacteria, especially coliforms and *Enterobacter sakazakii*. This may be due to the contaminants from dialysis water or a dairy plant hygiene problem. Gram-negative bacteria generally produce EPS and form biofilms, which are more resistant to biocides than those produced by Gram-positive bacteria (Tang et al., 2009). Therefore, the clean-in-place (CIP) procedures might not effectively sanitize these Gram-negative bacteria.

The operating temperature for membrane filtration in the dairy industry usually performs at a temperature that minimizes biological activities at 10 °C or higher than 50 °C. In the case of ultrafiltration of cheese whey, the permeate flux decreases with increasing temperature from 10 to 30 °C due to the decreasing solubility of calcium phosphate (Cheryan, 1998; Rice et al., 2009). However, at temperatures higher than 30 °C, the permeate flux increases because the low viscosity and high diffusivity effect are more pronounced than the negative impact of calcium phosphate precipitation (Cheryan, 1998). These operating parameters have to be considered when filtering milk through the membrane system in order to obtain the desired product.

2.4 INTEGRATED MEMBRANE BIOREACTOR

In recent years, the biocatalytic membrane reactor has received great attention because it is beneficial in the product inhibition case. In addition, the product is preconcentrated before going to further separation steps, making product recovery easy. The membrane bioreactor can be used in various processes such as fruit juice and wine clarification, syrup production, meat tenderization, production of food additives, lactose removal from milk or whey for human consumption (Giorno and Drioli, 2000; Jochems et al., 2011), the production of amino acids, antibiotics, anti-inflammatories, anticancer drugs, and vitamins (Charcosset, 2006). The biocatalysts (e.g. enzymes, cells) can be operated in two forms: (1) freely suspended in a solution and circulating on the retentate side and (2) immobilized on to the membrane surface or inside the membrane matrix itself, as illustrated in Figure 2.4.

Table 2.3 lists the advantages and disadvantages of the integrated membrane bioreactor. Each membrane bioreactor configuration has its own properties and advantages. For instance, the configuration of the

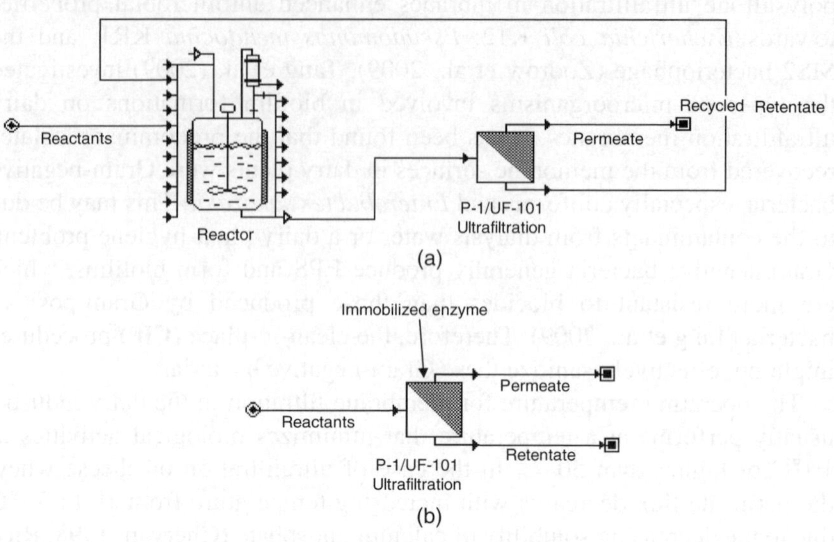

Figure 2.4 Configuration types of (a) a membrane bioreactor reactor coupled with a membrane separation unit and (b) an integrated membrane bioreactor in which the membrane serves as a catalytic and separation unit.

Table 2.3 Advantages and disadvantages of the integrated membrane bioreactor (modified from Garcia et al., 1999; Rios et al., 2004).

Advantages	**Disadvantages**
• Retention and reuse of biocatalystsvs	• Heterogeneous of reaction conditions between the bulk solution and membrane surface
• Minimize processing time	• Decrease of enzyme activity as a function of time
• Integrated process (simultaneous reaction and separation)	• Concentration polarization and membrane fouling
• Reduce substrate or product inhibition continuously	• Difficulty in maintaining aseptic conditions (in the case of the cell membrane bioreactor)
• Economical and high productivity	
• Easy to control product properties by enzyme (specificity) and/or membrane (selectivity)	

reactor coupled with a membrane separation unit is suitable for reactions where the homogeneous catalyst distribution is the key concern. The membrane is used to retain the large molecules of enzymes or cells and substrate in the retentate while allowing small molecules of product to pass through the membrane. Examples of reactions for this configuration are the hydrolysis of starch, pectin, κ-casein, and collagen (Giorno and Drioli, 2000). The membrane reactor in which enzymes or cells are immobilized in the membrane module is suitable in the case where enzymes or cells are sensitive or can be deactivated by shear stress, such as for therapeutic production. Also, the product inhibition reaction is another application of this membrane module because the product is continuously removed from the reaction. The examples of the processes for this configuration include a bioartificial pancreas, a bioartificial liver, and the extracorporeal detoxification device (Giorno and Drioli, 2000). Besides functioning as a selective separation system, membranes also act as the support for enzymes or cells. The advantages of immobilizing enzymes on the membrane are to increase the reactor stability and productivity and to improve product purity and quality. However, the loss of enzyme activity over time is the main concern.

When the enzyme is immobilized within the membrane, the substrate has to be transported across the membrane to the biocatalyst and the product also has to be transported from the reaction site through the membrane (permeate side). In general, mass transport resistance is the main influence affecting the performance of these membrane systems. The parameter that can explain the comparison between the reaction rate and the diffusion rate is the Thiele modulus, which is illustrated by the following equation (Giorno and Drioli, 2000):

$$\varphi = L\left(\frac{V_{\max}}{D_{eff}K_m}\right)^{\frac{1}{2}} \tag{2.16}$$

where L is length, $V_{\max}$ is the maximum velocity of the reaction, K_m is the Michaelis–Menten constant, and D_{eff} is the effective diffusivity. The rate of enzymatic conversion may be limited by diffusion of the substrate or reaction depending on the value of the Thiele modulus. If $\phi \gg 1$, the diffusion rate is limiting. For $\phi \ll 1$, the system is controlled by the enzyme kinetics and the mass transfer limitation is negligible.

2.4.1 Enzyme immobilization on polymeric membranes

Immobilization of biocatalysts on membranes can be divided into three types: entrapment, gelification, and attachment by chemical bonding (Giorno and Drioli, 2000). Entrapment is a process in which biocatalysts are entrapped in a polymeric membrane during membrane preparation by phase-inversion. Alternatively, cells or enzymes can be entrapped within the polymeric membranes by filtration in the reverse direction (porous support is faced up), using a low operating pressure. In this immobilization technique, enzymes or cells are inside the membrane pores, so the substrate must transport across the membrane to reach the catalyst sites and the products will be filtered through the membrane. Gelification is performed by flushing the enzyme solution through the polymeric membrane. The enzymes or cells accumulate on the membrane surface and are deposited as a gel layer. Therefore, the substrate is hydrolyzed and the product is formed on the retentate side. If the enzyme is inhibited by the product, then the reactor performance is reduced. In attachment immobilization, biocatalyst immobilization can take place using ionic binding, van der Waals interaction, and covalent binding. Most commercial polymeric membranes such as polysulfone-based membranes have negative charges on the membrane surface. Therefore, the membranes have to be coated with a high cationic polymer such as polyethyleneimine (PEI) in order to bind with negative charges of enzymes or cells. PEI has been extensively used to coat membrane surfaces for enzyme immobilization (Albayrak and Yang, 2002; Sen et al., 2011). This immobilization technique provides high process stability; however, the enzyme itself may be denatured during the binding process.

2.4.2 Application of integrated membrane bioreactor for producing functional dairy products

2.4.2.1 Hydrolysis of lactose

Table 2.4 lists examples of integrated membrane bioreactors for producing functional dairy products. As mentioned earlier, β-galactosidase or lactase is the common enzyme used in the dairy industry. The main problem of using an enzyme to remove lactose is the increased sweetness of the dairy product. Therefore, the hydrolysis of lactose in a membrane bioreactor for the simultaneous removal of monosaccharides from the reaction mixture is the main advantage of this membrane bioreactor. The other advantages of this process are continuous operation at low pressure and high enzyme concentration.

Novalin, Neuhaus, and Kulbe (2005) presented a process using a hollow fiber membrane reactor to produce lactose-reduced skim milk by combining both ultrafiltration and enzymatic hydrolysis. The *Kluyveromyces lactis* β-galactosidasewere circulated abluminally during the luminal flow of skim milk in a hollow fiber module having an MWCO of 10 kDa. When skim milk is pumped through the hollow fiber, molecules that are smaller than the MWCO (salts, lactose, and small proteins) pass through into the shell side.

Table 2.4 Examples of integrated membrane bioreactors for the hydrolysis of lactose and GOS production.

Purpose	Enzyme sources	Substrate	Type of UF	MWCO	Results	References
Lactose hydrolysis	Lactozyme 3000L HP-G *Kluyveromyces fragilis*	Lactose	Hollow fiber module	5000 Da	The conversion in the hollow fiber system is lower than the stirred tank reactor	Jurado et al., 2004
	Maxilact 2000, Maxilact LX 5000 *Kluyveromyces lactis,*	Skim milk (51% lactose)	Hollow fiber module	10 kDa	78.11% conversion at 9.9 L/h of skim milk flow rate	Novalin, Neuhaus, and Kulbe, 2005
GOS production	*Kluyveromyces lactis*	Lactose	Continuous cross flow	10 kDa	6.4 mg GOS/U/h	Chockchaisa-wasdee et al., 2005
	Maxilact 2000 *Kluyveromyces lactis*	20% lactose	Continuous membrane chromatography membrane	0.8 µm	24% GOS yield at 82% lactose conversion	Engel, Ebrahimi, and Czermak, 2008
	Maxilact LX5000	Lactose, UF permeate, whey	Ceramic UF membrane	150 kDa	32.9 g/L GOS from 203.2 g/L lactose in UF permeate	Pocedicova et al., 2010
	Maxilact 2000 *Kluyveromyces lactis*	30% lactose	Ceramic membrane reactor	20 kDa	38% w/w GOS	Ebrahimi et al., 2010
	Biolacta FN5 Bacillus circulans	Lactose	Rotating disc membrane		30% yield	Sen et al., 2011

Then lactose is enzymatically converted to glucose and galactose and a small amount of GOS. The product of the reaction is then transferred back into the tube side because of the concentration gradient. The effects of skim milk concentration, enzyme activity and flow rates were studied. It was shown that the main loss of enzyme activity in hollow fiber reactors is due to the adsorption of enzyme on the membrane surface. To avoid this loss, a shell side of the membrane surface is covered with proteins by ultrafiltration of skim milk. The lactose conversion at 519 g/L h was achieved with a skim milk flow rate at 8.46 L/h and 63.99% lactose hydrolysis. Generally, productivity increases at a higher flow rate and higher skim milk concentrations. However, high concentration of skim milk may have an adverse effect due to fat and protein adsorption, leading to the pore blockage. In addition, the process with a high substrate concentration is more susceptible to microbial contamination than that with a lower substrate concentration, which may cause loss of enzyme activity due to digestion. An UV irradiation of the enzyme solution could minimize the microbial contamination. In comparison to a batch process, this integrated membrane has longer enzyme stability and no enzyme is released into the product.

2.4.2.2 Galacto-oligosaccharide production

Recently the market of functional food ingredients has been increasing and gaining much attention. It has been estimated that the global retail market for prebiotic products is 167 000 tons annually with 390 million from the euro market (Sen et al., 2011). A prebiotic is a nondigestible food ingredient that beneficially affects the host's health by selectively stimulating the growth of beneficial bacteria in the host (Gibson and Roberfroid, 1995). Galacto-oligosaccharide (GOS), a common prebiotic in dairy industry, can be produced from lactose by the reaction of β-galactosidase. They are nondigestible carbohydrates containing two to five molecules of galactose and one molecule of glucose connected through the glycosidic linkage. GOS occurs in natural products such as breast milk, garlic, onions, honey, and soybean. Besides its prebiotic effects, GOS also has low sweetness, low caloric value (1.73 kcal/g), and low cariogenicity (Sako, Matsumoto, and Tanaka, 1999; Park and Oh, 2010). GOS also exhibits good stability over high temperatures in an acidic condition; therefore, it can be applied in various processes of dairy, confectionary, baking, and soft-drink industries.

The enzyme β-galactosidase has not only the hydrolysis function but also the transgalactosylation reaction of lactose to synthesize GOS. However, the transgalactosylation reaction of β-galactosidase is favored at high substrate (lactose) concentration. During the initial rate of reaction, a hydrolysis reaction is dominant, resulting in released glucose from the enzyme–galactosyl complex. Then the enzyme–galactosyl complex is transferred to an acceptor containing a hydroxyl group (water or other saccharides). At low lactose concentrations, water is an acceptor; therefore galactose is formed. In contrast, at high lactose concentrations, lactose molecules act as the acceptor and bind the enzyme–galactosyl complex, repeatedly forming galactosyl-oligosaccharides (Mlichova and Rosenberg, 2006). The transgalactosylation reaction of β-galactosidase from lactose is always in equilibrium, with the hydrolysis reaction producing glucose and galactose. In general, the transgalactosylation reaction competes with the hydrolysis. As a result, the GOS yield is usually lower than 50%, but it also depends on the enzyme origin, temperature, and lactose concentration (Albayrak and Yang, 2002). The key problem on the GOS production is the formation of monosaccharides (glucose and galactose), which act as enzyme inhibititors. The continuous removal of the enzyme inhibitors, could therefore improve the product yield and product purity as well as the GOS productivity. Many studies on simultaneous production and separation of GOS using enzymes from different sources and different types of membrane have been performed in the biocatalytic membrane bioreactor. Boon, Riet, and Janssen (2000) reported a GOS yield improvement of around 23% in the case of a continuous product removal using activated carbon compared to the process without product removal. The immobilization of enzyme on the membrane with a suitable pore size or MWCO helps facilitate the separation of monosaccharides or product inhibiting reactions such that the reaction is improved. Enzyme immobilization also offers many advantages such as reusability, ease of operation in downstream processing, and increased stability over temperature, pH, and operation conditions. The membrane is an excellent enzyme support because it has a high surface for enzyme loading and can be operated in a convective flow mode (Bhattacharyya et al., 1996).

Other problems occurring in the biocatalytic membrane bioreactor are the pore blocking phenomena and the loss of enzyme activity in repeated uses. Palai and Bhattacharya (2013) studied the effect of *Bacillus circulans* β-galactosidase immobilization on pore blocking by measuring the membrane hydraulic resistance. The results indicated that the use of glutaraldehyde during the enzyme immobilization and the accumulation of products on or within the membrane cause the pore blocking, resulting in a reduced flux of around 74% after the third run. The hydraulic resistances of the membranes before immobilization, after the first run, secong run, and third run were 1.07, 1.73, 3.51, and 4.68 m^{-1}, respectively. The activity of enzyme also decreased to 57% of its initial value after the second run. The maximum GOS at 28% was obtained from 200 g/L lactose and 12 kU/L of enzyme activity. These indicated that the challenges of operating the biocatalytic membrane with a high operation stability should be addressed with further research and development efforts.

2.5 FUTURE AND CHALLENGES IN SEPARATING MILK SUGAR FOR A PRODUCTION OF LOW LACTOSE MILK

Although the current technology for the production of low lactose and lactose-free milk is already widely available by using combinations of enzymatic hydrolysis and membrane filtration, the improvement of new technology in enzyme and membrane materials could enhance both nutritional and sensory quality of the dairy products. However, the technology is rather limited to Europe and the United States. The major area of lactose intolerance is Asia and South America, but the availabilities of low-lactose or lactose-free dairy products in these areas are very low. The application of the membrane process together with the production of functional dairy products by an enzymatic reaction could make the process more attractive and feasible on an industrial scale. New advances in membrane materials and process operations to minimize the membrane fouling and enable enzyme stability and functionality in the integrated membrane bioreactor will broaden the applications of membranes in the dairy industry.

Nomenclature

A = filtration area
C_b = concentration in the bulk solution
C_p = solute concentration in the permeate
$C_{p,i}$ = concentration of component i in the permeate
$C_{r,i}$ = concentration of component i in the retentate
C_w = concentration at the membrane wall
D = solute diffusion coefficient
D_{eff} = effective diffusivity
J = permeate flux
K_m = Michaelis–Menten constant
ΔP = pressure drop across the membrane
R_f = resistance due to the fouling
R_i = rejection
R_m = resistance of flow through the membrane
$S_{i,j}$ = selectivity
t = filtration time
V_0 = initial feed volume
V_p = collected permeate volume
V_r = retentate volume
V_{max} = maximum velocity of the reaction
VCF = volume concentration factor

x = coordinate that is used to measure the boundary-layer thickness
$\Delta\pi$ = osmotic pressure across the membrane
μ_p = permeate viscosity
$\delta(x)$ = boundary-layer thickness
ϕ = Thiele modulus

References

Adhikari, K., Dooley, L.M., Chambers IV,, E., et al. (2010) Sensory characteristics of commercial lactose-free milks manufactured in the United States. *LWT – Food Science and Technology*, **43**, 113–118.

Albayrak, N. and Yang, S.T. (2002) Production of galacto-oligosaccharides from lactose by *Aspergillus oryzae* β-galactosidase immobilized on cotton cloth. *Biotechnology and Bioengineering*, **77**, 8–19.

Asatekin, A., Kang, S., Elimelech, M., et al. (2007) Anti-fouling ultrafiltration membranes containing
polyacrylonitrile-*graft*-poly(ethylene oxide) comb copolymer additives. *Journal of Membrane Science*, **298**, 136–146.

Atra, R., Vatai, G., Bekassy-Molnar, E., et al. (2005) Investigation of ultra- and nanofiltration for utilization of whey protein and lactose. *Journal of Food Engineering*, **67**, 325–332.

Bhattacharyya, D., Ganapathi, S., Vishwanath, S., et al. (1996) Immobilized enzyme reactions on beads and membranes, in Biofunctional Membrane (ed. D.A. Butterfield), Platinum Press, New York, pp. 117–129.

Boon, M.A., Riet, K.V., and Janssen, A.E. (2000) Enzymatic synthesis of oligosaccharides: product removal during a kinetically controlled reaction. *Biotechnology and Bioengineering*, **70**, 411–420.

Brans, G., Schroen, C.G.P.H., Sman, R.G.M., et al. (2004) Membrane fractionation of milk: state of the art and challenges. *Journal of Membrane Science*, **243**, 263–272.

Chiag, Y-C., Chang, Y., Chen, W.-Y., et al. (2012) Biofouling resistance of ultrafiltration membranes controlled by surface self-assembled coating with PEGylated copolymers. *Langmuir*, **28**, 1399–1407.

Charcosset, C. (2006) Membrane processes in biotechnology: an overview. *Biotechnology Advances*, **24**, 482–492.

Chen, C.-S., Hsu, C.-K., and Chiang, B.-H. (2002) Optimization of the enzymatic process for manufacturing low-lactose milk containing oligosaccharides. *Process Biochemistry*, **38**, 801–808.

Chen, Y., Zhang, Y., Zhang, H.Z., et al. (2013) Biofouling control of halloysite nanotubes-decorated polyethersulfone ultrafiltration membrane modified with chitosan-silver nanoparticles. *Chemical Engineering Journal*, **228**, 12–20.

Cheryan, M. (1998) Fouling and cleaning, in *Ultrafiltration and Microfiltration Handbook*, Taylor & Francis, Lancaster, pp. 237–291.

Cheryan, M., and Alvarez, J.R. (1995) Food and beverage industry applications, in *Membrane Separation Technology. Principles and Applications* (eds. R.D. Noble and S.A. Stern), Elsevier Science, pp. 415–465.

Chockchaisawasdee, S., Athanasopoulos, V.I., Niranjan, K., et al. (2005) Synthesis of galacto-oligosaccharides from lactose using β-galactosidase from *Kluyveromyces lactis*: studies on batch and continuous UF membrane-fitted bioreactors. *Biotechnology and Bioengineering*, **89**, 434–443.

Clark, T.E. and Heath, C.A. (1997) Ultrafiltration of skim milk in flat-plate and spiral-wound modules. *Journal of Food Engineering*, **33**, 373–383.

Cloidt, R. and Lehmann, H. (2007) Process for the preparation of lactose from whey. European patent EP1869984 A1.

Dahlqvist, A., Asp, N.-G., Burvall, A., et al. (1977) Hydrolysis of lactose in milk and whey with minute amounts of lactase. *Journal of Dairy Research*, **44**, 541–548.

Daufin, G., Escudier, J.P., Carrere, H., et al. (2001) Recent and emerging applications of membrane processes in the food and dairy industry. *Transactions of the Institute of Chemical Engineers*, **79**, 89–102.

de Wit, J.N. (2001) *Lecturer's Handbook on Whey and Whey Products*, European Whey Products Association, Brussels, p. 91.

Ebrahimi, M., Placido, L., Engel, L., et al. (2010) A novel ceramic membrane reactor system for the continuous enzymatic synthesis of oligosaccharides. *Desalination*, **250**, 1105–1108.

Engel, L., Ebrahimi, M., and Czermak, P. (2008) Membrane chromatography reactor system for the continuous synthesis of galactosyl-oligosaccharides. *Desalination*, **224**, 46–51.

Flynn, R.G., Bakal, A.I., and Snyder, M.A. (1994) Method of preparing lactose-hydrolysed milk with suppressed sweetness. US patent 5,334,399.

Garcia, A.A., Bonen, M.R., Ramirez-Vick, J., et al. (1999) *Bioseparation Process Science*, Blackwell Science, Massachusetts, p. 423.

Gibson, G.R. and Roberfroid, M.B. (1995) Dietary modulation of the human colonic microbiota: introducing the concept of prebiotics. *Journal of Nutrition*, **125**, 1401–1412.

Giorno, L. and Drioli, E. (2000) Biocatalytic membrane reactors: applications and perspectives. *Trends in Biotechnology*, **18**, 339–349.

Harju, M. (2004) Chromatographic and enzymatic removal of lactose from milk. *IDF Bulletin*, **389**, 4–8.

Harju, M. (1989) Process for the specific separation of lactose from milk. US patent 4,820,348.

Harju, M., Kallioinen, H. and Tossavainen, O. (2012) Lactose hydrolysis and other conversions in dairy products: technological aspects. *International Dairy Journal*, **22**, 104–109.

Hausmann, A., Sanciolo, P., Vasiljevic, T., et al. (2013) Fouling of dairy components on hydrophobic polytetrafluoroethylene (PTFE) membranes for membrane distillation. *Journal of Membrane Science*, **442**, 149–159.

Hertzler, S.R., Huynh, B.-C.L., and Savaiano, D.A. (1996) How much lactose is low lactose? *Journal of The American Dietetic Association*, **96**, 243–246.

Hilanddairy, http://hilanddairy.com/ (9 May 2014).

Husain, Q. (2010) β-Galactosidases and their potential applications: a review. *Critical Reviews in Biotechnology*, **30**, 41–62.

James, B.J., Jing, Y., and Chen, X.D. (2003) Membrane fouling during filtration of milk – a micro structural study. *Journal of Food Engineering*, **60**, 431–437.

Jelen, P. and Tossavainen, O. (2003) Low lactose and lactose-free milk and dairy products – prospects, technologies and applications. *The Australian Journal of Dairy Technology*, **58**, 161–165.

Jochems, P., Satyawali, Y., Roy, S.V., et al. (2011) Characterization and optimization of β-galactosidase immobilization process on a mixed-matrix membrane. *Enzyme and Microbial Technology*, **49**, 580–588.

Jurado, E., Camacho, F., Luzon, G., et al. (2004) Kinetic model for lactose hydrolysis in a recirculation hollow-fibre bioreactor. *Chemical Engineering Science*, **59**, 397–405.

Kessler, H.G. (2002) *Food and Bioprocess Engineering, Dairy Technology*, Verlag A. Kessler, Munich, Germany, pp. 56–96.

Kim, J.-H., and Lee, K.-H. (1998) Effect of PEG additive on membrane formation by phase inversion. *Journal of Membrane Science*, **138**, 153, http://hilanddairy.com/63.

Koh, L.L.A., Ashokkumar, M., and Kentish, S.E. (2013) Membrane fouling, cleaning and disinfection, in Membrane Processing: Dairy and Beverage Applications (ed. A.Y. Tamime), Blackwell Publishing, West Sussex, pp. 73–106.

Kulkarni, S.S., Funk, E.W., and Li, N.N. (2001) Introduction and definitions, in Membrane Handbook (eds. W.S. Winston Ho and K.K. Sirkar), Kluwer Academic Publishers, Massachusetts, pp. 393–397.

Lange, M. (2005) Process for making a lactose-free milk and milk so processed. US Patent 6,881,428 B2.
Limsawat, P. and Pruksasri, S. (2010) Separation of lactose from milk by ultrafiltration. *Asian Journal of Food and Agro-Industry*, **3**, 236–243.
Lipnizki, F. (2010) Cross-flow membrane applications in the food industry, in Membrane Technology, Vol. **3**, *Membranes for Food Applications* (eds. K.V. Peinemann, S.P. Nunes, and L. Giorno), Wiley-VCH, Weinheim, pp. 1–24.
Makardij, A., Chen, X.D., and Farid, M.M. (1999) Microfiltration and ultrafiltration of milk: some aspects of fouling and cleaning. *Transactions of the Institute of Chemical Engineers*, **77**, 107–113.
Mendoza, M.R., Olano, A., and Villamiel, M. (2005) Chemical indicators of heat treatment in fortified and special milks. *Journal of Agricultural and Food Chemistry*, **53**, 2995, http://hilanddairy.com/ 9.
Messia, M.C., Candigliota, T., and Marconi, E. (2007) Assessment of quality and technological characterization of lactose-hydrolyzed milk. *Food Chemistry*, **104**, 910–917.
Mlichova, Z. and Rosenberg, M. (2006) Current trends of β-galactosidase application in food technology. *Journal of Food and Nutrition Research*, **45**, 47–54.
Novalin, S., Neuhaus, W., and Kulbe, K.D. (2005) A new innovative process to produce lactose-reduced skim milk. *Journal of Biotechnology*, **119**, 212–218.
Organicmeadow, http://www.organicmeadow.com (9 May 2014).
Palai, T. and Bhattacharya, P.K. (2013) Kinetics of lactose conversion to galacto-oligosaccharides by β-galactosidase immobilized on PVDF membrane. *Journal of Bioscience and Bioengineering*, **115**, 668–673.
Park, A.R. and Oh, D.K. (2010) Galacto-oligosaccharide production using microbial beta-galactosidase: current state and perspectives. *Applied Microbiology and Biotechnology*, **85**, 1279–1286.
Parlapanova, M. (1990) The effect of different technological treatment upon the quality of milk protein. *Nahrung*, **34**, 953–958.
Pocedicova, K., Curda, L., Misun, D., et al. (2010) Preparation of galacto-oligosaccharides using membrane reactor. *Journal of Food Engineering*, **99**, 479–484.
Pouliot, Y. (2008) Membrane processes in dairy technology – From a simple idea to worldwide panacea. *International Dairy Journal*, **18**, 735–740.
Prescott, R. (2012) Lactose-free dairy market is booming, says new report. Available from: www.foodbev.com/news/lactose-free-dairy-market-is-booming-say (18 August 2013).
Rosenberg, M. (1995) Current and future applications for membrane processes in the dairy industry. *Trends in Food Science and Technology*, **6**, 12–19.
Rice, G., Barber, A., Connor, A.O., et al. (2009) Fouling of NF membranes by dairy ultrafiltration permeates. *Journal of Membrane Science*, **330**, 117–126.
Rios, G.M., Belleville, M.P., Paolucci, D., et al. (2004) Progress in enzymatic membrane reactors – a review. *Journal of Membrane Science*, **242**, 189–196.
Ruiz-Matute, A.I., Corzo-Martinez, M., Montilla, A., et al. (2012) Presence of mono-, di- and galactooligosaccharides in commercial lactose-free UHT dairy products. *Journal of Food Composition and Analysis*, **28**, 164–169.
Sako, T., Matsumoto, K., and Tanaka, R. (1999) Recent progress on research and applications of non-digestible galacto-oligosaccharides. *International Dairy Journal*, **9**, 69–80.
Sen, D., Sarkar, A., Gosling, A., et al. (2011) Feasibility study of enzyme immobilization on polymeric membrane: a case study with enzymatically galacto-oligosaccharides production from lactose. *Journal of Membrane Science*, **378**, 471–478.
Shakeel-Ur-Rehman (2009) Reduced lactose and lactose-free dairy products, in Advanced Dairy Chemistry, Vol. **3**, *Lactose, Water, Salts and Minor Constituents*, third edition (eds. P.L.H. McSweeney and P.F. Fox), Springer Science and Business Media, New York, pp. 98–104.
Smith, K. (2013) Development of membrane processes, in Membrane Processing: Dairy and Beverage Applications (ed A.Y. Tamime), Blackwell Publishing, West Sussex, pp. 1–16.

Souza, R.R., Bergamasco, R., Costa, S.C., et al. (2010) Recovery and purification of lactose from whey.
Chemical Engineering and Processing: Process Intensification, **49**, 1137–1143.

Tang, X., Flint, S.H., Brooks, J.D., et al. (2009) Factors affecting the attachment of micro-organisms isolated from ultrafiltration and reverse osmosis membranes in dairy processing plants. *Journal of Applied Microbiology*, **107**, 443–451.

Tossavainen, O. and Sahlstein, J. (2013) Lactose-free milk product and processes for producing the same. US patent 8,449,938B2.

Tuure, T. and Korpela, R. (2003) Lactose intolerance and low-lactose dairy products, in Handbook of Functional Dairy Products (eds. C. Shortt and J. O'Brien), CRC Press LLC, Florida, pp. 71–90.

Valio (2013). Available from: http://www.valio.com/solutions/valio-zero-lactose/ (18 August 2013).

Wang, J. (2005) Lactose-removed milk product and process for the preparation thereof. US patent 2005/0196508 A1.

Waninge, R., Walstra, P., Bastiaans, J., et al. (2005) Competitive adsorption between β-casein or β-lactoglobulin and model milk membrane lipids a toil-water interfaces. *Journal of Agricultural and Food Chemistry*, **53**, 716–724.

www.arla.com/ (9 May 2014).

www.lactaid.com/ (18August 2013).

www.landolakes.com/ (19 August 2013).

www.liddells.com.au/ (19August 2013).

www.prairiefarms.com/ (19 August 2013).

www.rajo.sk/ (9 May 2014).

www.gazi.de/ (9 May 2014).

www.greenvalleylactosefree.com/ (9 May 2014d).

Zadow, J.G. (1986) Lactose hydrolysed dairy products. *Food Technology in Australia*, **38**, 460–462.

Zodrow, K., Brunet, L., Mahendra, S., et al. (2009) Polysulfone ultrafiltration membranes impregnated with silver nanoparticles show improved biofouling resistance and virus removal. *Water Research*, **43**, 715–723.

3 Membrane fouling: a challenge during dairy ultrafiltration

Dharmesh Kanani

Natrix Separations Inc., Burlington, Ontario, Canada

3.1 DAIRY ULTRAFILTRATION

Membrane technology has found many applications in the dairy industry. Ultrafiltration (UF) is a very popular operation in whey processing and cheese making. It is also used in the production of skim milk. In almost all the dairy applications, UF membranes are used to remove water, lactose, and mineral salt while retaining proteins. The pore size of the UF membrane is selected based on the size of the molecules being separated. The key proteins in the milk are caseins (α, β, γ, κ), α-lactalbumin, and β-lactoglobulin and therefore the use of UF membrane having MWCO greater than 10 kDA would result in increased transmission or loss of valuable milk proteins. On the other hand, a UF membrane having a pore size rating less than 5 kDa would reduce the efficiency of the UF process due to a significant reduction in lactose transmission. Therefore UF membrane having a pore size rating of 10 kDa is quite popular in the dairy industry. Normally, milk and whey UF processes are operated at 2–4 atm pressure in a cross-flow mode, yielding on average 100 LMH (l/m^2 h) flux (Zeman and Zydney, 1996). In the dairy industry, UF systems are typically operated at either >50 °C or <20 °C to avoid bacterial contamination. From an economic point of view, high-temperature operation is favored in comparison to low-temperature operation. High-temperature operation gives higher flux and thereby reduces the requirement of a total membrane area. Besides, lower viscosity at a higher temperature reduces pumping costs and overall a 40% saving in annual cost has been reported for high-temperature UF operation (Rautenbach and Albrecht, 1989). Synthetic polymer-based UF membranes such as polysulfone (PS) and polyethersulfone (PES) membranes are more popular in comparison to cellulose acetate membranes.

Membrane Processing for Dairy Ingredient Separation, First Edition.
Edited by Kang Hu and James M. Dickson.

PS and PES membranes offer superior thermal stability, which facilitates high-temperature operation and better chemical stability to address the stronger cleaning requirements of the dairy industry. They are available in an appropriate pore size and withstand required operating pressures. They can better tolerate extreme pH and chlorine and are amicable to acidic and alkali solutions used for cleaning as well as hypochlorite solutions used as a disinfecting agent. There has been some interest in ceramic membranes due to their very high thermal and chemical stability but they are at least 3 to 6 times more expensive then polymeric membranes.

3.2 FLUX–PRESSURE RELATIONSHIP

UF is a size-based separation process and its performance is typically measured through selectivity and flux. The selectivity of the UF process provides an estimate of the actual separation and the flux gives an idea about the efficiency of the process. For any membrane processes, the flux (J_v) is typically defined as volumetric flow rate of the permeate or filtrate per unit membrane surface area. It characterizes the throughput of membrane process: higher the flux, better the throughput and therefore smaller is the requirement of membrane area. The flux in the UF process depends on the membrane properties, solvent properties, transmembrane pressure, solute concentration, and hydrodynamics in the membrane module. For a typical UF membrane, pure solvent or water flux increases linearly with applied transmembrane pressure and is significantly higher than the actual process flux at moderate to high pressure. In fact, the actual process flux of any UF process shows two distinct pressure areas, which are shown schematically in Figure 3.1.

At a relatively low pressure, the process flux increases with increasing pressure but soon becomes almost independent of pressure at higher pressures. In this high-pressure area, the flux increases only by changing other process parameters, such as the cross-flow rate (or system hydrodynamics), feed concentration, etc. This asymptotic pressure–flux relationship is the direct result of concentration polarization. Concentration polarization is the build-up of retained solute molecules in the solution near the membrane surface. The build-up of retained solutes near the membrane surface not only presents the extra hydraulic resistance to flow but also reduces the effective transmembrane pressure by offering osmotic pressure. The influence of the concentration polarization layer is minimal to moderate in a pressure-controlled area (see Figure 3.1), which is typically characterized by either low pressure, low feed concentration, or high cross-flow rate. At higher pressure, the flux–pressure relationship deviates from linearity due to the heavy influence of the polarized layer of the solute near the membrane surface and flux is mainly controlled by the back mass transport

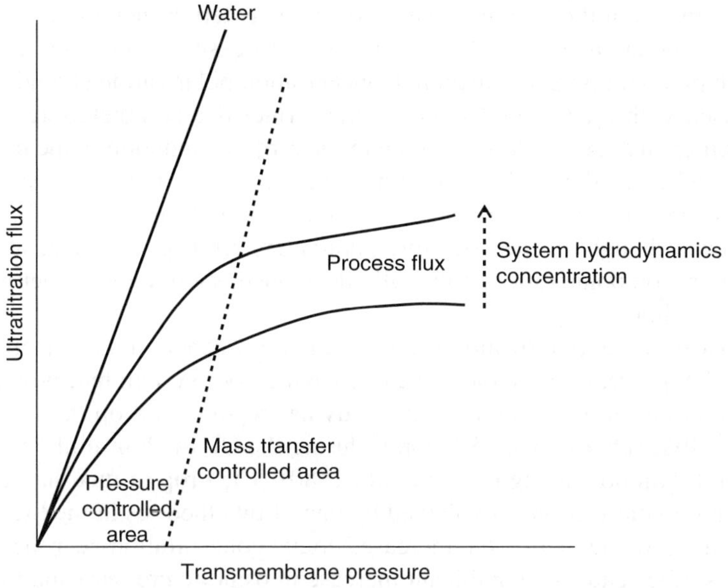

Figure 3.1 Typical correlation between UF flux and transmembrane pressure, indicating a pressure and mass transfer controlled area (Cheryan, 1998).

rate of accumulated solute molecules near the membrane surface. The pressure–flux relationship is further complicated by the presence of membrane fouling. Membrane fouling reduces the flux with time even though all the operating parameters including pressure and system hydrodynamics are constant. Both concentration polarization and membrane fouling are discussed further in more detail in the sections below.

3.3 CONCENTRATION POLARIZATION

Concentration polarization and membrane fouling are two major unavoidable phenomena associated with any commercial UF process. If the existence of a hydrodynamic boundary layer next to the membrane is assumed, which is parallel to the membrane surface, then concentration polarization layer is nothing but the concentration gradient of retained solute molecules in this hydrodynamic boundary layer. The transport of solute molecules into the boundary layer toward the membrane takes place by convection while their backtransport away from the boundary layer towards the bulk feed takes place by diffusion. Depending on the type of retained solute molecules, this layer could be fairly viscous and gelatinous, which would provide additional resistance to the filtrate flow.

Many studies in the literature have used various terms such as "gel layer", "cake", "polarization layer", or "CP layer" to describe the same concentration polarization phenomenon. Concentration polarization should not be confused with membrane fouling as the former is a completely reversible phenomenon arising solely from the mass transfer limitation in the module. On the other hand, membrane fouling is irreversible and requires some type of mechanical or chemical cleaning. Membrane fouling reduces flux with time even though all the operating parameters are kept constant, whereas concentration polarization, once fully developed, should not further reduce flux over time.

Concentration polarization can have a major impact on the performance of the UF process, particularly the one having poor system hydrodynamics or mass transfer. For example, one study has reported a drop in flux to 5% of its initial value in just 25 seconds during the UF of skim milk in a dead end cell with no stirring (Bruin et al., 1980). The drop in flux due to concentration polarization is explained by one of two theories in the literature. According to one theory, the increased solute concentration near the membrane surface causes a significant increase in osmotic pressure and thereby reduces the effective transmembrane pressure and flux. The alternate theory for lower flux in a concentration polarized limited system takes into account the additional hydrodynamic resistance of the boundary layer. This polarized layer is assumed to be dynamic in nature and any change in operating conditions, such as lowering pressure, increasing cross-flow rate, or reducing feed concentration, should increase flux by reducing the effect of concentration polarization.

3.4 MEMBRANE FOULING

Fouling is the major challenge for any membrane process. Membrane fouling refers to the adsorption and deposition of solute molecules on or inside the membrane. It increases the membrane resistance during the process, even when all the operating parameters, such as pressure, flow rate, temperature, and feed concentration, are kept constant. The increase in membrane resistance caused by fouling is irreversible in nature and requires some kind of mechanical or chemical cleaning. In a constant pressure membrane process, the flux decreases with time due to membrane fouling while in a constant flux membrane process, fouling results in increased transmembrane pressure in order to sustain a particular flux through the membrane (see Figure 3.2). Fouling reduces the productivity of the process, shortens the membrane life (often due to aggressive cleaning agents), and impairs the separation capability of the membrane.

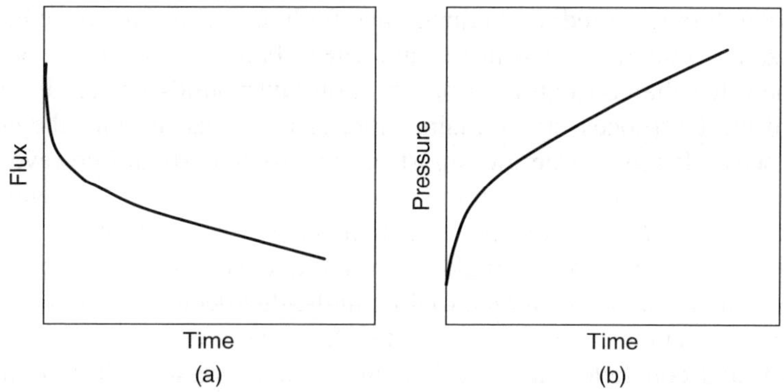

Figure 3.2 Loss of productivity due to membrane fouling in (a) constant pressure UF and (b) constant flux UF.

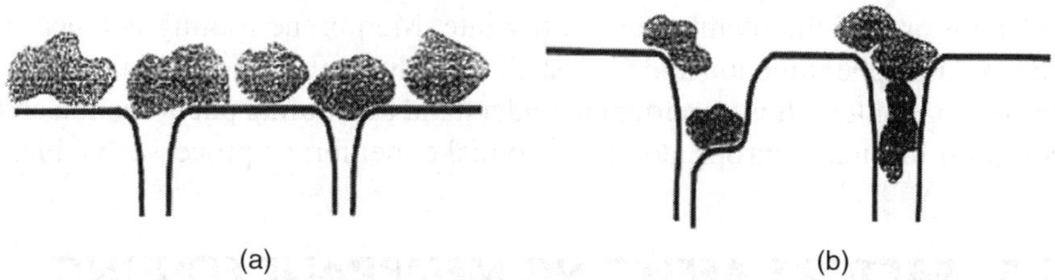

Figure 3.3 Mechanisms of membrane fouling: (a) external fouling and (b) internal fouling.

Figure 3.3 shows the different fouling mechanisms, which can be classified into two categories: (1) internal fouling and (2) external fouling. Internal fouling refers to the adsorption or deposition of solute molecules within the pores of the membrane. The three possible mechanisms by which membrane can be fouled internally are direct interception, inertial impaction, and diffusional transport of solutes to the pore walls (Ghosh, 2003). External membrane fouling refers to the adsorption/deposition of solutes on the external membrane surface. External fouling hinders the transmission of solutes by either increasing the effective membrane thickness or completely and/or partially blocking the pore entrance. In external fouling, solute molecules are transported to the membrane surface for adsorption/deposition either by diffusion or convection and sieving.

In constant pressure UF, flux decreases with time due to both concentration polarization and membrane fouling. At high transmembrane pressure, the flux decreases quite rapidly at the start of the process followed by a more gradual decline in flux. Many literature studies have attributed the initial rapid decline in flux to concentration polarization whereas the more gradual decline in flux was credited to membrane fouling at constant concentration polarization conditions (Aimar et al., 1988; Fane, 1983). These studies assumed that the concentration polarization layer builds up rapidly at the start of the UF process and remains more or less constant until the end of the process. If this is true then most of the flux loss should be reversible during a pure water flux test at the end of the process, since concentration polarization is a reversible phenomenon. However, experiments suggest that flux can be only restored after substantial cleaning, a clear indication of the greater role of membrane fouling in the flux decline process.

Kanani and Ghosh (2007) have studied flux loss in constant pressure protein UF and concluded that flux loss in constant pressure UF is dynamic in nature. Both concentration polarization and the membrane fouling rate decrease with declining flux due to a reduction in the convective transport of solute toward the membrane. Thus, flux loss in constant pressure UF is almost self-attenuating in nature as declining flux reduces concentration polarization and the membrane fouling rate. Membrane fouling is a nearly unavoidable phenomenon and almost all the solute molecules foul the membrane more or less. It is important to understand the fouling phenomenon and control it through appropriate means to make membrane processes viable.

3.5 FACTORS AFFECTING MEMBRANE FOULING

Membrane fouling is the direct result of specific interactions between the membrane and various components of the feed solution. Different components of the feed streams would interact with the membrane differently under different operating conditions and therefore it is quite difficult to make a universally applicable theory or rules about the nature and extent of membrane fouling. In general, the factors affecting membrane fouling can be grouped into three categories: (1) feed properties such as physicochemical characteristics of feed components, concentration, salt, and pH; (2) membrane properties such as hydrophobicity, surface charge of the membrane, pore size, porosity, and pore size distribution; and (3) operating parameters such as transmembrane pressure, cross-flow velocity, and temperature (Cheryan, 1998; Marshall, Munro, and Tragardh, 1993). It should be noted that different factors from these three categories can interact with each other and would give different effects in combination if they were studied individually or with model systems.

3.5.1 Feed properties

A good understanding of physicochemical characteristics of individual feed components is very helpful in optimizing membrane performance, as different components will foul membrane to different extents depending on their chemical nature and solution conditions. Dairy fluid is quite a complex feed containing a variety of materials including proteins, minerals, lactose, and fat. Among these, proteins and minerals are considered the major foulant during dairy UF (Zeman and Zydney, 1996; Cheryan, 1998). Proteins make up a large proportion of the foulant layer during dairy UF, which is not surprising considering their complex secondary and tertiary structure, multiplicity of functional groups, and varying degree of hydrophobicity and charge density. Proteins can interact with the membrane or other feed components through various types of interactions, such as electrostatic, hydrophobic, disulfide bonds, etc., and thereby act as a major source of membrane fouling (Zeman and Zydney, 1996; Kelly and Zydney, 1995).

Tong, Barbano, and Rudan (1988) have found α-lactalbumin and β-lactalglobulin as the major foulants during milk UF. In another study, they reported casein proteolysis products and α-lactalbulin as the major foulant during sweet whey UF (Tong, Barbano, and Jordan, 1989). It has been found that skim milk fouls less than whey despite having higher levels of foulants such as proteins and calcium. This may be due to the presence of large casein micelles in skim milk containing almost two-thirds of the protein and a majority of free calcium (Cheryan, 1998). The micelles would have a relatively lower propensity of membrane fouling due to their bigger size and would simply roll off the membrane surface under high shear forces typically present in the commercial module.

3.5.1.1 Concentration

Generally, permeate flux decreases as feed concentration increases during UF. Many studies have reported a flux reduction with the logarithm of concentration, as predicted by concentration polarization theory (Marshall, Munro, and Tragardh, 1993). Daufin et al. (1991a) have reported the effect of concentration on flux decline in terms of reversible (the one that can be removed by just water flushing) and irreversible fouling. They reported almost no effect of concentration on irreversible fouling whereas reversible fouling increased with concentration. This suggests that concentration has little effect on membrane fouling and flux decline is solely due to concentration polarization. Additionally, retention characteristics of the membrane do not change with concentration in spite of decreasing flux.

3.5.1.2 Salt and pH

Salt and pH have a significant influence on protein–protein and protein–membrane interaction and thus membrane fouling. Protein conformation and stability depend on pH and salt or ionic strength of the solution and any change in pH or salt concentration would change the tendency of protein to deposit on the membrane. Salt and pH can also change the porosity of the protein deposit and thus the fouling rate as the size of proteins also depends on solution pH and conductivity. Calcium and phosphate are considered major foulants in dairy streams due to their tendency to form insoluble salts. Their ionic form could also act as a catalyst or bridging agents in protein deposition on the membrane (Zeman and Zydney, 1996; Cheryan, 1998).

Muller, Hayes, and Griffin (1973) and Hayes et al. (1974) noted minimum flux around a pH of 4.5–5 during UF of hydrochloric acid casein whey. Daufin et al. (1989) reported a higher initial flux at pH 6.9 compared to pH 5.6 during UF of defatted whey protein concentrate. However, the flux drop was quite high at pH 6.9 and the final flux was lower than that observed at pH 5.6. The difference in fouling rate at different pH values was attributed to increased protein fouling on the surface and increased calcium phosphate fouling both on the surface and within the pores. Additionally, phosphate changes its form from sodium hydrogen phosphate to insoluble precipitate, hydroxyapatite, at pH 6.9. Renner and Abd El-Slam (1991) found that addition of NaCl increases transmission of β-lactoglobulin and α-lactalbumin during UF of whole milk. The level of calcium in the whey has a significant influence on membrane fouling and associated flux decline. Hayes et al. (1974) have found that addition of calcium to cheese whey reduced the flux at pH 6.2 whereas Patocka and Jelen (1987) have attributed the higher flux during UF of pretreated cottage cheese whey to removal of free calcium during the pretreatment.

3.5.2 Membrane properties

It is quite obvious that significant influence can be expected of membrane properties on the performance of the UF process. The influence of key membrane properties such as hydrophobicity, surface charge, pore size, porosity, and pore size distribution are summarized below.

3.5.2.1 Hydrophobicity

Hydrophobicity of membrane material is one of the important parameters and hydrophilic membranes typically give better performance with an aqueous feed stream. Hydrophobicity of different membrane materials can be compared through contact angle measurements, which provide an idea about the wettability of the surface. However, reliability of contact angle data is subjective due to its dependence on surface roughness, water purity, and measurement techniques.

Marshall, Munro, and Tragardh (1993) have cited several studies reporting higher permeate flux and less severe fouling with hydrophilic membranes. Rolchigo, Raymond, and Hildebrandt (1989) have studied protein UF using two 100 kDa membranes having mostly similar properties except hydrophobicity. Both membranes gave similar flux in a membrane system where concentration polarization was dominant, but flux of the hydrophilic membrane was much higher compared to other membranes in a membrane system having low concentration polarization due to improved mass transfer. Also, protein transmission was higher with a hydrophilic membrane, the difference being more pronounced in a membrane system with better mass transfer. Sheldon, Reed, and Hawes (1991) have reported the effect of hydrophobicity of a membrane on the nature of protein deposit during BSA UF. The study showed not only a 3 times increase in protein deposit but also a significant increase in the specific resistance of a 10 kDa polysulfone membrane in comparison to a regenerated cellulose membrane with a similar MWCO. Further analysis of protein deposits suggested protein denaturation on a polysulfone membrane compared to regenerated cellulose. Overall, the hydrophilic membrane fouls less compared to hydrophobic membranes and offer improved UF performance. However, the effect of hydrophobicity is not visible when concentration polarization and total protein deposits are high.

3.5.2.2 Surface charge on the membrane

The surface charge on the membrane is important during UF of charged molecules like proteins. The surface charge on the membrane depends on not only membrane material but also pH and ionic strength of the feed solution. Most of the membranes are negatively charged under normal process conditions. Generally membranes offer higher flux during the filtration of molecules having a similar charge. The charge on the membrane can also help to increase the selectivity of the UF membrane during protein fractionation by selecting an appropriate operating pH.

3.5.2.3 Pore size

The relative size of the pores and solute in the feed has a big influence on the performance of UF membranes. Membranes having a larger pore size rating give higher flux at the start of the process than membranes having a smaller pore size, but show rapid flux decline and eventually settle at a lower flux. Marshall, Munro, and Tragardh (1993) have reviewed many studies that indicate the existence of an optimum pore size for permeate flux, below which resistance of the membrane and surface fouling layer reduces flux and above which a higher rate of membrane internal fouling (deposition within the pores) reduces longer-term flux. Fane, Fell, and Waters (1983) have compared UF of BSA and lysozyme and noted that the initial permeate flux was higher with lysozyme but dropped rapidly and the final steady-state flux was lower than that of the totally retained BSA. Overall, it appears that membrane fouling is more severe with increasing pore size and there would be an optimum pore size below which membrane resistance dictates the flux and above membrane fouling.

3.5.2.4 Porosity and pore size distribution

UF membranes are typically characterized with low surface porosity and wide pore size distribution. Suki, Fane, and Fell (1984) have studied the effect of porosity on protein deposition and reported that the largest deposition or fouling of the membrane had the lowest porosity. This can be explained through concentration polarization theory whereby the membrane having the lowest porosity would experience relatively higher local concentration polarization (or higher local velocity normal to the membrane surface) and thereby more rapid fouling.

The effect of pore size distribution on flux can be realized by understanding the balance of flow through different pore sizes. According to the Hagen Poiseuille equation, the solvent flow is strongly biased to larger pores. Fane, Fell, and Waters (1981) calculated that 20–25% of pores on the larger-end side account for almost 50% of the flux through the UF membrane. Therefore, the overall flux is very sensitive to the population of these large pores and their loss through blockage by large feed components would result in a disproportionately high loss in the permeate flux. Membrane selectivity is also negatively affected by pore size distribution. Overall, the flux is more biased towards larger pores and therefore is very sensitive to fouling of larger pores, for example, by protein aggregates. Membrane fouling changes both porosity as well as pore size distribution and thereby membrane selectivity.

3.5.3 Operating parameters

The selection of operating parameters such as transmembrane pressure, cross-flow rate, temperature, and feed concentration is quite important during UF as they can have a great influence on membrane fouling.

3.5.3.1 Transmembrane pressure

Transmembrane pressure is one of the important operating parameters determining the total throughput of the UF process. Generally, a higher transmembrane pressure gives a higher flux but at the cost of increased concentration polarization and a higher membrane fouling rate (Kanani and Ghosh, 2007). Taddei et al. (1986, 1988) have reported increased membrane fouling with increasing transmembrane pressure during UF of cheese whey. Aimer et al. (1988) have noted that an initial flux decline was higher at higher pressure during UF of cheese whey. As an indirect proof of higher fouling at higher pressure, Nakanishi and Kessler (1985) have reported that rinsing was more effective to remove the deposited foulant layer when UF was carried out at reduced transmembrane pressure.

Higher fouling at higher pressure can be explained in two ways. According to one theory, a fouling layer forms during the early phase of UF and get further compressed at a higher pressure and becomes less permeable. On the other hand, few constant flux studies have reported a severe increase in pressure due to the formation of a tightly packed dynamic protein layer caused by convective protein deposition at a higher flux. Overall, selection of the optimum transmembrane pressure is very important for the UF process. A low-pressure process suffers from reduced productivity and demands a higher capital expense in terms of higher membrane area. On the other hand, a high-pressure process suffers from severe membrane fouling, which reduces the membrane life due to necessary stringent cleaning.

3.5.3.2 Cross-flow velocity

Cross-flow velocity is another important operating parameter, which can significantly improve the performance of any UF process. Higher cross-flow velocity improves the solute mass transfer from the polarization layer to the bulk feed. The improved mass transfer reduces concentration polarization and the membrane fouling rate, which in turn improves the throughput of the process. Nakanishi and Kessler (1985) have studied the effect of cross-flow velocity on flux during UF and recovery of flux during rinsing. Both process flux and flux recovery during rinsing increased with increasing cross-flow velocity during UF, indicating a positive effect on concentration polarization and membrane fouling.

3.5.3.3 Temperature

The effect of temperature on fouling is not well understood. It is generally believed that temperature has a positive effect on flux due to a reduction in permeate viscosity and an increase in solute diffusivity at elevated temperatures. Vetier, Bennasar, and Turodo de la Fuente (1988) have reported a positive effect of temperature on the UF of raw whole milk. They noted a considerable reduction in the thickness of the surface cake due to reduced viscosity of the milk at elevated temperatures. This resulted in increased flux and nitrogen transmission through the membrane. On the other hand, both Piot et al. (1984) and van Boxtel, Otten, and van der Linden (1991) reported the negative effect of temperature on flux decline with cheese whey. This was attributed to precipitation of calcium phosphate. Similarly, Vetier, Bennasar, and Turodo de la Fuente (1988) also reported an increase in calcium and phosphate fouling with skim milk at 50 °C compared to 20 °C. Hayes et al. (1974) reported a positive effect of prior short heat treatments (80–85 °C for 15 seconds) of cheese whey on permeate flux during UF. They proposed the formation of a complex between casein components and β-lactoglobulin due to heat treatment, which did not foul the membrane to the same extent. In another study, Meireles, Aimar, and Sanchez (1991) have suggested that BSA became denatured during UF at elevated temperatures, which further amplified the fouling problem.

It can be concluded that permeate flux generally increases with increasing temperature due to a reduction in viscosity and concentration polarization. However, protein denaturation or precipitation of calcium phosphate may increase the fouling rate, thereby giving a final flux lower than that at a lower temperature.

3.6 ENGINEERING MODELS FOR MEMBRANE FOULING

Membrane fouling is a complex process and depends on many factors, which has discouraged the development of mechanistic models. Most of the models in the literature are either empirical or semi-empirical in nature. They attempt to relate flux to the time or permeate volume in some kind of exponential relationship considering the shapes of the flux profile during the UF process (Cheryan, 1998):

$$J_t = J_0 t^{-b} \tag{3.1}$$

$$J_t = J_0 e^{-bt} \tag{3.2}$$

$$J_t = J_0 V^{-b} \tag{3.3}$$

$$J_t = J_{SS} + k e^{-bt} \tag{3.4}$$

where J_0 is the initial flux, J_t is the flux at any time t, J_{SS} is the steady-state flux, and V is the permeate volume. The constants k and b characterize the flux decline profile during UF. According to some of these models, flux will be zero at infinite time, which may not occur in reality.

In another approach, many researchers have modeled their flux decline data using one of the three classical fouling models or a combination of these: (1) complete pore blockage, (2) pore constriction, and (3) cake filtration. These models were originally developed for particle filtration (Hermia, 1982; Field, 1996; Jaffrin et al., 1997). The complete pore blockage model assumes that a fraction of the pores is completely blocked by particles or foulants and is proportional to the amount of permeate volume. The following equation describes the final form of this model:

$$J_t = J_0 \exp(-J_0 bt) \tag{3.5}$$

The pore constriction model assumes that the pore diameter could be reduced by the adsorption or deposition of the foulant inside the pores. This model is described by

$$J_t = J_0 \left(1 + \frac{J_0 bt}{\varepsilon \lambda}\right)^{-2} \tag{3.6}$$

where ε is the membrane porosity and λ is the thickness of the membrane. The cake filtration model assumes that the entire membrane surface is covered by a foulant layer and this layer continues to grow with the course of filtration. Therefore, cake resistance is proportional to the permeate volume and the model takes the form

$$J_t = J_0 \left(1 + \frac{2AJ_0 bt}{R_M}\right)^{-1/2} \tag{3.7}$$

where A is the membrane surface area and R_M is intrinsic membrane resistance or hydraulic resistance. With complex feed like dairy fluid, which contains foulants of varying size, the membrane can foul by more than one mechanism at the same time. Although these models are empirical or semi-empirical in nature, they still serve as useful tools to determine the strategy for minimizing fouling by correlating model parameters (k, b, or J_0) with operating variables.

3.7 CONTROL STRATEGIES

Membrane fouling cannot be completely eliminated during UF but it can be effectively controlled using one of the following strategies:

- Membrane pretreatment
- Feed pretreatment
- Improved system hydrodynamics.

3.7.1 Membrane pretreatment

The ideal UF membrane for dairy applications should be hydrophilic and homogeneously permeable. The membrane surface can be made more hydrophilic by introducing large numbers of hydrophilic functional groups such as hydroxyl groups (–OH) on its surface. This can be achieved through surface coating by adsorption, free radical, or radiation grafting of hydrophilic monomers, microwave or plasma treatment, chemical conversion of polymer side chains to hydrophilic groups, or coating using Langmuir–Blodgett methods (Hildebrandt, 1991). For PS and PVDF membranes, the simple static soak in a nonionic surfactant or water-soluble polymer followed by drying is a quite effective technique (Brink and Romjin, 1990; Kim, Fane, and Fell, 1988) although the life of the coating may be limited to only a limited number of cycles. Additionally, the coated membrane's tolerance to aggressive environments may be limited in comparison to a native membrane.

3.7.2 Feed pretreatment

Feed pretreatment can have a significant influence on the performance of the UF process. Rosenberg reported the positive effect of prior fat removal from whey on UF flux (Rosenberg, 1995). Prior removal of insoluble calcium salts from cheese whey by either filtration (Daufin, Michel, and Merin, 1992) or centrifugation (Kuo and Cheryan, 1983) was also reported to improve the UF flux. The method of pH adjustment prior to UF can potentially affect the process flux. Acid whey has typically almost twice the amount of calcium than cheddar (sweet) whey and the method of pH adjustment has been reported to have a significant influence on fouling by calcium phosphate (Cheryan, 1998). The rapid pH adjustment increases membrane fouling by calcium phosphate salts in comparison to slow pH adjustment, mainly due to the difference in the physical form of this foulant. Heat treatment of whey is another option to improve the flux of a UF membrane since it may cause the formation of less fouling casein–β-lactoglobulin complexes and/or calcium apatite complexes (Cheryan, 1998). All of these studies indicate the beneficial effect of feed pretreatment on reducing the membrane fouling propensity of feed components and thereby improved flux during the UF process.

3.7.3 Improved system hydrodynamics

Adjustment of operating conditions, particularly system hydrodynamics, is a most useful and practical approach to control both membrane fouling and concentration polarization. System hydrodynamics can be improved by increasing mass transfer and other back-transport mechanisms in the membrane module. Increasing cross-flow velocity reduces both concentration polarization and membrane fouling. Similarly, several studies (Belfort, Davis, and Zydney, 1994; Gupta et al., 1995; Winzeler and Belfort, 1993) have reported positive effects of turbulence promoters/inserts/baffles in the feed channel of a membrane module on flux due to increased mass transfer. Some studies (Bauser et al., 1982; Breslau et al., 1980) have also reported a positive effect of pulsating flow and backwash on membrane fouling.

Periodic backwash can be created by pumping filtrate back in to the feed channel to lift the deposited material off the membrane surface, though its effectiveness depends on the nature of the foulant.

Both pressure and flux have a negative impact on fouling as fouling generally increases at a higher pressure or flux. High-pressure processes initially give a higher flux but the average flux during the entire process can be lower due to severe membrane fouling. Besides, severe fouling at higher pressures normally demands more stringent cleaning, thereby reducing the membrane life. Therefore, it may be beneficial to operate the UF process at low to moderate pressure to minimize the intensity of membrane fouling. Fouling can be minimized quite dramatically by operating the process below a certain critical flux or above a critical shear stress/flux ratio. The concept of critical flux or critical shear stress/flux ratio is discussed in detail by Field et al. (1995) and Gesan, Daufin, and Merin (1995) and has to be determined experimentally.

3.8 MEMBRANE CLEANING AND SANITIZATION

The throughput of the UF process decreases with time due to membrane fouling and requires some kind of periodic membrane cleaning to maintain the efficiency of the process at desirable levels. Besides, membrane cleaning and sanitization is an important component of the dairy filtration process from a law and regulation point of view. Therefore, the selection of an appropriate membrane system for a given application should also involve ease of cleaning and sanitization as important criteria besides flux and rejection properties. Membrane cleaning involves the removal of foulant material from the membrane system whereas the purpose of membrane sanitization is to reduce microbial levels in the membrane system down to acceptable levels.

A membrane system can be cleaned either physically or using chemical cleaning agents (Zeman and Zydney, 1996; Cheryan, 1998). The physical cleaning of the system involves some kind of mechanical means, for example, circulation of the feed solution or water at a high shear rate and zero or minimal transmemrane pressure through the membrane system. Back-flushing is another important cleaning strategy to flush the foulants out and off the membrane surface, particularly with a hollow fiber or flat sheet membrane configuration. Back-flushing can be performed by applying negative transmembrane pressure during the actual process or cleaning with either rinse water or an appropriate chemical cleaning agent. Chemical cleaning agents are normally more effective than physical cleaning if selected appropriately based on the type of foulant and membrane. Their mode of action can involve a number of different mechanisms such as displacement of foulant through competitive adsorption of surface active agents, improving the solubility of foulant, and/or chemical modification of the foulant. There is variety of cleaning solutions in the market, which can be classified in to acids, alkalis, surfactants, sequestrants, and enzymes depending on their cleaning mechanisms. Acids are quite effective in removing calcium salts and metal oxides whereas alkalis are more effective for silica, inorganic colloids, and many biological/organic foulants. Surfactants are amphiphilic in nature and thus very effective at removing oils, fats, and other organic foulants from the membrane surface. Sequestrants are effective in removing metal cations due to their ability to form chelating complexes. Enzymes are expensive but quite effective in removing proteins (proteases), starches (amylases), fats and oils (lipases), and cellulose (cellulases).

For effective and safe membrane cleaning, a good understanding of the interactions among foulant, membrane, and operating conditions are quite helpful. Normally, membrane fouling during dairy UF is dominated by protein deposits and to a lesser extent by calcium precipitates. Therefore the cleaning cycle after dairy UF involved first alkali-based cleaner to remove proteins followed by acid cleaner to remove the calcium salts (McCray and Glater, 1985). However, during whey UF, calcium precipitates dominates the membrane fouling and therefore cleaning cycle involving acid cleaner followed by alkali solution is more appropriate (McCray and Glater, 1985). In another study, Daufin et al. (1991b) noted that nitric acid was more effective as a cleaning agent after whey UF and sodium hypochlorite at pH 11 was more effective after milk UF. Again, the difference in cleaning efficiency was attributed to the difference in the relative amount of protein and inorganic fouling in these systems. Kulozik and Kessler (1989) have studied the effect of the operating parameters on the cleaning of milk-fouled UF membranes. The effectiveness of the cleaning was lower when filtration was carried out at higher transmembrane pressure or lower cross-flow velocity due to the formation of thicker or denser protein deposits.

A typical cleaning/sanitization program consists of three steps: (1) rinsing with DI water, (2) cleaning with chemical agents, and finally (3) sanitization with appropriate disinfectant. In the dairy industry, the membrane systems are typically disinfected on a daily basis. Chlorine, hydrogen peroxide, and peracetic acid are the most common disinfectants. Chlorine is available as hypochlorites of sodium or chlorine gas. Sodium hypochlorite disinfection is typically carried out at pH 10 as hypochlorite is very corrosive at low pH values.

Membrane life is quite dependent on membrane cleaning and sanitization procedures. Therefore, it is prudent to pay careful attention to both cleaning and sanitization. Typically, the UF system is cleaned on a daily basis as prolonged processing fouls membranes more severely and makes membrane cleaning more difficult. The effectiveness of the cleaning process is typically measured through a water flux test. Cleaning solutions should be prepared with soft potable water that is free of sparingly soluble salts. Cleaning at a higher temperature is more effective, so the temperature of cleaning should be as high as dictated by the limitations of the membrane material or module. Chemical compatibility of the cleaning agent should be checked with not only membrane material but also with different parts of the module, such as gaskets, epoxy material, housing, etc. It is important to verify and conduct the necessary calibration of metering pumps, temperature, and pH probes on a regular basis.

References

Aimar, P., Taddei, C., Lafaille, J.P., and Sanchez, V. (1988) Mass transfer limitations during ultrafiltration of cheese whey with inorganic membranes. *Journal of Membrane Science*, **38**, 203–221.

Bauser, H., Chmiel, H., Stroh N., and Walitxa, E. (1982) Interfacial effects with microfiltration membranes. *Journal of Membrane Science*, **11**, 321–332.

Belfort, G., Davis, R.H., and Zydney, A.L. (1994) The behavior of suspensions and macromolecular solutions in cross flow microfiltration. *Journal of Membrane Science*, **96**, 1–58.

Breslau, B.R., Testa, A.J., Milnes, B.A., and Medjanis, G. (1980) Advances in hollow fiber ultrafiltration technology, in Ultrafiltration Membranes and Applications (ed. A.R. Cooper), Plenum Press, New York, pp. 109–128.

Brink, L.E.S. and Romjin, D.J. (1990) Reducing the protein fouling of polysulfone surfaces and polysulfone ultrafiltration membranes optimization of the type of presorbed layer. *Desalination*, **78**, 209–233.

Bruin, S., Kikkert, A., Weldring, J.A.G., and Hiddink, J. (1980) Overview of concentration polarization in ultrafiltration. *Desalination*, **35**, 223–242.

Cheryan, M. (1998) *Ultrafiltration and Microfiltration Handbook*, Technomic, Lancaster.

Daufin, G., Michel, F., and Merin, U. (1992) Ultrafiltration of defatted whey influence of some physicochemical characteristics. *Australian Journal of Dairy Technology,* **47**, 7–13.

Daufin, G., Labbe, J.P., Quemerais, A., Michel, F., Robert, F., and Fiaud, C. (1989) Spectroscopic analysis of fouling layers of inorganic ultrafiltration membranes, in Fouling and Cleaning in Food Processing (eds. H.G. Kessler and D.B. Lund), Technische Universitat, Munich, pp. 207–215.

Daufin, G., Labbe, J.P., Quemerais, A., and Michel, F. (1991a) Fouling of an inorganic membrane during ultrafiltration of defatted whey–protein concentrates. *Netherlands Milk and Dairy Journal*, **45**, 259–272.

Daufin, G., Merin, U., Labbe, J.P., Quemarais, A., and Kerherve, F.L. (1991b) Cleaning of inorganic membranes after whey and milk ultrafiltration. *Biotechnology and Bioengineering*, **38**, 82–89.

Fane, A.G. (1983) Factors affecting flux and rejection in ultrafiltration. *Journal of Separation Process Technology*, **4**, 15–23.

Fane, A.G., Fell, C.J.D., and Waters, A.G. (1981) The relationship between membrane surface pore characteristics and flux for ultrafiltration membranes. *Journal of Membrane Science*, **9**, 245–262.

Fane, A.G., Fell, C.J.D., and Waters, A.G. (1983) Ultrafiltration of protein solutions through partially permeable membranes – the effect of adsorption and solution environment. *Journal of Membrane Science*, **16**, 211–224.

Field, R.W. (1996) Mass transport and the design of membrane systems, in Industrial Membrane Separation Technology (eds. K. Scott and R. Hughes), Blackie Academic, London, UK, p. 67.

Field, R.W., Wu, D., Howell, J.A., and Gupta, B.B. (1995) Critical flux concept for microfiltration fouling. *Journal of Membrane Science*, **100**, 259–272.

Gesan, G., Daufin, G., and Merin, U. (1995) Performance of whey crossflow microfiltration during transient and stationary operating conditions. *Journal of Membrane Science*, **104**, 271–281.

Ghosh, R. (2003) *Protein Bioseparation Using Ultrafiltration: Theory, Applications and New Developments*, Imperial College Press, London.

Gupta, B.B., Howell, J.A., Wu, D., and Field, R.W. (1995) A helical baffle for cross-flow microfiltration. *Journal of Membrane Science*, **102**, 31–42.

Hayes, J.F., Dunkerley, J.A., Muller, L.L., and Griffin, A.T. (1974) Studies on whey processing by ultrafiltration, II. Improving permeation rates by preventing fouling. *Australian Journal of Dairy Technology*, **29**, 132–140.

Hermia, J. (1982) Constant pressure blocking filtration laws – application to power-law-non-Newtonian fluids. *Transactions of the Institute of Chemical Engineering*, **60**, 83–187.

Hildebrandt, J.R. (1991) Membranes for bioprocessing: design considerations, in Chromatographic and Membrane Processes in Biotechnology (eds. C.A. Costa and J.S. Cabral), Kluwer Academic Publishers, The Netherlands, pp. 363–378.

Jaffrin, M.Y., Ding, L.H., Couvreur, C., and Khari, P. (1997) Effect of ethanol on ultrafiltration of bovine albumin solutions with organic membranes. *Journal of Membrane Science*, **124**, 233–241.

Kanani, D.M. and Ghosh, R. (2007) A constant flux based mathematical model for predicting permeate flux decline in constant pressure protein ultrafiltration. *Journal of Membrane Science*, **290**, 207–215.

Kelly, S.T. and Zydney, A.L. (1995) Mechanisms for BSA fouling during microfiltration. *Journal of Membrane Science*, **107**, 115.

Kim, K.J., Fane, A.G., and Fell, C.J.D. (1988) The performance of ultrafiltration membranes pretreated by polymers. *Desalination*, **70**, 229–249.

Kulozik, U.M. and Kessler, H.G. (1989) Rinsing behavior of deposited layers in reverse osmosis and ultrafiltration, in Advances in Reverse Osmosis and Ultrafiltration (eds. T. Matsuura and S. Sourirajan), National Research Council of Canada, Ottawa, Canada, pp. 517–549.

Kuo, K.P. and Cheryan, M. (1983) Ultrafiltration of acid whey in a spiral-wound unit. Effect of operating parameters on membrane fouling. *Journal of Food Science*, **48**, 1113–1117.

Marshall, A.D., Munro, P.A., and Tragardh, G. (1993) The effect of protein fouling in microfiltration and ultrafiltration on permeate flux, protein retention and selectivity: a literature review. *Desalination*, **91**, 65–108.

McCray, S.B. and Glater, J. (1985) Effects of hydrolysis on cellulose acetate reverse osmosis transport coefficients, in Reverse Osmosis and Ultrafiltration (eds S. Sourirajan and S. Mastsuura), ACS Symposium Series, Washington, DC, pp. 141–151.

Meireles, M., Aimar, P., and Sanchez, V. (1991) Albumin denaturation during ultrafiltration: effects of operating conditions and consequences on membrane fouling. *Biotechnology and Bioengineering*, **38**, 528–534.

Muller, L.L., Hayes, J.F., and Griffin, A.T. (1973) Studies on whey processing by ultrafiltration, I. Comparative performance of various ultrafiltration modules on whey from hydrochloric acid casein and cheddar cheese. *Australian Journal of Dairy Technology*, **28**, 70–77.

Nakanishi, K. and Kessler, H.G. (1985) Stability and rinsing behaviour of deposited layers. *Journal of Food Science*, **50,** 1726–1731.

Patocka, J. and Jelen, P. (1987) Calcium chelation and other pretreatments for flux improvement in ultrafiltration of cottage cheese whey. *Journal of Food Science*, **52**, 1241–1244.

Piot, M., Maubois, J.L., Schaegis, P., Veyre, R., and Luccioni, M. (1984) Several types of microfiltration membranes have been tested with cheese whey. *Le Lait*, **64**, 102–120.

Rautenbach, R. and Albrecht, R. (1989) Membrane Processes, John Wiley & Sons, Inc., New York.

Renner, E. and Abd El-Slam, M.H. (1991) *Application of Ultrafiltration in the Dairy Industry*, Elsevier Applied Science, London.

Rolchigo, P.M., Raymond, W.A., and Hildebrandt, J.R. (1989) Improved control of ultrafiltration using vorticular hydrodynamics and hydrophilic membranes. *Process Biochemistry*, **24,** R3–R7.

Rosenberg, M. (1995) Current and future applications for membrane processes in the dairy industry. *Trends in Food Science and Technology*, **6**, 12–19.

Sheldon, J.M., Reed, I.M., and Hawes, C.R. (1991) The fine-structure of ultrafiltration membranes, II. Protein fouled membranes. *Journal of Membrane Science*, **62**, 87–102.

Suki, A., Fane, A.G., and Fell, C.J.D. (1984) Flux decline in protein ultrafiltration. *Journal of Membrane Science*, **21**, 269–283.

Taddei, C., Aimar, P., Daufin, G., and Sanchez, V. (1986) Mass transfer during sweet whey ultrafiltration on mineral membrane. *Le Lait*, **66**, 371–390.

Taddei, C., Aimar, P., Daufin, G., and Sanchez, V. (1988) Factors affecting fouling of an inorganic membrane during sweet whey ultrafiltration. *Le Lait*, **68**, 157–176.

Tong, P.S., Barbano, D.M., and Jordan, W.K. (1989) Characterization of proteinaceous membrane foulants from whey ultrafiltration. *Journal of Dairy Science,* **72**, 1435–1442.

Tong, P.S., Barbano, D.M., and Rudan, M.A. (1988) Characterization of proteinaceous membrane foulants and flux decline during the early stages of whole milk ultrafiltration. *Journal of Dairy Science*, **71**, 604–612.

van Boxtel, A.J.B, Otten, Z.E.H., and van der Linden, H.J.L.J. (1991) Evaluation of process models for fouling control reverse osmosis of cheese whey. *Journal of Membrane Science*, **58**, 89–111.

Vetier, C., Bennasar, M., and Tarodo de la Fuente, B. (1988) Study of the fouling of a mineral microfiltration membrane using scanning electron microscopy and physico-chemical analyses in the processing of milk. *Journal of Dairy Research*, **55**, 381–400.

Winzeler, H.B., and Belfort, G. (1993) Enhanced performance for pressure-driven membrane processes: the argument for fluid instabilities. *Journal of Membrane Science*, **80**, 35–47.

Zeman, L.J. and Zydney, A.L. (1996) *Microfiltration and Ultrafiltration Principles and Applications*, Marcel Dekker, New York.

4 Dairy protein fractionation and concentration using charged ultrafiltration membranes

Mark R. Etzel and Abhiram Arunkumar

Department of Chemical and Biological Engineering, Madison Wisconsin, USA

4.1 INTRODUCTION

Ultrafiltration is a common unit operation in the dairy industry for concentration of proteins to make protein-rich products like milk protein concentrate, whey protein concentrate, and whey protein isolate. While protein concentration deals with small-molecule clearance (e.g., water, lactose, minerals, and nonprotein nitrogen) and membranes are capable of performing this efficiently because all proteins must be retained, protein fractionation is more challenging because it requires selective permeation of one protein and simultaneous retention of another protein. Many studies have reported the use of conventional uncharged ultrafiltration membranes for fractionating individual dairy proteins, but most of these studies used ultrafiltration as a polishing step for concentrating the protein after a prepolishing step like thermal aggregation or chromatography that actually does the fractionation (Tolkach and Kulozik, 2005; Muller et al., 2003; Gesan-Guiziou et al., 1999; Toro-Sierra, Tolkach, and Kulozik, 2013; Xu et al., 2000; Bhattacharjee, Bhattacharjee, and Datta, 2006). This chapter reports on the use of positively charged ultrafiltration membranes to fractionate one dairy protein from another and the use of negatively charged ultrafiltration membranes to simply concentrate all dairy proteins while clearing small molecules.

4.1.1 Current practices in dairy protein fractionation using ultrafiltration membranes

Current practices in using ultrafiltration membranes for dairy protein fractionation base the separation on size differences alone between the different proteins. Because the predominant dairy proteins, α-lactalbumin

Membrane Processing for Dairy Ingredient Separation, First Edition.
Edited by Kang Hu and James M. Dickson.

(ALA) and β-lactoglobulin (BLG) have molecular masses of 14.4 kDa and 18.4 kDa, respectively, and because BLG is capable of existing as a mixture of monomers and dimers under conditions of milk processing (Mercadente et al., 2012), BLG should be mostly rejected by ultrafiltration membranes having a molecular weight cut-off (MWCO) of 30–50 kDa while ALA should be permeated more than BLG. Indeed, several researchers have reported the use of 30 kDa ultrafiltration membranes for fractionating ALA and BLG. Cheang and Zydney (2003, 2004) used a 30 kDa ultrafiltration membrane with constant-volume diafiltration to produce permeate fractions enriched in ALA and retentate fractions enriched in BLG, and a 100 kDa membrane in order to purify bovine-serum albumin (BSA). Metsämuuronnen and Nyström (2006, 2009) used both 30 kDa and 100 kDa membranes for fractionating ALA from diluted whey. While up to 88% pure ALA could be obtained from whey using 30 kDa membranes, the maximum purity that could be achieved using a 100 kDa membrane was only 35%. Moreover, there was little enrichment of BLG in the retentate because the 30 kDa membrane retained almost all of the BLG and permeated little ALA. Lucas et al. (1998) examined larger pore membranes functionalized with polyethleneimine, but they proposed that larger pore membranes would have to be modified in such a way to block the pores. Marella, Muthukumarappan, and Metzger (2011) examined ultrafiltration membranes for fractionation of ALA from Cheddar cheese whey using membranes of MWCO 50 kDa, 100 kDa, and 300 kDa. They could obtain at the most 63% pure ALA using membranes of 50 kDa with lower purities using the looser membranes. Konrad and Kleinschmidt (2008) examined the use of 100 kDa and 150 kDa ultrafiltration membranes to fractionate native ALA from whey and used tryptic treatment of BLG in the permeate to improve purity from 40% to more than 90%. The presence of the smallest whey protein, glycomacropeptide (GMP) was neglected in these studies. Several other studies have reported the use of purifying ALA from acid whey using tighter membranes (Muller et al., 2003; Gesan-Guiziou et al., 1999; Cheang and Zydney, 2003, 2004; Metsämuuronnen and Nyström, 2006, 2009; Lucas et al., 1998; Muller, Daufin, and Chaufer, 1999; Bottomley, 1991). Holland, Kackmar, and Corredig (2012) used a 100 kDa membrane for fractionating ALA from skim milk. They were able to obtain more than 80% pure ALA; however, this was at the cost of no enrichment of BLG or other components in the feed stream. The transmission of ALA itself was low, giving a permeate that contained only 10% of the total ALA in the feed stream. This means that the 100 kDa membrane mostly rejected ALA.

Xu et al. (2000) used anion-exchange chromatography to separate ALA, BLG, and BSA from whey and used ultrafiltration to concentrate immunoglobulins (IgGs). Approaches like cross-flow cation-exchange chromatography have been used to recover lactoferrin from whey (Saufi and Fee, 2013). The common method for isolating GMP from whey has been ion-exchange chromatography (Doultani, Turhan, and Etzel, 2003, 2004). Studies have been reported on the use of 20–50 kDa ultrafiltration membranes to fractionate GMP from whey (Tanimoto et al., 1991) or a combination of chromatography and ultrafiltration (Kawasaki and Dosako, 1994). A review of methods used for separation of GMP from whey was published in 2006 (Abd El-Salam, 2006).

With the exception of a few of the studies mentioned above (Cheang and Zydney, 2003, 2004), transmembrane pressure was held constant during ultrafiltration to fractionate proteins. Controlling the permeate flux during ultrafiltration is rare. However, as discussed in more detail in Section 4.2, permeate flux control is essential when it comes to protein fractionation because the extent of concentration polarization increases exponentially with increasing permeate flux, and this in turn controls the selectivity for transmission of individual proteins (Lebreton, Brown, and van Reis, 2008; Arunkumar and Etzel, 2014). The need for flux control in the proper use of large-pore charged ultrafiltration membranes for dairy protein fractionation is spelled out later in this chapter.

4.1.2 Current practices for protein concentration

Protein concentration involves the separation of dairy proteins from lactose, minerals, lactic acid, and nonprotein nitrogen in order to increase the ratio of proteins to solids in the final product stream. Typical products obtained from protein concentration are whey protein concentrates (WPCs) of varying protein contents (WPC 80 containing at least 80% protein and WPC 34 containing at least 34% protein), whey protein isolate (WPI) (>90% protein), and milk protein concentrate (MPC). While micellar casein concentrate (MCC) is made by microfiltration by separating the serum proteins from the casein micelles (Beckman et al., 2010; Hurt et al., 2010; Nelson and Barbano, 2005), WPC, WPI, and MPC are made using ultrafiltration membranes (Mistry, 2002). The composition and properties of whey protein concentrates can be found in Delaney (1976). Rinn et al. (1990) gave a procedure to produce whey protein concentrate in which a 10 kDa ultrafiltration membrane is used. The process involves a 24-fold volume reduction and diafiltration using three diafiltration volumes of deionized water per part of initial concentrate. Barba et al. (2001) described a two-stage diafiltration step for producing whey protein concentrates while Yee, Wiley, and Bao (2007) described the effects of operating conditions during the continuous ultrafiltration of whey. Rektor and Vatai (2004) describe the use of large-pore size ultrafiltration membranes of 100 kDa for the ultrafiltration of Mozarella whey, but the retention of proteins using this membrane is only 75%. Common modes of ultrafiltration for protein concentration almost always use constant pressure control, described in the literature mentioned in this section. The use of constant flux has not been reported to the knowledge of the authors; however, there have been studies that have investigated controlling the deposit layer during the ultrafiltration of milk and whey to improve small-molecule clearance (Kulozik and Kessler, 1990; Obermeyer, Kulozik, and Kessler, 1993).

Many studies have been conducted on the performance characteristics of ultrafiltration membranes in terms of the membrane resistance and fouling, but all of these works have focused on using tight membranes from 5 to 30 kDa for whey protein concentration (Cheryan and Kuo, 1984; Raubiller-Baudry et al., 2005; Da Costa, Fane, and Wiley, 1993; Yee, Wiley, and Bao, 2009; Davey et al., 2004). Indeed, there is an inherent trade-off between permeability (flux) and selectivity (protein retention) for protein concentration during ultrafiltration (Mehta and Zydney, 2005). According to this rule of thumb, more porous membranes offer a higher flux for protein concentration at the expense of lower protein retention and vice versa.

4.1.3 Use of charged ultrafiltration membranes for dairy protein fractionation and concentration

It is evident from Section 4.1 that use of tight membranes for fractionating whey proteins would require diafiltration along with other approaches like tryptic digestion of other proteins (causing protein loss) or chromatographic methods to increase the purity of different components in the product. This is because the inherent protein transmission of these membranes is low and large volumes of water must be used to flush the protein through these tight membranes. However, if a membrane could selectively and almost completely permeate one protein and selectively retain the other protein, then the separation could potentially be performed with just membranes and without the addition of water for diafiltration. For example, situations where the separation factor is high but transmission of both proteins is low (Cheang and Zydney, 2003, 2004; Cowan and Ritchie, 2007) result in no enrichment of the more retained protein and low yield of the purified protein when no water is added for diafiltration. Charged ultrafiltration membranes have been examined as alternatives to conventional membranes for high-resolution protein fractionation (van Reis et al., 1999). Charged ultrafiltration membranes work on the principle of electrostatic exclusion, wherein the protein that is charged the same as the membrane charge is electrostatically excluded from the membrane pores while proteins that carry a net charge of zero or are oppositely charged are permeated through the membrane (Arunkumar and Etzel, 2013, 2014; van Reis et al.,

1999, Rohani and Zydney, 2009). The principle of working of a charged ultrafiltration membrane is described later in this chapter.

While tight ultrafiltration membranes (e.g., 30 kDa) have been shown to increase the selectivity for dairy protein fractionation compared to the unmodified analog, the problem of low transmission of both proteins still exists (Arunkumar and Etzel, 2013; Bhushan and Etzel, 2009). In our previous work, we examined a 30 kDa positively charged membrane for fractionating ALA from a binary mixture of ALA and BLG in buffer. The 30 kDa positively charged membrane offered a selectivity boost of up to 125% for fractionation of ALA from BLG, but the transmission of ALA was less than 20% and that of BLG was less than 5% (Arunkumar and Etzel 2013).

In our more recent work (Arunkumar and Etzel, 2014), we have shown that imparting a positive charge on a 300 kDa ultrafiltration membrane and adjusting solution conditions such as pH and conductivity allows selective permeation of ALA (transmission >50%) and retention of BLG (transmission <20%). As described in Section 4.3, these large pore size membranes alone, without the need to add diafiltration water, can give products enriched in both ALA and BLG.

In the case of protein concentration, negatively charged membranes of 100 to 300 kDa MWCO retain proteins to the same extent as the standard uncharged 10 kDa membranes, but allow operation at a flux that was at least twofold higher. Negatively charged ultrafiltration membranes are also antifouling.

4.2 THEORY

4.2.1 Working principle

Protein transmission through the membrane has been modeled by transport of a charged spherical solute through a fluid-filled cylindrical pore (Smith and Deen, 1980, 1983; Pujar and Zydney, 1997, 1998; Burns and Zydney, 2001; Rohani and Zydney, 2010). The transmission of protein is described using the actual sieving coefficient ($S_a = C_{permeate}/C_{wall}$), which is theoretically given by

$$S_a = K_c\left(1 - \frac{r_s}{r_p}\right)^2 \exp\left(\frac{-\varphi_E}{kT}\right) \tag{4.1}$$

where r_s and r_p are the radii of the solute and the pore, K_c is the hindrance factor for convection through the pores, and φ_E is the energy of interaction between the spherical solute and the cylindrical pore. The energy of interaction can be evaluated in terms of the dimensionless surface charge density of the solute σ_s and the dimensionless surface charge density of the pore

σ_p by means of

$$\frac{\varphi_E}{kT} = \frac{A_s\sigma_s{}^2 + A_p\sigma_p{}^2 + A_{sp}\sigma_s\sigma_p}{A_{den}} \tag{4.2}$$

The first term in the numerator on the right-hand side of Equation (4.2) represents the energetic penalty for disrupting the double layer around the spherical solute due to the presence of the pore wall. This term exists even for an uncharged membrane. The second term in the numerator represents the energetic penalty for disrupting the double layer around the membrane pore walls as the charged solute convects through the pores. The last term represents the energetic penalty arising from the direct interaction of the charged solute and the charged pore. The coefficients A_s, A_p, A_{sp}, and A_{den} are positive functions that have a complex dependence on the solution pH, ionic strength, solute radius, and pore radius. More details on evaluating these functions may be found in Smith and Deen (1980) and Pujar and Zydney (1997).

At the protein isoelectric point (pI), $\sigma_s = 0$, which causes two out of the three terms to vanish. For an uncharged pore, even the term containing σ_p vanishes, making protein transmission based on a size-exclusion mechanism alone. Because of this, protein transmission is generally a maximum when the solution pH equals the pI of the protein.

When the protein and the membrane are of opposite charge, then protein transmission is determined by all three parameters in the numerator: the first and second terms control electrostatic exclusion, while the third term controls sieving. Burns and Zydney (2001) explain this balance of energies in detail for the transmission of ovotransferrin through positively and negatively charged polyethersulfone membranes.

The exponential dependence of solute sieving on the electrostatic interaction energy provides two advantages for using charged ultrafiltration membranes to separate proteins. First, proteins that are charged similarly to the membrane can be electrostatically excluded from passing through the membrane pores even when using a wide-pore size membrane, allowing operation at a higher flux, but at the same retention as tighter membranes. Second, a mixture of similarly sized proteins can be separated using a charged ultrafiltration membrane if the proteins differ somewhat in the isoelectric point. By adjusting the solution pH to the isoelectric point of one of the proteins, that protein will permeate the membrane while the other protein is retained by electrostatics.

For protein fractionation, charged ultrafiltration membranes have advantages over chromatography because membranes do not use the buffers, salts, and water required for chromatography. Furthermore, as shown by Rohani and Zydney (2009), protein transmission for charged membranes depends only on the net protein surface charge density and not on the presence of individual charged patches on the protein surface, as found for chromatography. For this reason, the effect of charge on chromatographic separations is less predictable than for charged membrane separations. Sections 4.3 and 4.4 in this chapter are case studies for demonstrating the use of charged ultrafiltration membranes for protein fractionation and protein concentration.

4.2.2 Concentration polarization and permeate flux

Proteins in the feed solution decrease the permeate flux (J_v) to less than the clean water flux because of the effect known as concentration polarization. As pressure drives the permeate through the membrane, convection transports proteins to the membrane surface, causing accumulation of proteins that do not pass through the membrane (Figure 4.1). The stagnant film model (SFM) first proposed by Michaels (1968) was developed to explain concentration polarization. It uses the following mass balance equation:

$$J_v C - J_v C_p = -D\frac{\mathrm{d}C}{\mathrm{d}x} \quad (4.3)$$

where C is the protein concentration upstream of the membrane at a distance x, C_p is the protein concentration in the permeate stream, and D is the protein diffusion coefficient. The first term on the left-hand side accounts for protein moving toward the membrane by convection, the second term on the left-hand side accounts for protein leaking through the membrane by convection, and the term on the right-hand side accounts for protein moving away from the membrane by diffusion according to Fick's law. At steady state the rate of protein transport by back-diffusion is exactly balanced by the net rate of forward protein transport by convection through the membrane.

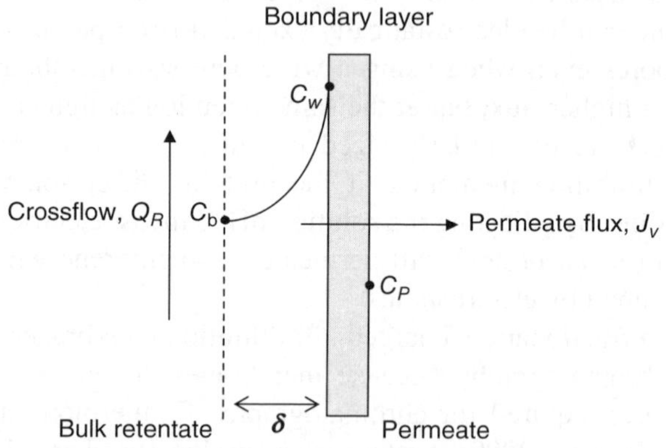

Figure 4.1 Schematic of the stagnant film model for concentration polarization.

When Equation (4.1) is integrated over the boundary layer thickness (i.e., from $C = C_w$ at $x = 0$ to $C = C_b$ at $x = \delta$), then

$$J_v = k \ln \frac{(C_w - C_p)}{(C_b - C_p)} \tag{4.4}$$

where the protein mass transfer coefficient (k) is equal to the ratio of the protein diffusion coefficient to the boundary layer thickness ($k = D/\delta$). The protein concentration is highest at the wall of the membrane (C_w) and falls to the bulk solution value (C_b) over the thickness of the concentration boundary layer (δ).

The observed sieving coefficient ($S_o = C_p/C_b$) is a fundamental measure of performance of an ultrafiltration membrane. From Equation (4.4), the concentration of the protein at the wall of the membrane, C_w, is determined by the observed sieving coefficient, S_o, and the protein concentration in the bulk solution (C_b) as follows (Arunkumar and Etzel, 2014):

$$C_w = C_b(S_o + (1 - S_o)\exp(J_v/k)) \tag{4.5}$$

where J_v is the permeate flux and k is the boundary-layer mass-transfer coefficient. The term J_v/k on the right-hand side of Equation (4.5) is called the polarization index (β) (Arunkumar and Etzel, 2014; Bakhshayeshi et al., 2012) and for cross-flow ultrafiltration membranes is given by

$$\beta = J_v/k = 1.225 J_v / \left(\frac{6Q_R}{hV_{HR}} D^2 \right)^{1/3} \tag{4.6}$$

where V_{HR} is the membrane hold-up volume, Q_R is the average recirculation flow rate, and h is the thickness of the feed-spacer material. Membrane manufacturers sometimes do not report V_{HR}, but it can be calculated using the relation $V_{HR} = hA_m$, where A_m is the membrane area.

The key adjustable parameters in Equation (4.6) for a given membrane module are J_v, and Q_R. In other words, by keeping the permeate flux (J_v) and recirculation flow rate (Q_R) constant, β also remains constant. Industrial ultrafiltration operations typically do not operate at constant β. That is because transmembrane pressure is held constant instead of J_v and J_v falls during ultrafiltration as a result of concentration polarization.

4.2.3 Constant flux (J_v) or constant pressure (ΔP)?

Existing industrial membrane operations take place at constant transmembrane pressure (ΔP) and permeate flux declines during operation, which means that β declines. Flux control is simply not done. The membrane skid cannot handle flux control because it has no pump or restriction to control the permeate flow rate. The problem with not having flux control is that the extent of concentration polarization on the wall of the membrane (C_w) is also not controlled. With uncontrolled flux, C_w can soar into the region of protein insolubility, aggregation, and precipitation. Furthermore, inevitable variations in membrane hydraulic permeability during manufacture result in different values of the flux at a given ΔP for each individual membrane. For example, a single membrane type often has a twofold difference in hydraulic permeability lot-to-lot (Lebreton, Brown, and van Reis 2008; van Reis et al., 1997). Therefore, each membrane element will have a different C_w at a given ΔP.

Operation at constant C_w as proposed by van Reis et al. (1997) is a more theoretically sound and rigorous approach to operation of an ultrafiltration system than pressure control. Constant C_w operation means the right-hand side of Equation (4.3) must remain constant as the protein bulk solution concentration (C_b) rises. Because S_o on the right-hand side does not change much during operation (Arunkumar and Etzel, 2013), β must decrease to keep the right-hand side constant. For protein concentration operations using a fully retentive membrane ($S_o \approx 0$), the protein concentration in the bulk tank increases from an initial value of C_0 at a batch volume of V_0 to a final value of $C_R = C_o V_o / V$ at a final concentrate volume of V. Defining the volume concentration ratio ($VCR = V_o/V$), the right-hand side of Equation (4.3) reduces to $C_o(VCR)\exp(\beta)$, which for a constant feed stream concentration C_o means $\ln(VCR) + \beta$ must remain constant. At time zero, $VCR_0 = 1$, and therefore at all other times we must have $\ln(VCR) + \beta = \beta_0$. For example, if $\beta_0 = 5$ and $VCR = 40$, then $\beta = 5 - \ln(40) = 1.3$. In other words, at the end of a 40-fold volume concentration of whey, the flux must decrease by 75% to keep C_w constant, and the flux must decline in proportion to the log of the VCR, not linearly.

Flux declines naturally during whey protein concentration using uncharged 10 kDa ultrafiltration membranes when operated at constant ΔP, although not in a controlled fashion. This decline in flux decreases β as the VCR increases and causes C_w to remain more constant than would otherwise occur for operation at constant flux. However, this occurs by happenstance not design. When we move to the large pore membranes (100 kDa to 1000 kDa) allowed by using a charged ultrafiltration membrane, then permeate flux is several-fold greater, as are values of β. This forces C_w to increase to very large values and can form a protein deposit on the membrane surface that can act as a finer filter than the membrane itself. Protein fractionation requires that one protein permeates the membrane more than another. The protein deposit participates in that permeation mechanism along with the membrane itself, and must be controlled to allow the fine control of sieving coefficients required to use charged ultrafiltration membranes for dairy protein fractionation. This means that flux control becomes an essential feature of the operation of large pore charged ultrafiltration membranes.

In our work, flux control was implemented by placing a pump on the permeate stream. This metered the permeate flow rate and held J_v and β constant. This is a key difference between our work and past work for dairy protein concentration and fractionation. It means that the present work described in Sections 4.3 and 4.4 is unique in that β was controlled to get a certain value of S_o. The charge on the membrane does not affect β in Equation (4.4), but β and the charge on the membrane do affect S_o. Placing a charge on the membrane and controlling β, S_o was also controlled to a target value. This allowed controlled permeation of one protein while retaining the other protein. In this way, the utility of ultrafiltration membranes was expanded beyond simple protein concentration to fractionation of proteins.

An example of flux control (constant β operation) illustrates the method of using Equation (4.4) for the Pellicon™ flat-sheet cross-flow membrane modules from EMD Millipore (Bedford, MA). For the Pellicon™ Mini with suspended screens, $Q_R = 1350\,\text{cm}^3/\text{min}$, $V_{HR} = 17\,\text{cm}^3$, $A_m = 1000\,\text{cm}^2$, and therefore $h = V_{HR}/A_m = 0.017\,\text{cm}$. Using protein diffusion coefficients of $D = 2 \times 10^{-7}$ cm²/s at 22 °C and $D = 3.2 \times 10^{-7}$ cm²/s at 40 °C, and $J_v = 36$ L/m² h (LMH), results in $\beta = 4.6$ at 22 °C and $\beta = 3.4$ at 40 °C. Decreasing the permeate flux to 12 LMH reduces β proportionately to $\beta = 1.1$.

In general, β decreases as J_v decreases, and as Q_R and D increase. For all membrane modules, β decreases as the spacer thickness (h) decreases, and the recirculation flow rate Q_R and temperature increase. In other words, operating at higher temperatures and higher recirculation flow rates, and using smaller spacer thicknesses and modules where the hold-up volume to membrane area ratio (V_{HR}/A_m) is lower, allows operation at higher flux without increasing β substantially. More details on the comparison between constant flux operation and constant ΔP operation were reported by Miller et al. (2014).

As described in Section 4.4, protein concentration using charged ultrafiltration membranes was performed by controlling the permeate flux (and hence β) to a smaller value than the maximum capacity of the membrane, but at a two-fold higher permeate flux than that of a 10 kDa membrane. This allowed a constant permeate flux to be maintained throughout the concentration operation at an overall higher average value than would have been obtained if the permeate flux was not controlled. In effect, limiting the flux initially meant that it was higher at the end, and greater on average, compared to the case of no flux control where the flux is high initially and falls to low values at the end. This strategy is analogous to pacing yourself during a marathon. Starting out too fast is a rookie mistake. Running at a slower but steady pace is more likely to win the race. We have found that limiting the flux to less than the maximum value at time zero, and controlling the flux at a constant value during ultrafiltration, resulted in a shorter overall time to process a fixed volume of protein solution.

4.3 CHARGED ULTRAFILTRATION MEMBRANES FOR PROTEIN FRACTIONATION

α-Lactalbumin (ALA) and β-lactoglobulin (BLG) are the predominant dairy proteins in milk serum and whey. While these proteins differ somewhat in isoelectric points (4.4 for ALA versus 5.2 for BLG), they do not differ much in size (14.4 kDa for ALA versus 18.4 kDa for BLG). Therefore, by basing the separation on size-based filtration alone, it is not possible to obtain a high yield of ALA in the permeate at high purity. One can either have nearly no yield of ALA in the permeate but high purity or vice versa, as explained in Section 4.1 of this chapter. However, by placing a positive charge on a large pore-size ultrafiltration membrane and adjusting the solution to pH 4.3, which is about the pI of ALA, it is possible to electrostatically reject BLG and simultaneously permeate ALA. This approach was illustrated by covalently attaching a positive charge (quaternary amine) to a 300 kDa ultrafiltration membrane and using the membrane to fractionate ALA and BLG from milk serum permeate.

4.3.1 Materials and methods

Cross-flow Pellicon™ XL ultrafiltration membranes (50 cm^2 area, EMD-Millipore) incorporating Ultracel™ PLC regenerated cellulose membranes of MWCO 300 kDa were charged using the procedure of van Reis (2006) and Bhushan and Etzel (2009). Pellicon XL modules had moderately coarse screens ("C" screens).

Milk serum permeate (MSP) was obtained from the microfiltration of skim milk using an 800 kDa microfiltration membrane at the Babcock Hall Dairy Plant, University of Wisconsin-Madison. The pH of MSP was adjusted to 4.3 using 1M H_3PO_4. The conductivity at this pH was about 8.8 mS/cm. MSP contained 1.2 g/L of ALA, 2.2 g/L of BLG, and traces of other proteins. Ultrafiltration was performed by recirculating MSP in the membrane module at an inlet pressure of 2 bar and a cross-flow rate of 50 mL/min (600 L/m^2 h). The permeate flow rate was controlled by placing a peristaltic pump on the permeate stream. The temperature was maintained at 40 °C. Permeate and retentate samples were recirculated back to the feed solution reservoir and samples were collected from the permeate and retentate tubing for analysis. Proteins were analyzed using SDS-PAGE and fluorescence laser densitometry as outlined by Arunkumar and Etzel (2013).

4.3.2 Results and discussion

4.3.2.1 Sieving coefficient (S_o) as a function of the polarization index (β)

Sieving coefficients of ALA and BLG and the selectivity for fractionating ALA from BLG were measured using positively charged and uncharged 300 kDa membranes (Table 4.1). Using the uncharged 300 kDa membrane, both ALA and BLG passed freely though the membrane; the sieving coefficient was high for both ALA ($S_o = 0.77$) and BLG ($S_o = 0.52$). The selectivity was $\psi = 1.5$, meaning there was essentially no fractionation. By placing a positive charge on the 300 kDa membrane, the sieving coefficient of ALA dropped slightly ($S_o = 0.52$), while BLG was substantially rejected ($S_o = 0.13$) compared to the uncharged membrane, increasing selectivity by 180% to $\psi = 4.2$. Also shown in Table 4.1 are the sieving coefficients obtained during the ultrafiltration of MSP using 10 kDa and 30 kDa uncharged membranes. While the 30 kDa membrane offers a selectivity of $\psi = 3$, both ALA and BLG are more than 90% rejected by the membrane. Using an even tighter membrane (10 kDa) further increases selectivity ($\psi = 21$), but at the cost of an even lower yield of ALA that is 97% rejected by the membrane. These data illustrate the classic permeability selectivity trade-off of uncharged ultrafiltration membranes mentioned above. In contrast, the 300 kDa charged membrane offers both high selectivity and high permeability (Table 4.1).

A flux excursion was performed using the charged 300 kDa membrane to examine the effects of concentration polarization and β on protein transmission and selectivity. As the permeate flux was reduced to 1.2 LMH ($\beta = 0.21$), the sieving coefficient of ALA and BLG both increased, reducing the selectivity from 4.2 at 36 LMH ($\beta = 6.4$) to 2.3 ($\beta = 0.21$). As the permeate flux was increased to 72 LMH ($\beta = 12.8$), the sieving

Table 4.1 Sieving coefficients of ALA and BLG and selectivity during the ultrafiltration of MSP at pH 4.3 using the charged and uncharged membranes. The reader is referred to Arunkumar and Etzel (2013) for more details.

Membrane	Permeate flux (L/m² h)	$\beta = J_v/k_m$	S_o ALA	S_o BLG	Selectivity ψ
300 kDa uncharged	36	6.4	0.77	0.52	1.5
300 kDa charged	1.2	0.21	0.72	0.32	2.3
	36	6.4	0.52	0.13	4.2
	72	12.8	0.58	0.17	3.5
30 kDa uncharged ($T = 22\,°C$)	18	4.4	0.09	0.03	3.0
10 kDa uncharged ($T = 22\,°C$)	7.2	1.8	0.033	0.0016	21

coefficients and selectivity were not statistically significantly different ($p > 0.05$) from 36 LMH ($\beta = 6.4$). This can be explained using the stagnant film model. As the permeate flux increases, the concentration of positively charged BLG on the membrane wall increases, which causes self-rejection of BLG. ALA, being at its isoelectric point, is less affected by the charged layer of BLG. When the polarization index is reduced 30-fold from $\beta = 6.4$ to $\beta = 0.21$, there is an exponential decrease in the wall concentration of BLG. The resulting decrease in self-rejection causes the sieving coefficient of BLG to increase by 146% as β decreases (Table 4.1). Therefore, for protein fractionation using charged ultrafiltration membranes, both the charge on the membrane and the polarization index, β, are critical for separation performance. The membrane charge does not affect β, but the charge and β do affect the sieving coefficient and selectivity.

4.3.2.2 Staging to increase purity of protein fractions

As discussed in Section 4.2, diafiltration is the most commonly used mode for enriching the permeate and retentate fractions. However, when tight membranes are used to fractionate ALA and BLG from MSP, the large amount of water that needs to be added for diafiltration to obtain high purity is problematic, and even then the purity of the protein fractions does not exceed 80%. One option to increase the purity of the protein fractions is to use staging in combination with diafiltration (Lightfoot, Root, and O'Dell, 2008; Mayani, Mohanty, and Ghosh, 2009). However, addition of diafiltration water can be avoided altogether when large pore-size charged ultrafiltration membranes are used in stages, and yet still produce pure protein fractions in the permeate and retentate. Analogous to distillation, the preferentially permeated protein is the "light" component and passes into the rectification (top) section and the preferentially retained protein is the "heavy" component and passes into the stripping (bottom) section. It follows that when $S_o > 0.5$, it can be considered a "light" component and when $S_o < 0.2$, it can be considered a "heavy" component. When we have a situation having one light component and one heavy component, an efficiently designed multistage system can give enriched fractions of both of these two components.

A schematic of two different three-stage systems is shown in Figure 4.2. Figure 4.2a shows a three-stage system with two rectification stages and one stripping stage. Briefly, the permeate from stage 1 (P_1) is fed to stage 2 and permeate P_2 is obtained as the enriched light product. The retentate R_1 from stage 1 is fed to stage 3, where it is stripped of the light component to give product R_3 enriched in the heavy component. Streams R_2 and P_3 have a composition similar to the feed stream and are recycled back to the feed stream. Using the sieving coefficients from Table 4.1 for ALA and BLG for the 300 kDa charged membrane at $\beta = 6.4$ (S_o ALA = 0.52, S_o

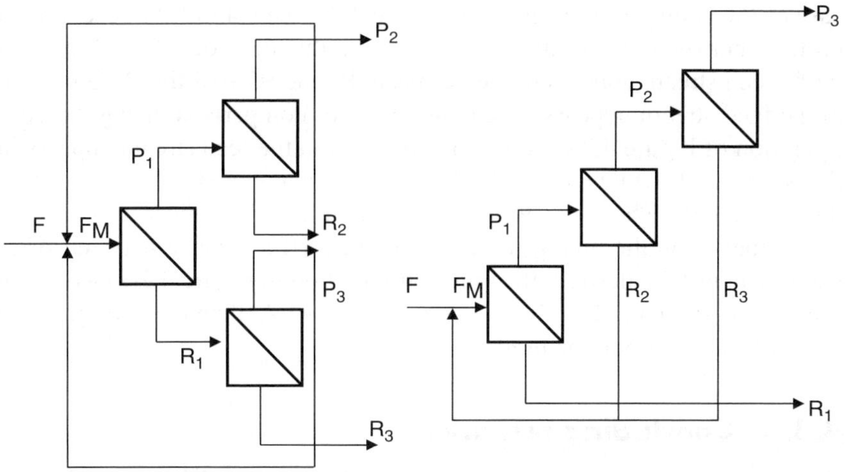

Figure 4.2 Different three-stage flow configurations: (a) three-stage system with two rectification stages and one stripping stage and (b) three-stage system with three rectification stages.

BLG = 0.13), feed concentrations of 1.2 g/L for ALA and 2.2 g/L for BLG (35% ALA and 65% BLG), and a volume concentration ratio (VCR) of 5 in each stage, the permeate P_2 is 87% pure in ALA and the retentate R_3 is 83% pure in BLG (Table 4.2) as calculated by mass balance for a continuous system (Arunkumar and Etzel, 2013).

Figure 4.2b shows a three-stage rectification system. In this system, permeate P_1 is fed to stage 2 and permeate P_2 from stage 2 is fed to stage 3 to obtain permeate P_3 as the light product. Retentate streams R_2 and R_3 are recycled back to the feed stream. Retentate R_1 from stage 1 is taken as the product stream enriched in the heavy component. This flow configuration gives 95% pure ALA in P_3 and 73% pure BLG in R_1. That compares to the feed stream that is 35% pure ALA and 65% pure BLG (Table 4.2). In the

Table 4.2 Purities of ALA and BLG in the permeates and retentates for different staging situations. The feed (MSP) contains 1.2 g/L ALA and 2.2 g/L BLG (35% ALA and 65% BLG). S_o ALA = 0.52, S_o BLG = 0.13, volume reduction of 5 times for each stage.

Staging configuration	**Purity of ALA in permeate (%)**	**Purity of BLG in permeate (%)**
One-stage system	63	78
Two-stage system with recycle	86	75
Three-stage system (two rectification stages and one stripping stage)	87	83
Three-stage rectification system	95	73

case of the configuration shown in Figure 4.2a, the split of the feed stream proteins between the product streams is such that 65% of the ALA present in the feed stream appears in the permeate P_2 and 95% of the BLG present in the feed stream appears in retentate R_3. For comparison, using the configuration of Figure 4.2(b), only 31% of ALA in the feed stream appears in the permeate P_3 while 99% of the BLG present in the feed stream appears in the retentate R_1. Therefore, the choice between using rectification and stripping depends on the final product requirement. If a balance between ALA and BLG purity is desired, then the configuration in Figure 4.2(a) would be used. If the primary objective is to obtain pure ALA, then the configuration in Figure 4.2(b) would be used.

4.3.3 Concluding remarks

Charged 300 kDa ultrafiltration membranes that have a MWCO 15–20 times larger than the proteins of interest (ALA and BLG) can be used effectively to fractionate these proteins at a 180% increased selectivity compared to an uncharged 300 kDa membrane. Using large-pore size charged ultrafiltration membranes in a staged configuration produces pure dairy protein fractions without the need for addition of water for diafiltration.

4.4 NEGATIVELY CHARGED ULTRAFILTRATION MEMBRANES FOR PROTEIN CONCENTRATION

Current standards for concentration of dairy protein require the use of tight membranes (10 kDa to 30 kDa). While these membranes offer high protein retention, they have the problem of low permeate flow rates and fouling due to pore blocking. Larger pore size ultrafiltration membranes offer higher flow rates but come at the expense of low protein retention. Charged ultrafiltration membranes can ameliorate that problem.

At near neutral pH (pH 6–7), the major dairy proteins carry a net-negative charge because the average pI is about 4.6. Therefore, a large-pore size negatively charged ultrafiltration membrane should be possible to reject these proteins. This approach was tested using negatively charged 100 kDa and 300 kDa ultrafiltration membranes.

4.4.1 Materials and methods

4.4.1.1 Membrane modification

Millipore membranes that are available commercially were modified to add a negative charge. Two negatively charged sulfonic acid ligands were evaluated: 3-bromopropane sulfonic acid (Bromo-S) and 2-aminoethane sulfonic acid (taurine). For Bromo-S, the bromine moiety reacts directly with the hydroxyl moieties on the cellulose to form a permanent covalent bond that will not leach off (van Reis, 2006). For the taurine, the membranes were first reacted with allyl glycidyl ether and *N*-bromosuccinimide to place the bromine moiety directly on the cellulose (Riordan et al., 2009; Rohani and Zydney, 2012). Then taurine was attached to the membrane via its free primary amine using two ligand densities (low caustic (LC) and high caustic (HC).

The amount of negative charge on the membrane was determined by measuring the number of protons that bind to the negatively charged membrane after treating it with an excess of strong acid (0.1 *M* HCl). The hydrogen ions were desorbed using 1 *M* KNO_3 and the eluate titrated using 0.02 *M* NaOH. The ionic capacity (I_c) of the membrane was calculated according to the formula: ionic capacity = mmol H^+ per m^2 membrane area = $C_{NaOH} x V_{OH} / A_m$, where C_{NaOH} = concentration of NaOH (*M*), V_{OH} = volume of NaOH at the equivalence point (mL), and A_m = membrane area (m^2). One mmol H^+ equals one milliequivalent. Just 1 *M* of KNO_3 by itself required small volumes (0.15 to 0.2 mL) of 0.02 *M* NaOH for titration to the equivalence point, corresponding to I_c = 0.60 to 0.80 mmol/m^2.

4.4.1.2 Ultrafiltration experiments using milk serum permeate (MSP)

Protein rejection for milk serum permeate (MSP) was measured using tangential-flow ultrafiltration membranes (Pellicon XL, Ultracel PLC, EMD Millipore, Bedford, MA) containing either the negatively charged ligand or no ligand at all for the uncharged, unmodified membrane. An unmodified, uncharged 10 kDa membrane was also tested for comparison purposes. The goal was to achieve about the same protein sieving coefficient (S_o) as the 10 kDa membrane, but at a higher milk serum permeate flux (J_v) and higher hydraulic permeability (L_p).

4.4.1.3 Concentration of Swiss cheese whey

An industrial process for producing WPC 80 (Figure 4.3) was simulated using the 10 kDa unmodified membrane, 100 kDa unmodified membrane, 100 kDa charged membrane, and 300 kDa charged membrane. Membrane modules used were the Pellicon-2™ Mini modules that had a membrane area of 1000 cm^2 and incorporated suspended screens ("V" screens). 5L of Swiss cheese whey at pH 6.8 was used as the feed stream. The first step of the process was a 10-fold volume reduction from 5 L to 0.5 L producing 4.5 L of permeate. Following volume reduction, the retentate was diafiltered by addition of 2.4 volumes of diafiltration water (1.2 L of DI water).

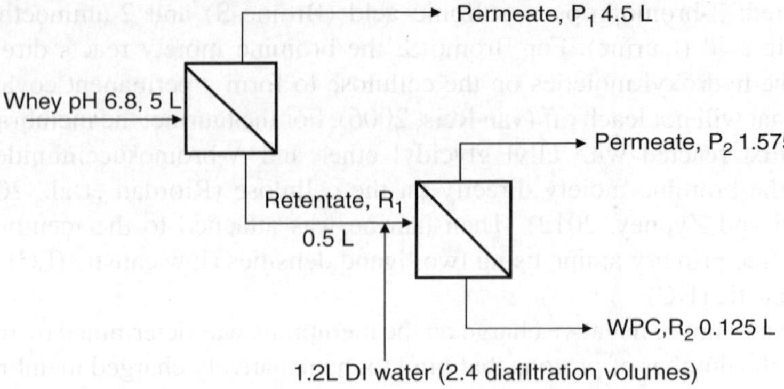

Figure 4.3 Schematic diagram of the process used to concentrate Swiss cheese whey by 40 times to manufacture WPC 80.

During diafiltration, water was added to the feed tank at the same rate as permeate removal, keeping the level of the tank constant. Following diafiltration, the retentate was further concentrated by a factor of 4 from 0.5 L to 0.125 L. At the end of this process, the membrane retentate hold-up volume was flushed with 70 mL of DI water to displace the residual concentrate held in the membrane. Samples of the permeate and retentate were collected for later analysis. Permeate flux was measured periodically during the ultrafiltration and diafiltration process.

Analysis of permeate and retentate samples was by HPLC. Samples were analyzed separately for glycomacropeptide (GMP) and the other whey proteins (OWP) using the method of Bhushan and Etzel (2009). Briefly, the OWP (primarily ALA and BLG) were bound to a cation-exchanger (1 mL HiTrap SPFF®) at pH 4 and eluted using 1 *M* NaCl. The signal at 280 nm was used to quantify the peaks for concentration using a known calibration curve. In the case of GMP, the OWP were precipitated using 7.2% trichloroacetic acid, centrifuged at 4000g for 1 h, and the desalted supernatant was injected into a BioSil SEC® (Bio-Rad, Hercules, CA) size-exclusion column. The area at 214 nm was used to quantify GMP using a calibration curve. Total whey protein (TWP) was the sum of OWP and GMP.

4.4.2 Results and discussion

4.4.2.1 Measures of membrane performance

There are two measures used to characterize the permeability of the membrane. The first measure is the permeability to pure water called the hydraulic permeability (L_p). L_p was determined by measuring the flux of deionized water at 22 °C versus pressure drop, and taking the slope. The second measure is the permeability using the protein mixture, such as whey or milk serum, and is the permeate flux (J_v). L_p is generally greater than J_v because, when using a protein mixture, a boundary layer of rejected protein builds up on the surface of the membrane and restricts flow. L_p is more a characteristic of the membrane itself, whereas J_v depends also on the solution characteristics such as the solution viscosity, protein concentration, protein diffusion coefficient, boundary-layer thickness, fluid shear rate, flow path length, and feed channel spacer height (Section 4.2.2). In a protein concentration process, J_v determines throughput. The goal was to achieve about the same sieving coefficient (S_o) as the 10 kDa membrane, but at a higher MSP or whey flux (J_v) and a higher hydraulic permeability (L_p) than the 10 kDa membrane.

4.4.2.2 Sieving coefficients during the ultrafiltration of MSP using negatively charged membranes

After the ultrafiltration of MSP described in Section 4.4.1, samples were analyzed for ALA and BLG using SDS-PAGE and fluorescence laser densitometry. Using the 300 kDa negatively charged membranes, it was desired to obtain a higher permeate flux (J_v) and the same protein retention as the 10 kDa standard membrane, where $S_o = 0.01$ for total protein (sum of ALA and BLG), $J_v = 6$ litres per square meter per hour (LMH) at a pressure drop of 2 bar, and $L_p = 50$ LMH/bar (Table 4.3).

As shown in Table 4.3, all three negatively charged membranes exceeded the flux target for J_v by 6 times, but the amount of negative charge provided by the Bromo S ligand (3.3 meq/m^2) was insufficient to reject enough protein

Table 4.3 Performance characteristics of unmodified and negatively charged ultrafiltration membranes during the ultrafiltration of MSP. Experiments were performed at 2 bar.

Membrane	Unmodified 10 kDa	Unmodified 300 kDa	Bromo S 300 kDa	LC Taurine S 300 kDa	HC Taurine S 300 kDa
Ionic capacity (mmol/m^2)	1.5	1.1	3.3	4.7	15.7
Hydraulic permeability (L_p) (LMH/bar)	50	250	200	190	180
Sieving coefficient	0.01	0.56	0.39	0.12	0.04
MSP flux (J_v) (LMH/bar)	3	18	18	18	18

(61% rejection, S_o = 0.39), although rejection was nevertheless better than the uncharged 300 kDa membrane (44% rejection, S_o = 0.56). LC taurine was better than the Bromo S because it had more charge (4.7 meq/m^2) and more protein rejection (88% rejection, S_o = 0.12), but the high caustic (HC) taurine chemistry was better because it had the most charge (15.7 meq/m^2) and the most protein rejection (96% rejection, S_o = 0.04). The difference between the 99% rejection found using the uncharged 10 kDa membrane and the 96% rejection found using the negatively charged 300 kDa HC taurine membrane was acceptable given that L_p was 3.6 times greater, and J_v was 6 times greater for the 300 kDa HC taurine membrane compared to the uncharged 10 kDa membrane.

There was a trade-off between L_p and S_o with increasing I_c for the 300 kDa membranes (Table 4.3). As I_c increased, both S_o and L_p decreased. The net result was that gaining from a higher recovery (smaller S_o) required suffering from a lower L_p as I_c increased. The proper balance between gaining recovery at the expense of losses in L_p will depend on the application. Nevertheless, in all cases, the negatively charged 300 kDa membrane was a three- to fourfold improvement over the L_p of the uncharged 10 kDa membrane (L_p = 50 LMH/bar) used presently to concentrate dairy proteins.

4.4.2.3 Scale-up to Pellicon-2™ Mini for producing WPC 80

An industrial process for producing WPC 80 (Figure 4.3) was simulated using the 10 kDa unmodified membrane, 100 kDa unmodified membrane, 100 kDa charged membrane, and 300 kDa charged membrane. The membranes used for this experiment were Pellicon-2™ Mini modules that had a membrane area of 1000 cm^2 and suspended screens. These modules were thus 20-fold larger in membrane area than the Pellicon XL modules mentioned earlier. Permeate flux was controlled throughout the ultrafiltration and there was no flux decline. As described in Section 4.4.1, samples were analyzed for concentrations of OWP, GMP, and TWP using HPLC.

Recoveries of OWP, GMP, and TWP in the WPC 80 product (retentate stream R_2 in Figure 4.3) were measured and compared to the feed stream. Values are plotted in Figure 4.4. Recovery of OWP was not different ($p > 0.05$) between the 10 kDa unmodified and 100 kDa S (negatively charged) membranes. However, the 100 kDa S membrane had a 28% higher recovery for GMP compared to the 10 kDa unmodified membrane and a 21-fold higher recovery for GMP compared to the 100 kDa unmodified membrane. The 100 kDa S membrane allowed a flux enhancement of 2.8 times compared to the 10 kDa unmodified membrane (Table 4.4). Recovery of OWP and TWP for the 300 kDa S (negatively charged) membrane was

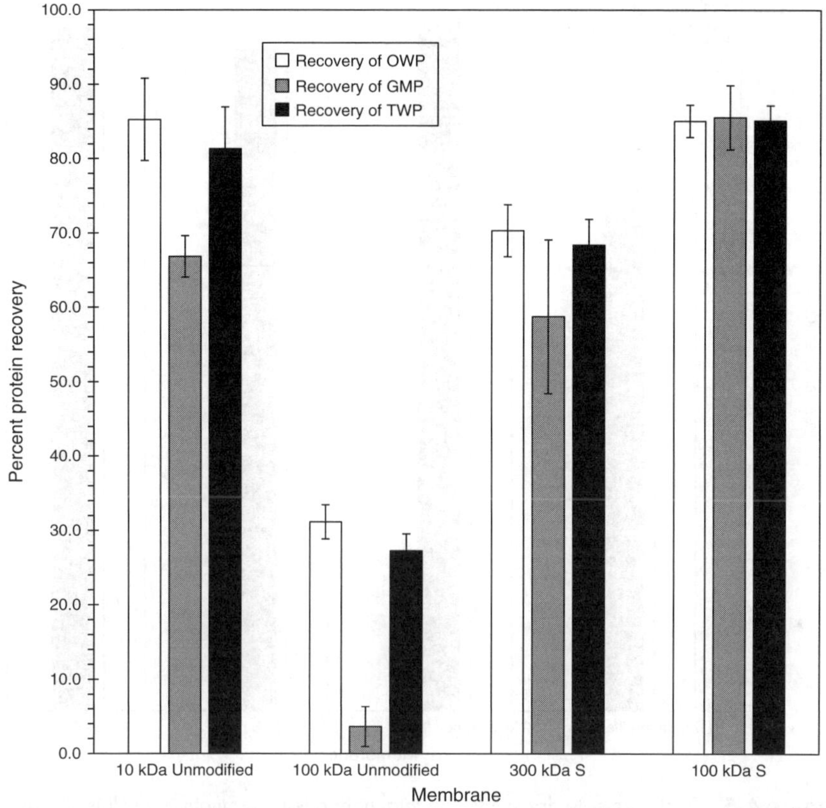

Figure 4.4 Protein recovery (%) and for the WPC 80 process in terms of other whey proteins (OWP), glycomacropeptide (GMP), and total whey protein (TWP).

Table 4.4 Permeate flux of different membranes during the ultrafiltration of Swiss cheese whey. Experiments for the unmodified membranes were performed at 2 bar and those for the charged membrane were performed at 1.4 bar.

Membrane	**Unmodified 10 kDa**	**Unmodified 100 kDa**	**Negatively charged 100 kDa S**	**Negatively charged 300 kDa S**
Permeate flux of whey (J_v) (LMH/bar)	6.0	12.0	17.4	34.8
Flux enhancement factor (compared to 10 kDa)	1.0	2.0	2.8	5.8

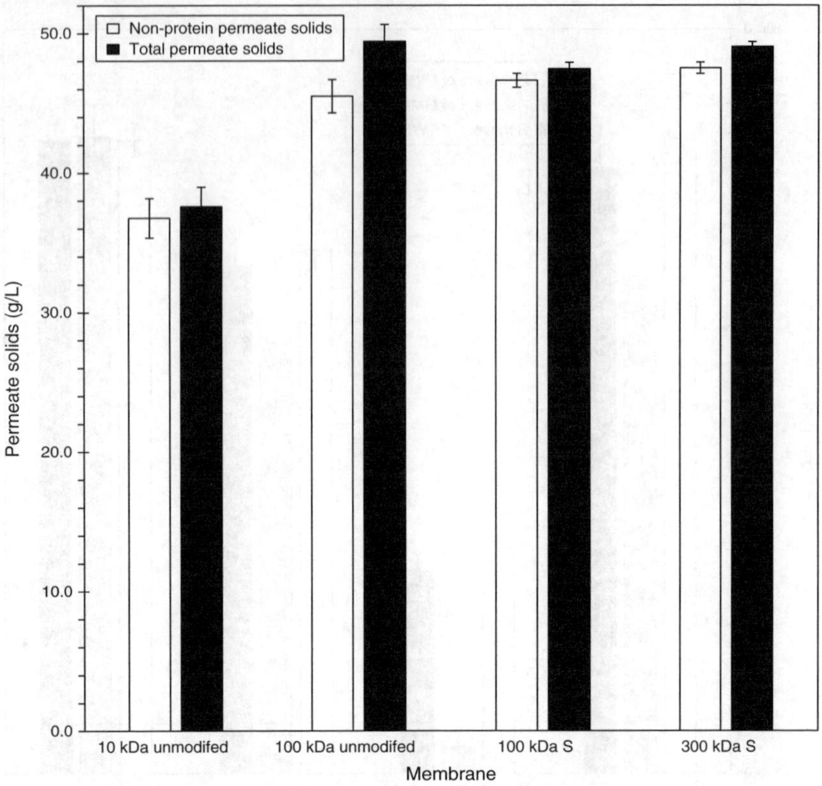

Figure 4.5 Total permeate dry solids and total nonprotein permeate dry solids measured from the composite permeate and diafiltrate streams.

17% lower than the 10 kDa unmodified membrane ($p < 0.05$), while that of GMP was not different ($p > 0.05$). However, the 300 kDa S membrane offered a flux that was 5.8 times higher than the 10 kDa unmodified membrane throughout the concentration process (Table 4.4).

As shown in Figure 4.5, transmission of nonprotein permeate solids (mainly lactose, minerals, and nonprotein nitrogen) was 27% higher for the 100 kDa S and 29% higher for 300 kDa S negatively charged membranes compared to the 10 kDa unmodified membrane. The nonprotein permeate solids were calculated by subtracting the protein dry solids from the total dry solids in the permeate. This result is significant because less water would be required for diafiltering out the nonprotein permeate solids using the negatively charged membranes compared to the 10 kDa unmodified membrane to get the same protein content in the retentate. Less added water would mean less permeate generation and greater sustainability of the process.

4.4.2.4 Membrane fouling

The extent of membrane fouling was measured by means of the normalized water permeability. The ultrafiltration membrane was rinsed with up to 100 L/m^2 of DI water after ultrafiltration of a dairy feed stream and the hydraulic permeability was remeasured without chemically cleaning the membrane first. Normalized water permeability (NWP) is the ratio of the permeability before ultrafiltration to the permeability after ultrafiltration, but before cleaning. A higher NWP means less fouling. It was found that even after the 40-fold concentration of cheese whey, the normalized water permeability was a 100% for the negatively charged membranes (data not shown) compared to 60% for the unmodified membranes. This means that the negatively charged membranes were antifouling. Therefore, they could be cleaned faster and with less cleaning chemicals compared to the unmodified membranes that are used currently.

4.4.3 Concluding remarks

Negatively charged 100 kDa ultrafiltration membranes can be used to concentrate whey at a flux that is 2.8 times higher than the 10 kDa industry standard membrane, but at the same protein recovery and higher transmission of small molecules in whey like lactose and nonprotein nitrogen. Especially in the case of large-pore size membranes, it is necessary to control the permeate flux rather than the trans-membrane pressure to ensure stable performance. Negatively charged membranes are antifouling, which has implications for sustainability.

Nomenclature

Symbols

A_m = membrane surface area (m^2)
C = protein concentration (g/L)
C_b = protein concentration in bulk feed solution (g/L)
C_p = instantaneous protein concentration in permeate (g/L)
C_r = instantaneous protein concentration in retentate (g/L)
C_w = concentration of protein on the membrane wall (g/L)
D = protein diffusion coefficient (m^2/s)
h = feed channel spacer height (m)
J_v = permeate flux (L/m^2 h or m/s)
k = boundary-layer mass transfer coefficient (L/m^2 h or m/s)
k_B = Boltzmann constant (J/K)
K_c = hindrance factor for solute convective transport

L_p = membrane hydraulic permeability (L/m^2 h bar)
Q_P = permeate flow rate (mL/min)
Q_R = retentate recirculation rate (mL/min)
r_s = radius of solute (m)
r_p = radius of pore (m)
S_a = actual sieving coefficient
S_o = observed sieving coefficient
T = absolute temperature (K)
V_{HR} = retentate hold-up volume (mL)
VCR = volume concentration ratio
x = axial coordinate (m)
ΔP = transmembrane pressure (TMP)

Greek letters

β = polarization index
δ = thickness of concentration polarization boundary layer (m)
σ_p = dimensionless surface charge density of membrane pore
σ_s = dimensionless surface charge density of solute (protein)
ψ = selectivity for protein fractionation
ψ_E = energy of interaction between solute and pore (J)

Abbreviations

ALA = α-lactalbumin
BLG = β-lactoglobulin
GMP = glycomacropeptide
MSP = milk serum permeate
MWCO = molecular weight cut-off
NWP = normalized water permeability
OWP = other whey proteins (all whey proteins other than GMP)
pI = isoelectric point
TWP = total whey protein

References

Abd El-Salam, M.H. (2006) Separation of casein glycomacropeptide from whey: methods of potential industrial application. *International Journal of Dairy Science*, **1**, 93–99.

Arunkumar, A. and Etzel, M.R. (2013) Fractionation of α-lactalbumin from β-lactoglobulin using positively charged tangential flow ultrafiltration membranes. *Separation and Purification Technology*, **105**, 121–128.

Arunkumar, A. and Etzel, M.R. (2014) Fractionation of α-lactalbumin and β-lactoglobulin from bovine milk serum using staged, positively charged, tangential flow ultrafiltration membranes. *Journal of Membrane Science*, **454**, 488–495.

Bakhshayeshi, M., Teella, A., Zhou, H., Olsen, C., et al. (2012) Development of an optimized dextran retention test for large pore size hollow fiber ultrafiltration membranes. *Journal of Membrane Science*, **421**, 32–38.
Barba, D., Beolchini, F., Cifoni, D., and Veglio, F. (2001) Whey protein concentrate production in a pilot scale two-stage diafiltration process. *Separation Science and Technology*, **36** (4), 587–603.
Beckman, S.L., Zulewska, J., Newbold, M., and Barbano, D.M. (2010) Production efficiency of micellar casein concentrates using polymeric spiral-wound microfiltration membranes. *Journal of Dairy Science*, **93**, 4506–4517.
Bhattacharjee, S., Bhattacharjee, C., and Datta, S. (2006) Studies on the fractionation of β-lactoglobulin from casein whey using ultrafiltration and ion-exchange membrane chromatography. *Journal of Membrane Science*, **275**, 141–150.
Bhushan, S. and Etzel, M.R. (2009) Charged ultrafiltration membranes increase the selectivity of whey protein separations. *Journal of Food Science*, **74**, E131–E139.
Bottomley, R.C. (1991) Process for obtaining concentrates having a high α-lactalbumin content from whey. US Patent **5**,008,376.
Burns, D.B. and Zydney, A.L. (2001) Contributions to electrostatic interactions on protein transport in membrane, *American Institute of Chemical Engineers Journal*, **47**, 1101–1114.
Cheang, B. and Zydney, A.L. (2003) Separation of α-lactalbumin and β-lactoglobulin using membrane ultrafiltration. *Biotechnology and Bioengineering*, **83**, 201–209.
Cheang, B. and Zydney, A.L. (2004) A two-stage ultrafiltration process for the fractionation of whey protein isolate. *Journal of Membrane Science*, **231**, 159–167.
Cheryan, M. and Kuo, K.P. (1984) Hollow fibers and spiral wound modules for the ultrafiltration of whey: energy consumption and performance. *Journal of Dairy Science*, **67**, 1406–1413.
Cowan, S. and Ritchie, S. (2007) Modified polyethersulfone (PES) ultrafiltration membranes for enhanced filtration of whey proteins. *Separation Science and Technology*, **42**, 2405–2418.
Da Costa, A.R., Fane, A.G., and Wiley, D.E. (1993) Ultrafiltration of whey protein solutions in spacer-filled flat channels. *Journal of Membrane Science*, **76**, 245–254.
Davey, M., Lindman, K., Perera, J.M., Stevens, G.W., et al. (2004) Measurement and prediction of ultrafiltration of whey protein. *American Institute of Chemical Engineers Journal*, **50** (7), 1431–1437.
Delaney, R.A. (1976) Composition, properties and uses of whey protein concentrates, *Journal of Society of Dairy Technology*, **29**, 91–101.
Doultani, S., Turhan, K.N., and Etzel, M.R. (2003) Whey protein isolate and glycomacropeptide recovery from whey using ion-exchange chromatography. *Journal of Food Science*, **68**, 1389–1395.
Doultani, S., Turhan, K.N., and Etzel, M.R. (2004) Fractionation of proteins from whey using cation exchange chromatography. *Process Biochemistry*, **39**, 1737–1743.
Gesan-Guiziou, G., Daufin, G., Timmer, M., Allersma, D., and Van der Horst, C. (1999) Process steps for the preparation of purified fractions of α-lactalbumin and β-lactoglobulin from whey protein concentrates. *Journal of Dairy Research*, **66**, 225–236.
Holland, B., Kackmar, J., and Corredig, M. (2012) Isolation of a whey fraction rich in α-lactalbumin from skim milk using tangential flow filtration. *Journal of Dairy Science*, **95**, 5604–5607.
Hurt, E., Zulewska, J., Newbold, M. and Barbano, D.M. (2010) Micellar casein concentrate production with a 3×, 3-stage, uniform transmembrane pressure ceramic membrane process at 50 °C. *Journal of Dairy Science*, **93**, 5588–5600.
Kawasaki, Y. and Dosako, S. (1994) Process for producing κ-casein glycomacropeptide. US Patent 5,278,288.
Konrad, G. and Kleinschmidt, T. (2008) A new method for isolation of native α-lactalbumin from sweet whey. *International Dairy Journal*, **18**, 47–54.

Kulozik, U. and Kessler, H.G. (1990) Effects of salt ions and deposit formation on the permeation of organic molecules in complex media in reverse osmosis. *Journal of Membrane Science*, **54**, 339–354.

Lebreton, B., Brown, A., and van Reis, R. (2008) Application of high-performance tangential flow filtration (HPTFF) to the purification of a human pharmaceutical antibody fragment expressed in *Escherichia coli*. *Biotechnology and Bioengineering*, **100**, 964–974.

Lightfoot, E.N., Root, T.W., and O'Dell, J.L. (2008) Emergence of ideal membrane cascades for downstream processing. *Biotechnology Progress*, **24**, 599–605.

Lucas, D., Rabiller-Baudry, M., Millesime, L., Chaufer, B., and Daufin,G. (1998) Extraction of α-lactalbumin from whey protein concentrate with modified inorganic membranes. *Journal of Membrane Science*, **148**, 1–12.

Marella, C., Muthukumarappan, K., and Metzger, L.E. (2011) Evaluation of commercially available wide-pore size ultrafiltration membranes for production of alpha-lactalbumin enriched whey protein concentrate. *Journal of Dairy Science*, **94**, 1165–1175.

Mayani, M., Mohanty, K., and Ghosh, R. (2009) Continuous fractionation of plasma proteins HSA and HIgG using cascade ultrafiltration systems. *Separation and Purification Technology*, **70**, 231–241.

Mehta, A. and Zydney, A.L. (2005) Permeability-selectivity analysis for ultrafiltration membranes. *Journal of Membrane Science*, **249**, 245–249.

Mercadante, D., Melton, L.D., Norris, G.E., Loo, T.S., Williams, M.A., Dobson, R.C., et al. (2012) Bovine β-lactoglobulin is dimeric under imitative hysiological conditions: dissociation equilibrium and rate constants over the pH range of 2.5–7.5. *Biophysical Journal*, **103**, 303–312.

Metsämuuronnen, S. and Nyström,M. (2006) Evaluation of six flat-sheet ultrafiltration membranes for fractionation of whey proteins. *Desalination*, **200**, 290–291.

Metsämuuronnen, S. and Nyström,M. (2009) Enrichment of α-lactalbumin from diluted whey with polymeric ultrafiltration membranes. *Journal of Membrane Science*, **337**, 248–256.

Michaels, A.S. (1968) New separation technique for the CPI. *Chemical Engineering Progress*, **64**, 31–43.

Miller, D.J., Kasemset, S., Paul, D.R., and Freeman, B.D. (2014) Comparison of membrane fouling at constant flux and constant transmembrane pressure conditions. *Journal of Membrane Science*, **454**, 505–515.

Mistry, V.V. (2002) Manufacture and application of high milk protein powder. *Lait*, **82**, 515–522.

Muller, A., Chaufer, B., Merin, U., and Daufin, G. (2003) Purification of α-lactalbumin from pre-purified whey acid whey: ultrafiltration or precipitation. *Lait*, **83**, 439–451.

Muller, A., Daufin, G., and Chaufer, B. (1999) Ultrafiltration modes for the operation for the separation of α-lactalbumin from acid casein whey. *Journal of Membrane Science*, **153**, 9–21.

Nelson, B.K. and Barbano, D.M. (2005) A microfiltration process to maximize removal of serum proteins from skim milk before cheese making. *Journal of Dairy Science*, **88**, 1891–1900.

Obermeyer, J.H., Kulozik, U., and Kessler, H.G. (1993) Controlled deposit formation to influence the retention of solutes in reverse osmosis and ultrafiltration. *Desalination*, **90**, 161–172.

Pujar, N.S. and Zydney, A.L. (1997) Charge regulation and electrostatic interactions for a spherical particle in a cylindrical pore. *Journal of Colloid Interface Science*, **192**, 338–349.

Pujar, N.S. and Zydney, A.L. (1998) Electrostatic effects on protein partitioning in size-exclusion chromatography and membrane ultrafiltration. *Journal of Chromatography A*, **796**, 229–238.

Raubiller-Baudry, M., Gesan-Guiziou, G., Roldan-Calbo, D., Beaulieu, S. and Michel, F. (2005) Limiting flux in skimmed milk ultrafiltration: impact of electrostatic repulsion due to casein micelles. *Desalination*, **175**, 49–59.

Rektor, A. and Vatai, G. (2004) Membrane filtration of Mozarella whey. *Desalination*, **162**, 279–286.

Rinn, J.C., Morr,C.V., Seo, A., and Surak, J.G. (1990) Evaluation of nine semi-pilot scale whey pretreatment modifications for producing whey protein concentrate. *Journal of Food Science*, **55**, 510–515.

Riordan, W.T., Heilmann, S., Brorson, K., Seshadri, K., He, Y., et al. (2009) Design of salt-tolerant membrane adsorbers for viral clearance. *Biotechnology and Bioengineering*, **103**, 920–929.

Rohani, M.M. and Zydney, A.L. (2009) Effect of surface charge distribution on protein transport through semipermeable ultrafiltration membranes. *Journal of Membrane Science*, **337**, 324–331.

Rohani, M.M. and Zydney, A.L. (2010) Role of electrostatic interactions during protein ultrafiltration. *Journal of Colloid Interface Science*, **160**, 40–48.

Rohani, M.M. and Zydney, A.L. (2012) Protein transport through zwitterionic ultrafiltration membranes. *Journal of Membrane Science*, **397**, 1–8.

Saufi, S.M. and Fee, C.J. (2013) Recovery of lactoferrin from whey using cross-flow cation-exchange mixed matrix membrane chromatography. *Separation and Purification Technology*, **77**, 68–75.

Smith, F.G. and Deen, W.M. (1980) Electrostatic double-layer interactions for spherical colloids in cylindrical pores. *Journal of Colloid and Interface Science*, **78**, 444–465.

Smith, F.G. and Deen, W.M. (1983) Electrostatic effects on the partitioning of spherical colloids between dilute bulk solution and cylindrical pores. *Journal of Colloid and Interface Science*, **91**, 571–590.

Tanimoto, Y., Kawasaki, Y., Shinmoto, H., Dosako, S., and Tomizawa, A. (1991) Process for producing kappa-casein glycomacropeptide. US Patent 5,075,424.

Tolkach, A. and Kulozik, U. (2005) Fractionation of whey proteins and caseinomacropeptide by enzymatic crosslinking and membrane separation techniques. *Journal of Food Engineering*, **67**, 13–20.

Tolkach, A., Steinle, S., and Kulozik, U. (2005) Optimization of thermal pretreatment conditions, for the separation of native α-lactalbumin from whey protein concentrates by means of selective denaturation of β-lactoglobulin. *Journal of Food Science*, **70**, E557–E566.

Toro-Sierra, J., Tolkach, A., and Kulozik, U. (2013) Fractionation of α-lactalbumin and β-lactoglobulin from whey protein isolate using selective thermal aggregation, an optimized membrane separation procedure and resolubilization techniques and pilot plant scale. *Food Bioprocess Technology*, **6**, 1032–1043.

van Reis, R. (2006) Charged filtration membranes and uses therefore. US Patent 7,001,550 B2.

van Reis, R., Goodrich, E.M., Yson, C.L., Frautschy, L.N., Whiteley, R., et al. (1997) Constant C_{wall} ultrafiltration process control. *Journal of Membrane Science*, **130**, 123–140.

van Reis, R., Brake, J.M., Charkoudian, J., Burns, D.B., and Zydney, A.L. (1999) High performance tangential flow filtration using charged membranes. *Journal of Membrane Science*, **159**, 133–142.

Xu, Y., Sleigh, R., Hourigan, J., and Johnson, R. (2000) Separation of bovine immunoglobulin G and glycomacropeptide from dairy whey. *Process Biochemistry*, **36**, 393–399.

Yee, K.W.K., Wiley, D.E., and Bao, J. (2007) Whey protein concentrate production by continuous ultrafiltration: operability under constant operating conditions. *Journal of Membrane Science*, **290**, 125–137.

Yee, K.W.K., Wiley, D.E., and Bao, J. (2009) A unified model of the time dependence of flux decline for the long-term ultrafiltration of whey. *Journal of Membrane Science*, **332**, 69–80.

5 Demineralization of dairy streams and dairy mineral recovery using nanofiltration

Sandra E. Kentish[1] and G. Rice[2]

[1] *ARC Dairy Innovation Hub, Department of Chemical and Biomolecular Engineering, The University of Melbourne, Victoria, Australia*
[2] *Beca Pty Ltd, Melbourne, Victoria, Australia*

5.1 INTRODUCTION

Nanofiltration (NF) is the membrane filtration operation that fits between ultrafiltration and reverse osmosis. Thus, the membrane retains lactose and other species of molecular weight greater than around 200 daltons, but it freely permeates many smaller molecules and ions, in particular, sodium and chloride. Many multivalent salts, such as calcium, magnesium, and phosphate, can be at least partially retained. However, the rejection of such salts relies upon subtle balances between the chemical nature of the divalent compounds, their electric charge, and their hydrated size. These balances are discussed further in Section 5.3.

The primary application of nanofiltration in dairy operations is to remove water and salt from dairy fluids. These fluids can be whole milk (Matsui et al., 2006), whey (Gernigon et al., 2011), or whey ultrafiltration permeate (Cuartas-Uribe et al., 2009; Leruyet and Maugas, 2001). Nanofiltration of milk and whey concentrates upstream of evaporators protects these equipment items from unwanted scaling and reduces the salt content of the final spray-dried product. More importantly, nanofiltration of whey ultrafiltration permeate allows for the profitable use of the retained concentrated lactose solution. The world dairy lactose market has grown substantially over the last two decades and currently sits at around 0.9 million tonnes (Paterson, 2011). This lactose is used in baby and infant formulations, as human milk contains more lactose than bovine milk. It is also widely employed in adult food products such as chocolate, confectionary, bakery, and meat products (Paterson, 2011). Further uses are as a fermentation medium and as a substrate or carrier for drugs in the pharmaceutical industry. Lactose derivatives, such as lactulose, lactitol, and lactobionic acid, are also potential products

Membrane Processing for Dairy Ingredient Separation, First Edition.
Edited by Kang Hu and James M. Dickson.

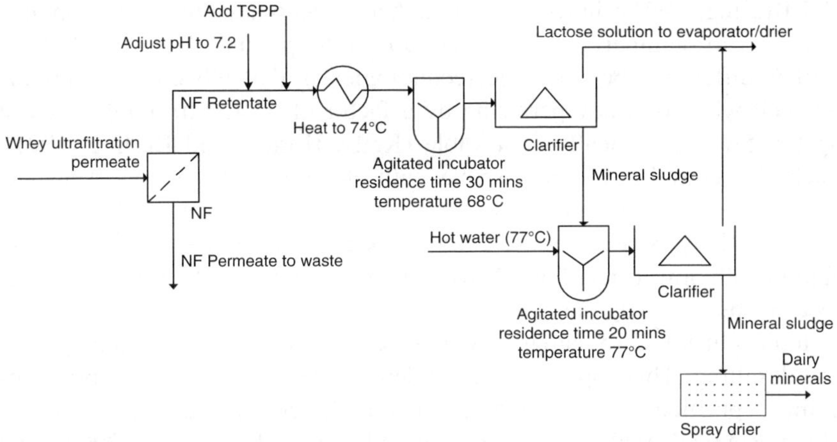

Figure 5.1 A process flowsheet showing how dairy minerals can be recovered from an ultrafiltration whey permeate stream following nanofiltration (based on Vembu and Rathinam, 1997).

with uses in pharmaceuticals, as well as in sweeteners and in animal feed (Horton, 1995).

In desalting applications, nanofiltration competes with reverse osmosis, electrodialysis, and ion exchange (Gernigon et al., 2011). Electrodialysis is covered in more detail in Chapter 4. A major distinction is that NF is more selective toward removal of the monovalent sodium ions, while the more nutritionally relevant divalent ions, calcium, zinc, and magnesium, can be retained (Rice et al., 2005; Van der Bruggen, Koninckx, and Vandecasteele, 2004). This means that calcium and magnesium "fortified" dairy ingredients can be prepared using this approach. Further, it is possible to recover these important dairy minerals directly from the NF retentate for sale as dietary supplements. In one such process (Figure 5.1), the dairy minerals are precipitated from the nanofiltration retentate by pH adjustment to around 7.2, addition of pyrophosphate or tetrasodium pyrophosphate (TSPP), and incubation at 62–80 °C (Vembu and Rathinam, 1997). Traditionally, such mineral supplements are derived from mining, from bone meal, and from oyster shells, but these sources can contain heavy metal impurities such as lead and arsenic (Vembu and Rathinam, 1997). The use of a dairy source eliminates such concerns over toxicity. Further, as the purifed minerals are in the same form in which they occur in milk (phosphates and citrates) they can be more easily absorbed by humans than the calcium carbonate of traditional calcium supplements (Vembu and Rathinam, 1997).

However, demineralization is less extensive than with other technologies. In NF, only 30% demineralization is achievable, with a 70% reduction in monovalent ions (Greiter et al., 2002). Electrodialysis can achieve demineralization of up to 90% (Hoppe and Higgins, 1992; Kelly, Horton, and Burling, 1991); however, ion exchange becomes the most economic approach to demineralization above 50% (Hoppe and Higgins, 1992). A further difference between the technologies is that NF concentrates the retained whey or whey permeate by a factor of 3–4, with total solids of up to 25% economically achievable (Kelly, Horton, and Burling, 1991). Such concentration is not possible with electrodialysis or ion exchange. In general, both operating and membrane costs are lower for NF relative to electrodialysis (Rice et al., 2005). Process water requirements are also significantly lower than for ED, with estimates of 0.05–0.1 litre per litre of feed (Hoppe and Higgins, 1992).

Ion exclusion chromatography may also be used for lactose purification and desalting. This approach is considerably more expensive but produces a more purified lactose stream, suitable directly for a wide range of pharmaceutical applications (Durham, Sleigh, and Hourigan, 2004). Ion exchange-based processes can also be more selective, for instance allowing the selective recovery of particular dairy sugars such as 3'-sialyl lactose (Scott and Krishnapillai, 2010).

Nanofiltration has also been often considered as a processing step in the purification and recovery of dairy wastewaters (Balannec et al., 2002; Frappart et al., 2006; Luo and Ding, 2011; Turan, 2004; Zhang et al., 2012). NF is able to concentrate the organic contaminants for disposal. However, the permeate water is still too salty for reuse and downstream reverse osmosis is required (Balannec et al., 2002; Vourch et al., 2003, 2005). NF is also often considered in combination with ultrafiltration (Gong, Zhang, and Cheng, 2012; Luo, et al., 2011, 2012) and/or microfiltration in these applications (Zhang et al., 2008).

NF can be used to recover cleaning agents such as caustic soda and detergents (Fernandez et al., 2010; Gesan-Guiziou, Boyaval, and Daufin, 2002; Gesan-Guiziou et al., 2007) from clean-in-place solutions. In this case, membranes that can accommodate the extremes in pH are required. Again, NF can be considered in combination with microfiltration or reverse osmosis (Merin et al., 2002; Rasanen et al., 2002).

More minor applications include the purification of lactic acid (Patel, Sandry, and Seguin, 2008) or acetic acid (Nayak and Pal, 2013) following production from dairy ultrafiltration permeate, the recovery of sialyloligosaccharides (Roth, Weil, and Spade, 2001; Spade, Weil, and McHale, 2003), and purification of galacto-oligosaccharides (Samhaber, 2005) manufactured from lactose hydrolysis.

5.2 MEMBRANE OPERATIONS

The membranes used commercially for nanofiltration are usually spiral wound polyamide membranes (Table 5.1). The polyamide layer is formed

Table 5.1 Current NF membranes sold for whey demineralisation and lactose Concentration. (Data taken from Manufacturer Specification Sheets.)

Supplier	Type	Molecular weight cut-off (daltons)	$MgSO_4$ rejection (%)	NaCl rejection (%)	pH range
Koch	SR3D	200	>99		1.8–11.0
Dow Filmtec	NF245	300			
	NF				
Synder	NFX	150–300	99	60	2.0–11.0
	NFW	300–500	98	40	3.0–10.5
Sepro	NF2A		98.5	35	
	NF3A		99.5	40	
	NF6		80	10	
GE	Dairy DK	150–300	98		3.0–9.0
	Dairy DL	150–300	96		3.0–9.0

interfacially and is only around 200 nm in thickness. It is thus supported by a more porous polysulfone interlayer and then a nonwoven polyester backing material. The polyamide layer can be based on either the classical chemistry created by the reaction between trimesoyl chloride and 1,3-benzenediamine or may use piperazine as the amine reagent, creating a polypiperazine amide structure (Tang, Kwon, and Leckie, 2009a). These polypiperazine amide membranes are more hydrophilic (Tang, Kwon, and Leckie, 2009b) and less prone to fouling (Tang and Leckie, 2007; Vrijenhoek, Hong, and Elimelech, 2001) than the standard formulation and thus are the type usually selected for dairy applications. The membranes can be treated with organic acids or alcohols at high concentrations and high temperatures to further improve performance (Mattila and Koivikko, 2011).

The presence of both carboxyl groups (–COOH) and amine groups ($-NH_2$) within the polyamide layer (Elimelech, Chen, and Waypa, 1994) means that it will usually carry a charge. At low solute concentrations, this charge can be determined from streaming potential (or zeta potential) measurements. These measurements can be performed both along the surface of flat-sheet polymeric membranes (Burns and Zydney, 2000; Childress and Elimelech, 1996; Kim et al., 1996; Peeters, Mulder, and Strathmann, 1999; Tay, Liu, and Delai Sun, 2002) and through the pores (Teixeira, Rosa, and Nyström, 2005).

Typically, the membrane carries a positive charge below a pH of around 3–4, but is negatively charged across the usual range of operation, i.e. above pH = 5. However, the net charge is also a function of the solution in contact with the membrane, as charged solutes will interact with the membrane charges (Afonso, Hagmeyer, and Gimbel, 2001; Childress and Elimelech, 1996; Shim et al., 2002; Tay, Liu, and Delai Sun, 2002). Thus for example, calcium decreases the magnitude of the zeta potential as it

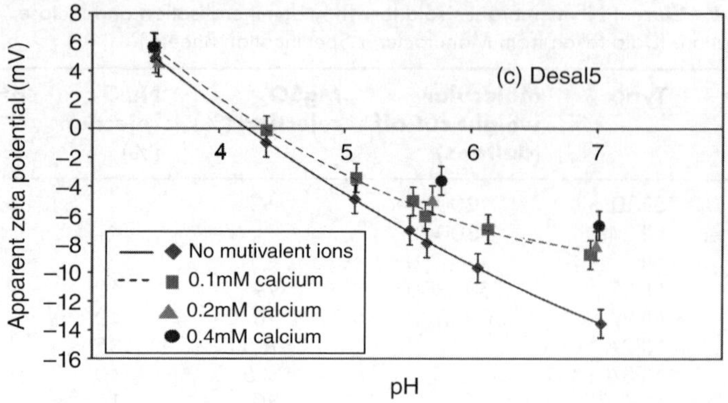

Figure 5.2 Effect of calcium on the measured zeta potential of a GE Osmonics Desal 5 membrane. All experiments were performed with a background electrolyte of 1 mM KCl. Error bars signify ±1 mV (Rice et al., 2011a). Reproduced with permission from Elsevier.

readily adsorbs to the membrane surface (see Figure 5.2). This adsorbed calcium can be difficult to remove during the cleaning process (Rice et al., 2011a). Conversely, anions such as phosphate and citrate have a minimal effect on the membrane charge (Rice et al., 2011a).

Nanofiltration feed pressures can vary from 5 to 20 bar. The nanofiltration operation can be run in a batch mode, where the feed is continuously recycled across the membrane until sufficient salt is removed and/or the lactose is sufficiently concentrated (Figure 5.3a). However, a feed and bleed

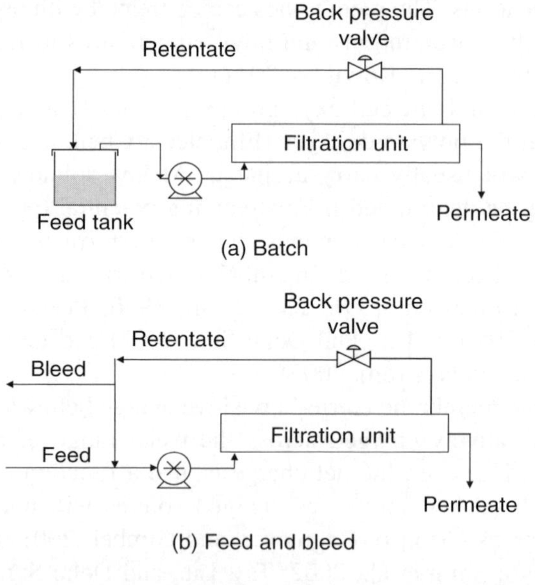

Figure 5.3 Operational modes of a typical dairy membrane operation.

operation is common (Figure 5.3b), where a feed is continuously supplied and a product stream is continuously removed in addition to the permeate. Diafiltration may also be used, where clean water is added to the feed and the process continued. In this mode the contents of the recycling retentate are effectively "washed" of further salt, as sodium chloride continues to permeate the membrane.

5.3 THEORY OF SEPARATION

The driving force for the flow of water across a nanofiltration membrane is predominantly the pressure difference between the feed and permeate sides (ΔP). However, this flow is somewhat restricted by the second law of thermodynamics, which states that fluids will tend to move to minimize concentration differences, rather than to create them. This concept is usually expressed in terms of the osmotic pressure (π).

As an example, consider a lactose solution placed on one side of an NF membrane, with pure water on the other side and no applied pressure difference between the two (Figure 5.4a). As the lactose cannot flow into the water to even out the concentration differences, water will instead flow to dilute the lactose. This flow will continue until the lactose solution reaches a height where the gravitational forces balance the osmotic pressure – the need for water to try and balance these concentration differences. In nanofiltration (and also in reverse osmosis), we apply a pressure difference between the two solutions that exceeds this osmotic pressure. This causes the situation to reverse and for water to flow in the opposite direction to that of constant concentration (Figure 5.4b).

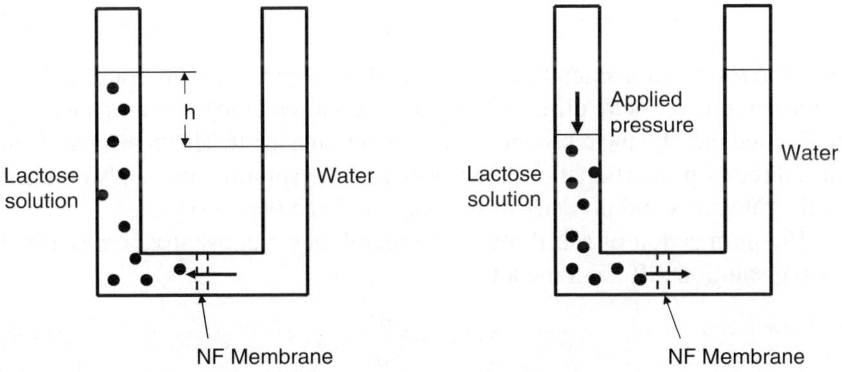

Figure 5.4 The concept of osmotic pressure (a) causes water to flow into a salt solution until it reaches a certain height to even out the concentration difference – this is osmosis. (b) An applied pressure in excess of this osmotic pressure is required to drive flow in the opposite direction – this is reverse osmosis.

The osmotic pressure (π) can usually be approximated by the Van't Hoff relationship, which is similar in form to the ideal gas law, although the two concepts are totally unrelated:

$$\pi = cRT \tag{5.1}$$

where c is the total solute concentration in mol/m^3, R is the universal gas constant (8.314 J/mol K), and T is the absolute temperature in Kelvin. It is important to note that this equation refers to dissociated solutes. Thus, for 1 mol/litre of lactose, c = 1 mol/litre, while for 1 mol/litre of NaCl, c = 2 mol/litre of ions. For 1 mol/litre of $MgCl_2$, c = 3 mol/litre. As salts have a lower molecular weight than lactose, they are usually the largest contributor to osmotic pressure.

While Equation (5.1) is the most widely used approach, Seki et al. (2012) find that for their whey system, the equation proposed by Morse (Morse, Frazer, and Rogers, 1907) showed better coincidence with experimental data than the van't Hoff equation:

$$\pi = m\rho^{\circ}RT \tag{5.2}$$

where m is the total solute molar concentration (mol/kg) while ρ° is the density of the pure solvent (water) at the same temperature.

For solutions with high ionic strength, neither approach is likely to be effective, as they do not take into account ionic interactions and activities (Van der Bruggen et al., 2001). In this case, more complex equations are required, such as the Pitzer model (Pitzer, 1991) or the electrolyte NRTL approach (Chen and Song, 2004).

The total volumetric flux (flow rate per unit membrane area, in m^3/m^2 s or litres/m^2 h) through a nanofiltration membrane (J_v) is then given by

$$J_v = \frac{\Delta P - \Delta\pi}{R_t} \tag{5.3}$$

where ΔP is the transmembrane pressure drop across the membrane and $\Delta\pi$ is the osmotic pressure difference. In an ideal system, the total resistance (R_t) is dictated only by the resistance of the membrane itself. This resistance is in turn directly proportional to the viscosity of the solution and is also related to the thickness and porosity of the polyamide active layer.

The permeation of salt through the membrane can usually be described by the solution diffusion model:

$$J_s = -B\frac{\mathrm{d}C}{\mathrm{d}x} \tag{5.4}$$

where B is the salt permeance, C is the concentration of salt in the bulk solution, and x is the thickness of the membrane.

However, for small and/or noncharged solutes, convective forces also become significant. Many authors describe this situation using the

Spielger–Kedem (Kedem–Katchalsky) model (Hoon and Jae-Hong, 2006; Hyung and Kim, 2006; Kedem and Katchalsky, 1961; Mariñas and Selleck, 1992; Spiegler and Kedem, 1966; Taniguchi, Kurihara, and Kimura, 2001). This model describes the water flux (Equation (5.5)) and solute flux (Equation (5.6)) as follows:

$$J_v = -A\left(\frac{dP}{dx} - \sigma\frac{d\pi}{dx}\right) \tag{5.5}$$

$$J_s = -B\frac{dC}{dx} + (1-\sigma)J_v\overline{C} \tag{5.6}$$

In this case, the effect of the osmotic pressure difference is corrected by a reflection coefficient (Staverman reflection coefficient, σ) that represents the correction for nonideality in a semi-permeable membrane $\left(\sigma \equiv \left(\frac{\Delta P}{\Delta\pi}\right)_{J_v=0}\right)$. For an ideal semi-permeable membrane, which passes water but has 100 % rejection of solutes, $\sigma = 1$, while for a completely nonselective porous membrane $\sigma = 0$ (Spiegler and Kedem, 1966). Equation (5.6) represents the solute flux (J_s), combining the contribution of a diffusion term (embedded in the solute permeance, B) as well as the convection of the bulk solution.

However, a more accurate description of performance in such cases requires the extended Nernst–Planck equation, which considers the flux of each solute or ion, J_i, to be controlled by convection, diffusion, and the electrical potential (Bowen, Mohammad, and Hilal, 1997; Bowen and Cao, 1998; Bowen and Welfoot, 2002; Bowen et al., 2004):

$$\frac{J_i}{A_k} = -D_{ip}\frac{dc_i}{dx} - \frac{K_{ic}c_iJ_v}{A_k} - D_{ip}\frac{z_ic_iF}{RT}\frac{d\psi}{dx} \tag{5.7}$$

In this expression, A_k is the porosity of the membrane, c_i is the concentration of the species at any position x within the membrane, K_{ic} is the convective hindrance factor, D_{ip} is the diffusion coefficient of the species within the membrane pores, z_i is the species charge, F is the Faraday constant (96487 C/mol), and $d\psi/dx$ is the electric potential gradient.

In real systems, both the water flux and the solute flux can be further modified by concentration polarization (Section 5.3.2) and by membrane fouling (Section 5.5).

5.3.1 Salt rejection and the Donnan exclusion mechanism

The intrinsic rejection of salt from the membrane is given by

$$\text{Intrinsic rejection:}\quad \mathfrak{R}_i = 1 - \frac{C_p}{C_m} \tag{5.8}$$

where C_m is the salt concentration in the fluid on the retentate side of the membrane and C_p that on the permeate side.

The rejection of any particular ion is a function of both its size and charge. This can be summarized in the partition coefficient (or equilibrium relationship) at the membrane–solution interface, which summarizes the relationship between the concentration of the species (i) in the solution (C_{im}) with that in the membrane phase (c_{im}):

$$\frac{c_{im}}{C_{im}} = \Phi_i \exp\left(-\frac{z_i F}{RT}\Delta\psi_D\right) \tag{5.9}$$

The steric partition coefficient Φ_i is a function of the hydrated ionic radius of the ion (r_i, in Table 5.2) and the radius of the pores within the membrane (r_p):

$$\Phi_i = (1 - r_i/r_p)^2 \tag{5.10}$$

while the Donnan potential difference ($\Delta\psi_D$) is a function of the fixed charge in the membrane and reflects the difference in charge between the solute and the membrane itself (Bowen and Mukhtar, 1996). As a nanofiltration membrane is negatively charged under most circumstances, it will repel negatively charged anions while attracting positively charged cations. Equation (5.9) shows that a high valence anion will be more strongly rejected than a weakly charged anion if they are of the same size. Similarly, strongly charged divalent cations permeate in preference to monovalent cations.

This Donnan exclusion mechanism explains the decrease in sodium rejection that is sometimes observed after the addition of a multivalent

Table 5.2 Hydrated ionic radii for some ions commonly present in dairy solutions.

Ion	Hydrated ionic radius (Å)		
	Marcus (1994)	**Kiriukhin and Collins (2002)**	**Conway (1981)**
H^+	–	1.96	–
H_3O^+	1.90	–	–
OH^-	2.12	2.41	–
Na^+	2.18	1.78	3.58
K^+	2.12	2.01	3.31
Cl^-	2.24	1.95	3.32
Ca^{2+}	2.71	2.60	4.12
Mg^{2+}	2.99	3.00	4.28
$H_2PO_4^-$	2.59	3.02	–
HPO_4^{2-}	2.78	3.27	–
PO_4^{3-}	–	3.39	–

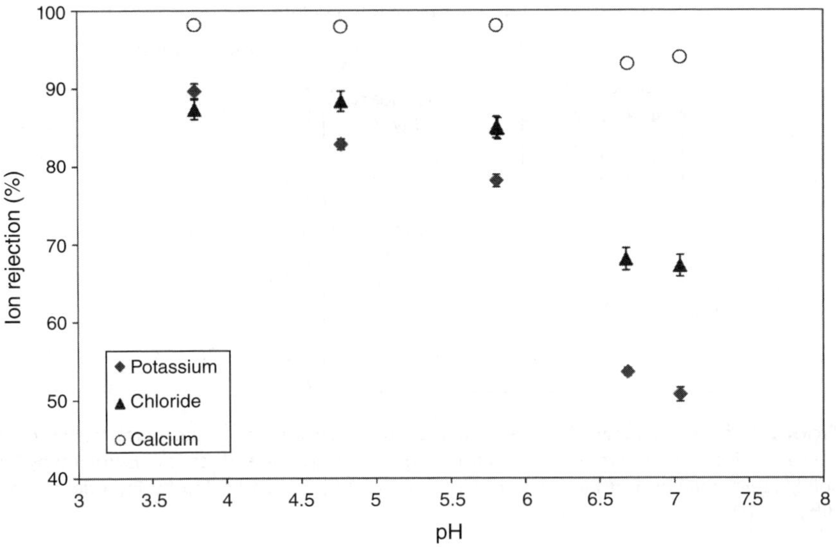

Figure 5.5 Rejection of ions during nanofiltration of a solution containing 10 mM KCl and 2 mM $CaCl_2$, measured at 15 bar TMP, 0.45 m/s cross-flow velocity and 25 °C. Error bars represent ±2 standard deviations (Rice et al., 2011b). Reproduced with permission from Elsevier.

co-ion such as calcium (Balannec et al., 2002). However, the balance between size and charge is subtle and, in many other cases, the larger size of calcium can lead to stronger rejection of this species (Garcia-Aleman and Dickson, 2004; Labbez et al., 2003; Rice et al., 2011b; Schaep et al., 1999) (Figure 5.5). The lower rejection of monovalent counter-ions can also often be attributed to charge inversion of the membrane due to adsorption of the divalent counter-ion (see Section 5.2) (Garcia-Aleman and Dickson, 2004; Rice et al., 2009, 2011b).

5.3.2 Concentration polarization

Concentration polarization is the accumulation of membrane rejected solutes at the membrane surface. This phenomenon arises when the mass transfer resistance within the boundary layer of fluid above the membrane itself becomes significant. Overall, concentration polarization leads to lower fluxes and lower rejections than would otherwise occur and should be eliminated as much as possible.

The increased concentration of solute at the membrane surface leads to back-diffusion according to Fick's law. At steady state, the rate of back-diffusion of solute must equal the difference between the molar

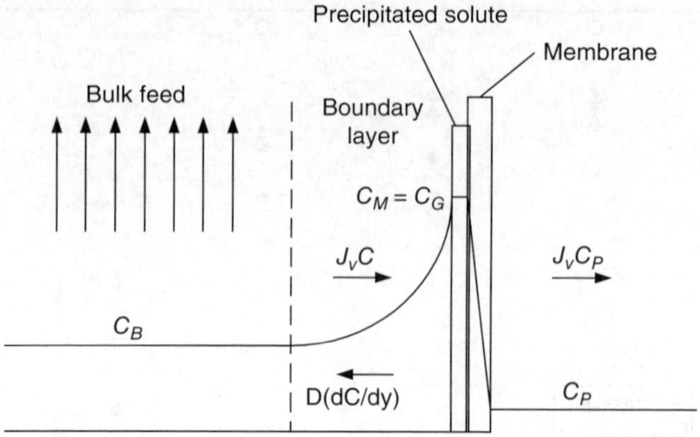

Figure 5.6 The impact of concentration polarisation on membrane performance once solute precipitation occurs. The concentration at the membrane surface is constrained by the solid/liquid equilibria of the solute at C_G. The flux (J_v) is then limited by the rate of back diffusion ($D\,dC/dy$).

flux towards the membrane and the rate of solute leakage through the membrane:

$$J_V C - J_V C_p = -D\frac{dC}{dy} \tag{5.11}$$

In the absence of a fouling cake, integration of this expression leads to

$$J_V = k_{CP} \ln\left(\frac{C_m - C_p}{C_B - C_p}\right) \approx k_{cp} \ln\left(\frac{C_m}{C_B}\right) \tag{5.12}$$

where C_m is the concentration at the surface of the membrane and k_{cp} the mass transfer coefficient in the boundary layer (D/δ). However, when this concentration reaches a point where solute precipitation occurs (C_G) then (Figure 5.6)

$$J_V = k_{CP} \ln\left(\frac{C_G - C_p}{C_B - C_p}\right) \approx k_{cp} \ln\left(\frac{C_G}{C_B}\right) \tag{5.13}$$

where C_G is known as the gel concentration. The concentration at the membrane surface cannot exceed C_G and hence the flux cannot increase beyond this limiting value.

When concentration polarisation occurs, the observed rejection, as measured by the bulk concentration C_b (Equation (5.14)), falls relative to the intrinsic value (Equation (5.8)), as C_b is always lower than C_m:

$$\text{Observed rejection:} \quad \Re_o = 1 - \frac{C_p}{C_b} \tag{5.14}$$

This observed rejection can be related to the observed flux via the concentration polarisation ratio (C_m/C_b):

$$\frac{C_m}{C_b} = (1 - \mathfrak{R}_o) + \mathfrak{R}_o \exp\left(\frac{J_v}{k_{cp}}\right) \tag{5.15}$$

Experimentally, the mass transfer coefficient, k_{CP}, is determined by obtaining J_v at varying C_B; a plot of J against ln C_B gives a gradient of $-k_{CP}$. Alternatively, k_{CP} can be obtained by using mass transfer correlations. These correlations take the general form of

$$Sh = aRe^b Sc^c \left(\frac{d_H}{L}\right)^e \tag{5.16}$$

where

$$Sh = \frac{k_{CP}d_H}{D},\ Re = \frac{\rho v d_H}{\mu_F},\ \text{and } Sc = \frac{\mu_F}{\rho d_H} \tag{5.17}$$

where Sh is the Sherwood number, Re is the Reynolds number, Sc is the Schmidt number, d_H is the hydraulic diameter (m), L is the length of the module (m), ρ is the density of the feed fluid (kg m^{-3}), μ_F is the dynamic viscosity of the feed fluid (Pa /s), v is the cross-flow velocity (m/s), and a, b, c, and e are empirical constants. The hydraulic diameter is defined as

$$d_H = \frac{4 \times \text{cross} - \text{sectional area of flow}}{\text{wetted perimeter of channel}} \tag{5.18}$$

The hydraulic diameter for a slit-shaped channel is equal to twice the channel spacing, $2H$. For a membrane module with a feed side spacer in the channel, the hydraulic diameter is a function of spacer porosity, ε, and specific surface area, s (Zeman and Zydney, 1996):

$$d_H = \frac{4\varepsilon}{\frac{2}{H} + (1 - \varepsilon)s} \tag{5.19}$$

Typical values of the empirical constants a, b, c, and e in Equation (5.16) have been tabulated by Prudich et al. (2008). The values for slit-shaped channels and channels containing spacers is as shown in Table 5.3. For spacer-filled modules, the constant a in Equation (5.16) is dependent on the geometry of the spacer and the term L is taken as the spacer mesh size (Da Costa, Fane, and Wiley, 1994; Zeman and Zydney, 1996).

Concentration polarisation can be reduced by increasing the mass transfer coefficient, k_{CP}, and this can be achieved by increasing the cross-flow velocity, by increasing turbulence through a judicious choice of membrane spacers, or by shear enhancement. In this latter case, rotating or vibrating disc membranes are used to induce turbulence within the boundary layer (Luo et al., 2012). Ultrasound has also been considered for this purpose (Muthukumaran et al., 2005).

Table 5.3 Typical values of the parameters *a*, *b*, *c*, and *e* for use in Equation (5.16) for relevant membrane modules (Prudich et al., 2008).

Geometry	Flow	*a*	*b*	*c*	*e*	Reference
Slit	Laminar	1.86	0.33	0.33	0.33	Lévêque, 1928
	Turbulent	0.023	0.80	0.33	–	Gekas and Hallström, 1987
Spacer	–	0.664	0.50	0.33	0.50	Da Costa, Fane, and Wiley, 1994

5.4 DAIRY SALTS AND SALT EQUILIBRIUM

An understanding of the NF operation in dairy processing requires an understanding of the salts that are present and how they interact with each other.

The total content of inorganic salts present in milk can be determined by measuring the ash content, which generally remains fairly constant in bovine milk between 0.7 and 0.8% (Fox and McSweeney, 1998). These salts are present either as soluble free ions, dissolved neutral salts, or within insoluble colloids (Table 5.4). The insoluble colloidal salts are generally associated with casein micelles and hence are readily removed during cheese making or microfiltration/ultrafiltration operations.

Holt, Dalgleish and Jenness (1981) developed a more detailed understanding of the specific salts that are expected to be present in the aqueous phase of skim milk at room temperature and natural pH (Table 5.5). In our own work (Rice et al., 2010), we determined the calcium:phosphate:citrate speciation in solutions typical of ultrafiltration whey permeate and predicted how these compositions might change with respect to pH and temperature. These results were consistent with Holt in showing that $H_2PO_4^-$ was the most abundant phosphate species present across the pH range 3–7, but above pH 7, HPO_4^{2-} became more prevalent (Figure 5.7). H_2Cit^- was the most

Table 5.4 Distribution of milk salts between soluble and insoluble colloidal fractions in cow's milk (Fox and McSweeney, 1998).

Component	Average Conc. (mg/L)	Soluble % (w/w)	Colloidal % (w/w)
Calcium	1000–1400	33.5	66.5
Magnesium	100–150	67	33
Sodium	350–600	92	8
Potassium	1350–1550	92	8
Phosphorus (total)	750–1100	43	57
Chloride	800–1400	100	0
Citrate (as citric acid)	1750	94	6

Table 5.5 Calculated concentration (mM) of ions and complexes in the aqueous phase of milk (Holt, Dalgleish, and Jenness, 1981).

		Cation complex			
Anion	**Free ion**	**Ca^{2+}**	**Mg^{2+}**	**Na^{+}**	**K^{+}**
H_2Cit^-	+	+	+	+	+
$HCit^{2-}$	0.04	0.01	+	+	+
Cit^{3-}	0.26	6.96	2.02	0.03	0.04
$H_2PO_4^-$	7.50	0.07	0.04	0.10	0.18
HPO_4^{2-}	2.65	0.59	0.34	0.39	0.52
PO_4^{3-}	+	0.01	+	+	+
GLC-1-HPO_4^-	0.50	+	+	0.01	0.01
GLC-1-PO_4^{2-}	1.59	0.17	0.07	0.10	0.14
H_2CO_3	0.11	–	–	–	–
HCO_3^-	0.32	0.01	+	+	+
CO_3^{2-}	+	+	+	+	+
Cl^-	30.90	0.26	0.07	0.39	0.68
HSO_4^-	+	+	+	+	+
SO_4^{2-}	0.96	0.07	0.03	0.04	0.10
RCOOH	0.02	–	–	–	–
$RCOO^-$	2.98	0.03	0.02	0.02	0.04
Free ion	–	2.00	0.81	20.92	36.29

+ < 0.005 μM; – not estimated, GLC glucose.

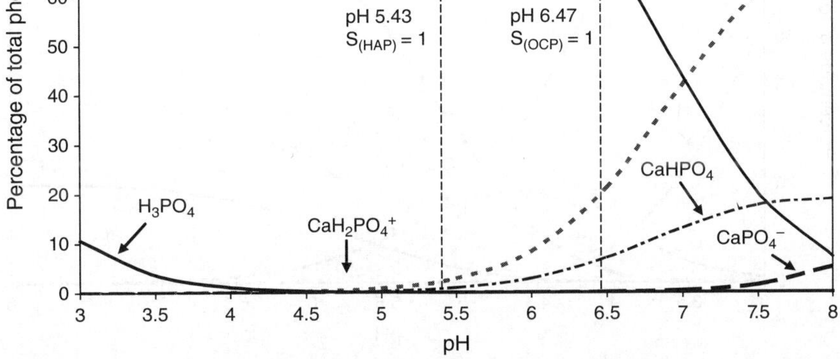

Figure 5.7 Predicted distribution of phosphate species in an aqueous phase of a 10 mM KCl +2 mM $CaCl_2$ +2 mM KH_2PO_4 system as a function of pH. The ionic strength was 0.020 ± 0.003 M (Rice et al., 2010). Reproduced with permission from Elsevier.

abundant citrate species present below pH 4.5. Following this, $HCit^{2-}$ was most abundant between 4.5 and 5, after which $CaCit^{-}$ became most prevalent in the system (Figure 5.8). Ionic calcium levels dropped significantly as the pH was raised due to the formation first of CaHCit and then $CaCit^{-}$. Only small quantities of calcium phosphate species were observed, with less than 10% of the total calcium being associated with phosphate (Figure 5.9). The predicted ionic calcium concentration was only slightly influenced by temperature within the range of 10–50 °C (Rice et al., 2010).

The calcium equilibrium will also be influenced by additional minerals such as sulphate and carbonate while Holt, Dalgleish, and Jenness (1981) also suggested the presence of phosphate esters that bind small quantities of minerals.

An understanding of whether salts will remain soluble or precipitate as concentration is increased during nanofiltration is also important. This depends upon the intrinsic solubility product of the salt (Table 5.6), the ionic strength, and the pH. Another significant factor can be the presence of even trace levels of protein, which has been shown to be present even in ultrafiltration permeate (Rice et al., 2010). Sequestration of both calcium and phosphate by protein molecules (in particular casein) will reduce the likelihood of precipitation.

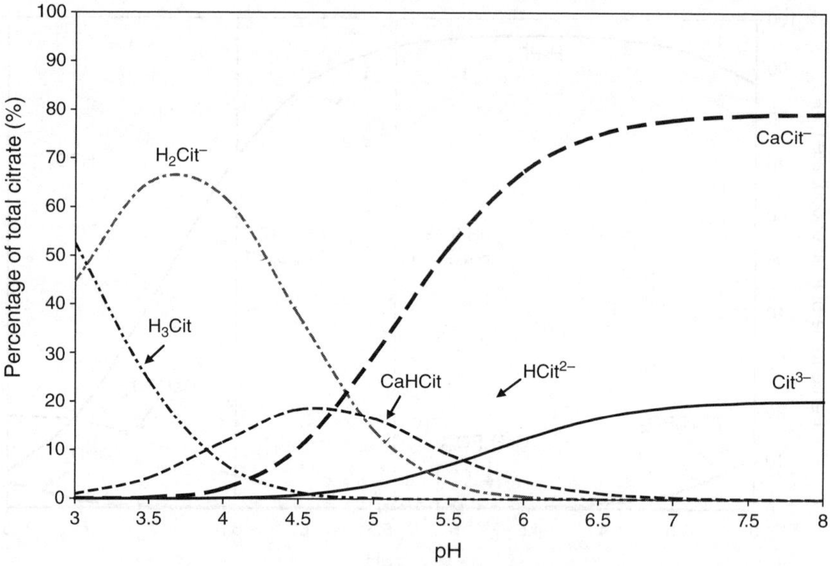

Figure 5.8 Predicted distribution of citrate species in a 10 mM KCl + 2 mM $CaCl_2$ + 2 mM KH_2Cit system as a function of pH. The ionic strength was 0.018 ± 0.0004 M (Rice et al., 2010). Reproduced with permission from Elsevier.

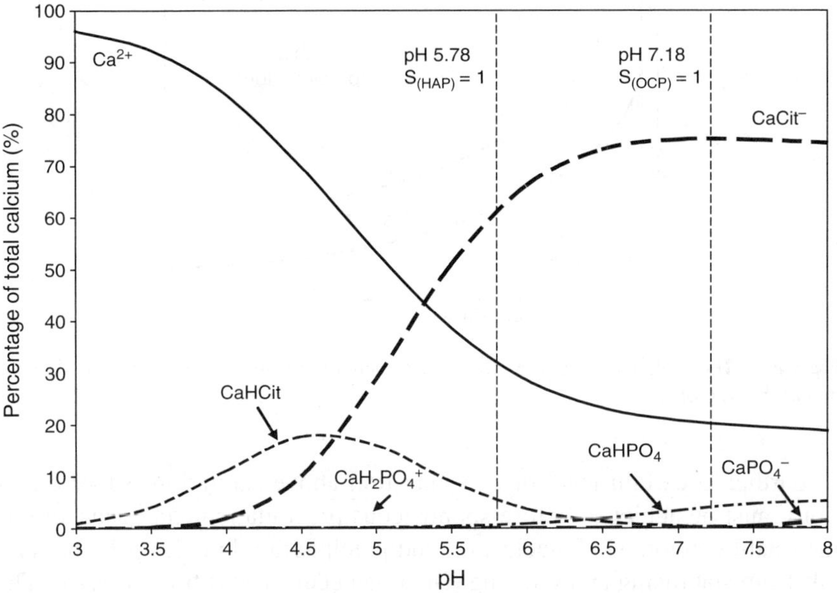

Figure 5.9 Predicted distribution of calcium in an aqueous phase of a 10 mM KCl + 2 mM $CaCl_2$ + 2 mM KH_2Cit + 2 mM KH_2PO_4 system as a function of pH. The ionic strength is 0.021 ± 0.003 M (Rice et al., 2010). Reproduced with permission from Elsevier.

Table 5.6 Intrinsic solubility products for some dairy salts (Marshall and Daufin, 1995; Walstra and Jenness, 1984).

Salt name	Formula	Solubility product (K_{sp}°, 20 °C)
Tricalcium citrate	Ca_3Cit_2	2.34×10^{-18}
Dicalcium phosphate (monetite)	$CaHPO_4$	1.26×10^{-7} (25 °C) (Martell, Smith, and Motekaitis, 2002)
Dicalcium phosphate dihydrate (DCPD) (brushite)	$CaHPO_4.2H_2O$	2.6×10^{-7}
Octacalcium phosphate (OCP)	$Ca_4H(PO_4)_3$	1.2×10^{-47}
Tricalcium phosphate	$Ca_3(PO_4)_2$	1.15×10^{-29}
Hydroxyapatite (HAP)	$Ca_5OH(PO_4)_3$	1.8×10^{-58} (25 °C)
Magnesium phosphate	$MgHPO_4.3H_2O$	$\approx 1.0 \times 10^{-4}$

It is also apparent from Table 5.6 that a variety of insoluble salts can precipitate, each with a different solubility product. In particular, calcium phosphate can precipitate in five different crystal habits, although three are more common: dicalcium phosphate dihydrate (DCPD, or brushite), octacalcium phosphate (OCP), and hydroxyapatite (HAP).

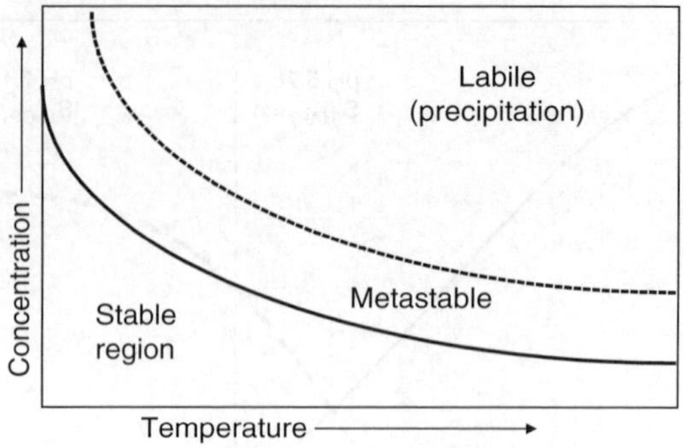

Figure 5.10 Solubility diagram for a salt such as calcium phosphate, showing the metastable region.

Further, a certain level of concentration above saturation (supersaturation) must be achieved before spontaneous precipitation can occur. In this metastable region, slow nucleation and precipitation kinetics prevent crystals from stabilising and growing (Boerlage et al., 1999) (Figure 5.10). The size of this metastable zone is dependent upon a large range of factors including the temperature and the degree of mixing in the solution.

The supersaturation ratio (S) of a precipitate is commonly used as an indication of the scaling propensity of a solution. In general, to avoid scaling, membrane manufacturers recommend maintaining a salt supersaturation ratio below 1. However, some authors have shown that due to the metastable region of saturation and long precipitation induction times, a higher critical supersaturation ratio is usually required before precipitation is observed (Boerlage et al., 1999; van de Lisdonk et al., 2001). The supersaturation ratio may be calculated using the following equations, where K_{sp} is the apparent solubility product:

$$A_xB_{y(s)} \leftrightarrow xA^+{}_{(aq)} + yB^-{}_{(aq)} \tag{5.20}$$

$$S = \left(\frac{[A^+]^x [B^-]^y}{K_{sp}} \right)^{\frac{1}{x+y}} \tag{5.21}$$

Supersaturation of calcium phosphate salts can be reached very readily, despite very low concentrations of the reacting species. For example, in our analysis of a solution containing 2 mM $CaCl_2$, 2 mM KH_2PO_4 and 10 mM KCl (with and without the addition of 2mM potassium citrate), saturation of hydroxyapatite (HAP) was reached between pH 5.43 and 5.78 and supersaturation ratios for this salt reached values above 35 at pH 8. Octacalcium phosphate (OCP) saturation was reached at pH 6.47 to 7.18

while dicalcium phosphate dihydrate (DCPD) was not found to be saturated at any point. We find a high likelihood that it is indeed HAP that precipitates first as a membrane surface scale (Rice et al., 2009, #521) from an ultrafiltration dairy permeate. However, other authors suggest that the formation of OCP or DCPD generally precedes HAP formation (Rosmaninho and Melo, 2006; Schmidt and Both, 1987; Spanos et al., 2007). In fact, van Kemenade and de Bruyn (1987) observed that HAP precipitation was always preceded by formation of another salt in calcium phosphate solutions between pH 6 and pH 8, despite very high HAP supersaturation ratios. It was found that at these concentrations, the saturation ratio of tricalcium citrate (TCC) remained below 1 at all times, despite increasing slightly from pH 3 to pH 6.5 due to increasing concentrations of Cit^{3-} in the system. However, at around pH 6 the saturation ratio reached a maximum and then began to decrease slowly due to loss of Ca^{2+} to form $CaCit^{-}$.

5.5 MEMBRANE FOULING

It has become widely accepted that proteins and minerals are the two major contributors to dairy membrane fouling (Marshall and Daufin, 1995). If dairy fat is also present, this can also play a role (Luo et al., 2012). However, both Nyström, Kaipia, and Luque (1995) and van der Horst (1995) have shown that uncharged highly rejected components such as lactose do not contribute substantially to fouling of NF membranes.

Milk proteins foul the membrane by adsorbing to and depositing on to the membrane surface in the form of a gelatinous cake layer. This protein adsorption has been shown to be influenced by electrostatic effects between the charged protein and charged membrane, and also by the hydrophobicity of the membrane (Marshall and Daufin, 1995). Calcium is able to contribute to this protein cake by forming protein–protein and protein–membrane bridges. During protein cake layer fouling, an initial rapid decline in flux is observed, which gradually levels out to a limiting flux.

Protein fouling is usually removed from the membrane by cleaning with an alkali solution. Sodium hydroxide or caustic soda is the most commonly used alkali because of its low cost and high alkalinity. Although sodium hydroxide is an effective saponification agent, it is a poor buffer (Zeman and Zydney, 1996). Sodium hydroxide also has a poor rinsing power: addition of appropriate surfactants to the cleaning agent is required so as to allow easy and rapid removal of sodium hydroxide from surfaces (Watkinson, 2008). Sodium carbonate or soda ash is the cheapest alkaline cleaning agent. It has a high alkali reserve: a low concentration of sodium carbonate is required for saponification and neutralisation. Sodium carbonate forms an insoluble salt with calcium, improving soap solubility but increasing the possibility of calcium carbonate deposit formation on the membrane surface (Zeman and Zydney, 1996). Silicates and phosphates can also be used to remove protein deposits.

Meanwhile, salts can also foul the membrane by precipitating once their solubility is exceeded, forming a scale within the pores or on the surface of the membrane (see Section 5.4 above). There can be a considerable nonfouling induction time before scaling occurs and then a very sudden and drastic flux decrease (Bhattacharjee and Johnston, 2002; Lee and Lee, 2000). In our own work, we find that the fouling behaviour can vary quite dramatically for different batches of ultrafiltration permeate tested under identical conditions (Figure 5.11) (Rice et al., 2009). In particular, the induction period prior to calcium salt scaling was quite inconsistent. This variation may reflect differing balances between protein cake fouling and calcium phosphate scaling, reflecting slight differences in both composition and cross-flow velocity (Rice et al., 2009).

Calcium phosphate scaling is of particular concern in the dairy industry due to the supersaturation of these salts in the aqueous phase of most dairy fluids (see Section 5.4). Calcium phosphate has a reverse solubility, meaning that the solubility decreases with an increase in temperature. This means that scaling is a greater issue at high temperatures (Rice et al., 2009). An increase in pH has also been linked to calcium phosphate precipitation, with low pH and low temperatures being utilised in the industry to reduce the

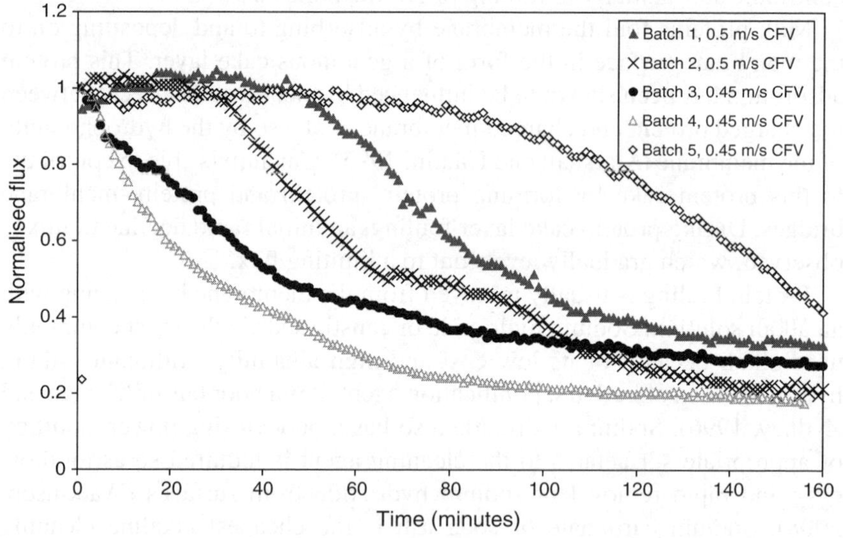

Figure 5.11 Variation in flux decline during nanofiltration of different batches of UF permeate at 30 °C and 15 bar transmembrane pressure (Rice et al., 2009). Reproduced with permission from Elsevier.

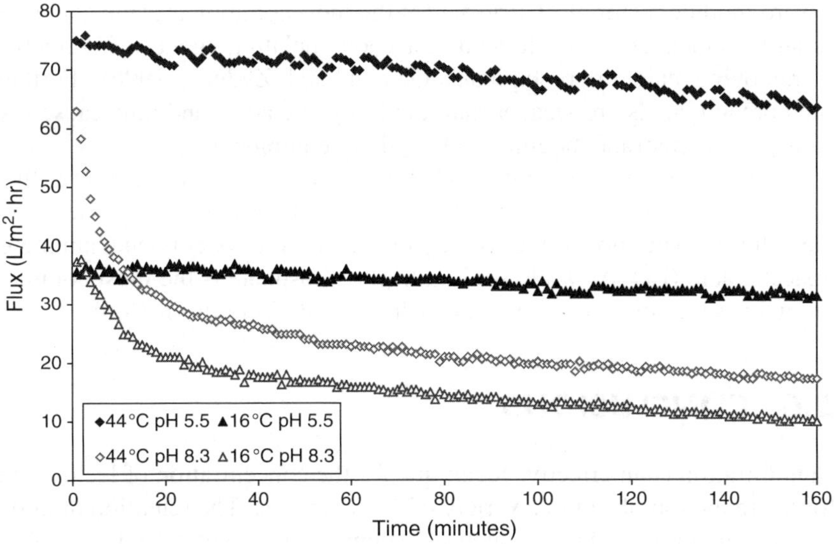

Figure 5.12 Flux decline during NF of UF permeate at various operating temperatures and modified feed pH values, 15 bar TMP, 0.45 m/s cross-flow velocity (Rice et al., 2009). Reproduced with permission from Elsevier.

likelihood of calcium phosphate scaling (Marshall and Daufin, 1995; Pouliot and Jelen, 1995). Our own work showed that the best approach might be to use a combination of low pH and high temperatures. In this instance, the calcium scaling is eliminated by the use of a low pH, but the high operating temperature provides greater net fluxes due to greater membrane permeability (Rice et al., 2009) (Figure 5.12). Under these conditions of low pH, both protein and salt contribute to the fouling process, even when the protein content in the feed is less than 0.2 wt%, as indicated by the need for both alkaline and acid cleaning to restore performance.

Acids are generally used to remove mineral scales. Inorganic acids, such as hydrochloric and sulfuric acids, are the cheapest and simplest to use. However, the corrosive properties of hydrochloric and sulfuric acids on membrane modules and stainless steel make them unfavorable for cleaning. The most common inorganic acid used is phosphoric acid. It is less corrosive and a strong chelating agent, facilitating the removal of metal cations. Phosphoric acid is also a good buffer, providing good pH control during cleaning (Zeman and Zydney, 1996). These properties make phosphoric acid a good acid cleaner but it is expensive compared to other inorganic acids. Nitric acid is effective for cleaning of organic and biological foulants as it is a strong oxidant, which causes nitration of organic compounds (Zeman and Zydney, 1996). It is commonly used in dairy cleaning but care must be taken as nitric acid is also corrosive toward module materials. Citric acid is the most common organic acid used as an acid cleaner. As citric acid is a good chelating agent, it is effective at removing calcium precipitates (Zeman and Zydney, 1996). Further, such organic acids are weaker than the inorganic acids and thus cause less damage to membrane modules and ancillary equipment.

Sequesterants are commonly added to such acid or alkali cleaners as they are effective in removing insoluble metal cations of fats and proteins from the solution. The most effective organic sequestrant is ethylenediaminetetraacetic acid (EDTA) while sodium tripolyphosphate is the most common inorganic sequestrant (Zeman and Zydney, 1996; Watkinson, 2008).

5.6 CONCLUSIONS

Nanofiltration is an effective technique for the concentration of lactose and for the reduction of salt in a variety of dairy streams. The retention of nutritionally important minerals such as calcium, magnesium, and zinc is an additional advantage. The effective development of a nanofiltration operation requires careful consideration of the concentration, size, and charge of the species present in the feed stream and the charge and pore size of the membrane. However, operation will always be less than the theoretical limit due to both membrane fouling and concentration polarisation.

Nomenclature

A = water permeance (m^4/kg s)
A_k = membrane porosity
B = salt permeance (m^2 s)
c = concentration within the membrane (mol/m^3)
C = concentration in the solution (mol/m^3)
d_H = hydraulic mean diameter (m)
D = diffusion coefficient (m^2 s)
D_p = diffusion coefficient in the membrane pores (m^2 s)
F = Faraday constant (96 487 C/mol)
H = channel spacing (m)
J = molar flux (mol/m^2 s)
J_v = volumetric flux (m^3/m^2 s or L/m^2 h)
k = mass transfer coefficient (m/s)
K_c = convective hindrance factor
K_{sp} = apparent solubility product
L = length of the module (m)

m = total solute molar concentration (mol/kg)
P = pressure (Pa)
r = radius (m)
r_p = pore radius (m)
R = universal gas constant (8.314 J/mol K)
Re = Reynolds number
R_t = total resistance (kg s/m^3)
$\mathfrak{R}$ = rejection
s = specific surface area of spacer (m^{-1})
S = supersaturation ratio
Sc = Schmidt number
Sh = Sherwood number
T = absolute temperature (K)
v = cross-flow velocity (m/s)
x = distance within the membrane thickness (m)
y = distance within the boundary layer (m)
z = species charge

Greek letters

π = osmotic pressure (Pa)
ρ = density (kg/m^3)
Δ = change in a variable
σ = Staverman reflection coefficient
ψ = electric potential (V)
Φ = steric partition coefficient
δ = film thickness (m)
ε = porosity of the spacer
ψ_D = Donnan potential (V)
μ = viscosity (kg/m s)

Subscripts

s = salt
i = ion or solute species
p = permeate
m = membrane
b = bulk
cp = concentration polarisation boundary layer
G = gel
o = observed
F = feed

Superscripts

o = pure solvent (water)

References

Afonso, M.D., Hagmeyer, G., and Gimbel, R. (2001) Streaming potential measurements to assess the variation of nanofiltration membranes surface charge with the concentration of salt solutions. *Separation and Purification Technology*, **22-23**, 529–541.

Balannec, B., Gesan-Guiziou, G., Chaufer, B., et al. (2002) Treatment of dairy process waters by membrane operations for water reuse and milk constituents concentration. *Desalination*, **147**, 89–94.

Bhattacharjee, S. and Johnston, G.M. (2002) A model of membrane fouling by salt precipitation from multicomponent ionic mixtures in crossflow nanofiltration. *Environmental Engineering Science*, **19**, 399–412.

Boerlage, S.F.E., Kennedy, M.D., Witkamp, G.J., et al. (1999) $BaSO_4$ solubility prediction in reverse osmosis membrane systems. *Journal of Membrane Science*, **159**, 47–59.

Bowen, W.R. and Mukhtar, H. (1996) Characterisation and prediction of separation performance of nanofiltration membranes. *Journal of Membrane Science*, **112**, 263–274.

Bowen, W.R., Mohammad, A.W., and Hilal, N. (1997) Characterisation of nanofiltration membranes for predictive purposes – use of salts, uncharged solutes and atomic force microscopy. *Journal of Membrane Science*, **126**, 91–105.

Bowen, W.R. and Cao, X. (1998) Electrokinetic effects in membrane pores and the determination of zeta-potential. *Journal of Membrane Science*, **140**, 267–273.

Bowen, W.R. and Welfoot, J.S. (2002) Modelling the performance of membrane nanofiltration – critical assessment and model development. *Chemical Engineering Science*, **57**, 1121–1137.

Bowen, W.R., Cassey, B., Jones, P., and Oatley, D.L. (2004) Modelling the performance of membrane nanofiltration – application to an industrially relevant separation. *Journal of Membrane Science*, **242**, 211–220.

Burns, D.B. and Zydney, A.L. (2000) Buffer effects on the zeta potential of ultrafiltration membranes. *Journal of Membrane Science*, **172**, 39–48.

Chen, C.-C. and Song, Y. (2004) Generalized electrolyte-NRTL model for mixed-solvent electrolyte systems. *American Institute of Chemical Engineers Journal*, **50**, 1928–1941.

Childress, A.E. and Elimelech, M. (1996) Effect of solution chemistry on the surface charge of polymeric reverse osmosis and nanofiltration membranes. *Journal of Membrane Science*, **119**, 253–268.

Conway, B.E. (1981) *Ionic Hydration in Chemistry and Biophysics*, Elsevier Scientific Publishing Company, Amsterdam, The Netherlands.

Cuartas-Uribe, B., Alcaina-Miranda, M.I., Soriano-Costa, E., et al. (2009) A study of the separation of lactose from whey ultrafiltration permeate using nanofiltration. *Desalination*, **241**, 244–255.

Da Costa, A.R., Fane, A.G., and Wiley, D.E. (1994) Spacer characterization and pressure drop modelling in spacer-filled channels for ultrafiltration. *Journal of Membrane Science*, **87**, 79–98.

Durham, R.J., Sleigh, R.W., and Hourigan, J.A. (2004) Pharmaceutical lactose: a new whey with no waste. *Australia Journal of Dairy Technology*, **59**, 138–141.

Elimelech, M., Chen, W.H., and Waypa, J.J. (1994) Measuring the zeta (electrokinetic) potential of reverse osmosis membranes by a streaming potential analyzer. *Desalination*, **95**, 269–286.

Fernandez, P., Riera, F.A., Alvarez, R., and Alvarez, S. (2010) Nanofiltration regeneration of contaminated single-phase detergents used in the dairy industry. *Journal of Food Engineering*, **97**, 319–328.

Fox, P.F. and McSweeney, P.L.H. (1998) *Dairy Chemistry and Biochemistry*, Blackie Academic & Professional, London.

Frappart, M., Akoum, O., Ding, L.H., and Jaffrin, M.Y. (2006) Treatment of dairy process waters modelled by diluted milk using dynamic nanofiltration with a rotating disk module. *Journal of Membrane Science*, **282**, 465–472.

Garcia-Aleman, J. and Dickson, J.M. (2004) Permeation of mixed-salt solutions with commercial and pore-filled nanofiltration membranes: membrane charge inversion phenomena. *Journal of Membrane Science*, **239**, 163–172.

Gekas, V. and Hallström, B. (1987) Mass transfer in the membrane concentration polarization layer under turbulent cross flow: I. Critical literature review and adaptation of existing Sherwood correlations to membrane operations. *Journal of Membrane Science*, **30**, 153–170.

Gernigon, G., Schuck, P., Jeantet, R., and Burling, H. (2011) Whey processing/demineralization, in *Encyclopedia of Dairy Sciences*, second edition (Editor-in-Chief: W.F. John), Academic Press, San Diego, CA, pp. 738–743.

Gesan-Guiziou, G., Boyaval, E., and Daufin, G. (2002) Nanofiltration for the recovery of caustic cleaning-in-place solutions: robustness towards large variations of composition. *Journal of Dairy Research*, **69**, 633–643.

Gesan-Guiziou, G., Alvarez, N., Jacob, D., and Daufin, G. (2007) Cleaning-in-place coupled with membrane regeneration for re-using caustic soda solutions. *Separation and Purification Technology*, **54**, 329–339.

Gong, Y.-W., Zhang, H.-X., and Cheng, X.-N. (2012) Treatment of dairy wastewater by two-stage membrane operation with ultrafiltration and nanofiltration. *Water Science and Technology*, **65**, 915–919.

Greiter, M., Novalin, S., Wendland, M., et al. (2002) Desalination of whey by electrodialysis and ion exchange resins: analysis of both processes with regard to sustainability by calculating their cumulative energy demand. *Journal of Membrane Science*, **210**, 91–102.

Holt, C., Dalgleish, D.G., and Jenness, R. (1981) Calculation of the ion equilibria in milk diffusate and comparison with experiment. *Analytical Biochemistry*, **113**, 154–163.

Hoon, H. and Jae-Hong, K. (2006) A mechanistic study on boron rejection by sea water reverse osmosis membranes. *Journal of Membrane Science*, **286**, 269–278.

Hoppe, G.K. and Higgins, J.J. (1992) Demineralization, in *Whey and Lactose Processing* (ed. J.G. Zadow), Elsevier Science Publishers Ltd, Essex, pp. 91–133.

Horton, B.S. (1995) Commercial utilisation of minor milk components in the health and food industries. *Journal of Dairy Science*, **78**, 2584–2589.

Hyung, H. and Kim, J.-H. (2006) A mechanistic study on boron rejection by sea water reverse osmosis membranes. *Journal of Membrane Science*, **286**, 269–278.

Kedem, O. and Katchalsky, A. (1961) A physical interpretation of the phenomenological coefficients of membrane permeability. *The Journal of General Physiology*, **45**, 143–179.

Kelly, P.M., Horton, B.S., and Burling, H. (1991) Partial demineralisation of whey by nanofiltration. *International Dairy Federation [*Special Issue*] S.I.*, **9201**, 130–140.

Kim, K.J., Fane, A.G., Nyström, M., et al. (1996) Evaluation of electroosmosis and streaming potential for measurement of electric charges of polymeric membranes. *Journal of Membrane Science*, **116**, 149–159.

Kiriukhin, M.Y. and Collins, K.D. (2002) Dynamic hydration numbers for biologically important ions. *Biophysical Chemistry*, **99**, 155–68.

Labbez, C., Fievet, P., Szymczyk, A., et al. (2003) Retention of mineral salts by a polyamide nanofiltration membrane. *Separation and Purification Technology*, **30**, 47–55.

Lee, S. and Lee, C.-H. (2000) Effect of operating conditions on $CaSO_4$ scale formation mechanism in nanofiltration for water softening. *Water Research*, **34**, 3854–3866.

Leruyet, P. and Maugas, J.-J. (2001) Patent WO2001093689A1.

Lévêque, M.D. (1928) Les lois de la transmission de chaleur par convection. *Ann. Mines*, **13**, 201.

Luo, J. and Ding, L. (2011) Influence of pH on treatment of dairy wastewater by nanofiltration using shear-enhanced filtration system. *Desalination*, **278**, 150–156.

Luo, J., Ding, L., Qi, B., et al. (2011) A two-stage ultrafiltration and nanofiltration process for recycling dairy wastewater. *Bioresource Technology*, **102**, 7437–7442.

Luo, J., Cao, W., Ding, L.H., et al. (2012) Treatment of dairy effluent by shear-enhanced membrane filtration: the role of foulants. *Separation and Purification Technology*, **96**, 194–203.

Marcus, Y. (1994) A simple empirical model describing the thermodynamics of hydration of ions of widely varying charges, sizes, and shapes. *Biophysical Chemistry*, **51**, 111–127.

Mariñas, B.J. and Selleck, R.E. (1992) Reverse osmosis treatment of multicomponent electrolyte solutions. *Journal of Membrane Science*, **72**, 211–229.

Marshall, A. and Daufin, G. (1995) Physico-chemical aspects of membrane fouling by dairy fluids. *International Dairy Federation [*Special Issue*] S.I.*, **9504**, 8–35.

Martell, A.E., Smith, R.M., and Motekaitis, R.J. (2002) NIST critically selected stability constants of metal complexes, in *NIST Standard Reference Database 46*, NIST, College Station, TX, USA.

Matsui, H., Mizota, Y., Sumi, M., et al. (2006) Studies on the sensory characteristics and physicochemical properties of UHT processed milk, part VI. Sensory characteristics of membrane treated-milk. *Nippon Shokuhin Kagaku Kogaku Kaishi*, **53**, 644–650.

Mattila, J. and Koivikko, H. (2011) Patent WO2011154604A1.

Merin, U., Gesan-Guiziou, G., Boyaval, E., and Daufin, G. (2002) Cleaning-in-place in the dairy industry: criteria for reuse of caustic (NaOH) solutions. *Lait*, **82**, 357–366.

Morse, H.N., Frazer, J.C.W., and Rogers, F.M. (1907) The osmotic pressure of glucose solutions in the vicinity of the freezing point of water. *American Chemical Journal*, **37**, 558.

Muthukumaran, S., Kentish, S.E., Ashokkumar, M., and Stevens, G.W. (2005) Mechanisms for the ultrasonic enhancement of dairy whey ultrafiltration. *Journal of Membrane Science*, **258**, 106–114.

Nayak, J. and Pal, P. (2013) Transforming waste cheese-whey into acetic acid through a continuous membrane-integrated hybrid process. *Industrial and Engineering Chemistry Research*, **52**, 2977–2984.

Nyström, M., Kaipia, L., and Luque, S. (1995) Fouling and retention of nanofiltration membranes. *Journal of Membrane Science*, **98**, 249–262.

Patel, R., Sandry, M., and Seguin, R. (2008) Patent US20080254165A1.

Paterson, A.H.J. (2011) Lactose and oligosaccharides/lactose: production, applications, in *Encyclopedia of Dairy Sciences*, second edition (Editor-in-Chief: W.F. John), Academic Press, San Diego, CA, pp. 196–201.

Peeters, J.M.M., Mulder, M.H.V., and Strathmann, H. (1999) Streaming potential measurements as a characterization method for nanofiltration membranes. *Colloids and Surfaces A: Physicochemical and Engineering Aspects*, **150**, 247–259.

Pitzer, K.S. (1991) Activity Coefficients in Electrolyte Solutions, CRC Press, Boca Raton, FL.

Pouliot, Y. and Jelen, P. (1995) Pretreatments of dairy fluids to minimize long-term membrane fouling. *International Dairy Federation [*Special Issue*] S.I.*, **9504**, 80–92.

Prudich, M.E., Chen, H., Gu, T., et al. (2008) Alternative separation processes, in *Perry's Chemical Engineers' Handbook* (ed. R.H. Green), McGraw-Hill, New York, pp. 36–71.

Rasanen, E., Nystrom, M., Sahlstein, J., and Tossavainen, O. (2002) Purification and regeneration of diluted caustic and acidic washing solutions by membrane filtration. *Desalination*, **149**, 185–190.

Rice, G., Kentish, S., Vivekanand, V., et al. (2005) Membrane-based dairy separation: a comparison of nanofiltration and electrodialysis. *Development in Chemical Engineering and Mineral Processing*, **13**, 43–54.

Rice, G., Barber, A., O'Connor, A., et al. (2009) Fouling of NF membranes by dairy ultrafiltration permeates. *Journal of Membrane Science*, **330**, 117–126.

Rice, G., Barber, A., O'Connor, A., et al. (2010) A theoretical and experimental analysis of calcium speciation and precipitation in dairy ultrafiltration permeate. *International Dairy Journal,* **20**, 694–706.

Rice, G., Barber, A., O'Connor, A., et al. (2011a) The Influence of dairy salts on nanofiltration membrane charge. *Journal of Food Engineering*, **107**, 164–172.

Rice, G., Barber, A.R., O'Connor, A.J., et al. (2011b) Rejection of dairy salts by a nanofiltration membrane. *Separation and Purification Technology*, **79**, 92–102.

Rosmaninho, R. and Melo, L.F. (2006) The effect of citrate on calcium phosphate deposition from simulated milk ultrafiltrate (SMUF) solution. *Journal of Food Engineering*, **73**, 379–387.

Roth, S.A., Weil, J.K., and Spade, M. (2001) Patent WO2001060171A1.

Samhaber, W.M. (2005) Uses and setting of tasks of nanofiltration in the food industry. *Chemie Ingenieur Technik*, **77**, 583–588.

Schaep, J., Vandecasteele, C., Mohammad, A.W., and Bowen, W.R. (1999) Analysis of the salt retention of nanofiltration membranes using the Donnan-steric partitioning pore model. *Separation Science and Technology*, **34**, 3009–3030.

Schmidt, D.G. and Both, P. (1987) Studies on the precipitation of calcium phosphate I. Experiments in the pH range 5.3 to 6.8 at 25 °C and 50 °C in the absence on additives. *Netherlands Milk and Dairy Journal*, **41**, 105–120.

Scott, S.N. and Krishnapillai, A. (2010) Patent WO2010106320A2.

Seki, N., Kinoshita, K., Saitoh, H., et al. (2012) Analysis of flux change on nanofiltration of dairy whey. *Kagaku Kogaku Ronbunshu*, **38**, 90–101.

Shim, Y., Lee, H.-J., Lee, S., et al. (2002) Effects of natural organic matter and ionic Sspecies on membrane surface charge. *Environmental Science and Technology*, **36**, 3864–3871.

Spade, M.E., Weil, J.K., and McHale, M.S. (2003) Patent US6623954B1.

Spanos, N., Patis, A., Kanellopoulou, D., et al. (2007) Precipitation of calcium phosphate from simulated milk ultrafiltrate solutions. *Crystal Growth and Design*, **7**, 25–29.

Spiegler, K.S. and Kedem, O. (1966) Thermodynamics of hyperfiltration (reverse osmosis): criteria for efficient membranes. *Desalination*, **1**, 311–326.

Tang, C.Y. and Leckie, J.O. (2007) Membrane independent limiting flux for RO and NF membranes fouled by humic acid. *Environmental Science Technology*, **41**, 4767–4773.

Tang, C.Y., Kwon, Y.N., and Leckie, J.O. (2009a) Effect of membrane chemistry and coating layer on physiochemical properties of thin film composite polyamide RO and NF membranes I. *FTIR and XPS characterization of polyamide and coating layer chemistry. Desalination*, **242**, 149–167.

Tang, C.Y., Kwon, Y.N., and Leckie, J.O. (2009b) Effect of membrane chemistry and coating layer on physiochemical properties of thin film composite polyamide RO and NF membranes. *II. Membrane physiochemical properties and their dependence on polyamide and coating layers. Desalination*, **242**, 168–182.

Taniguchi, M., Kurihara, M., and Kimura, S. (2001) Boron reduction performance of reverse osmosis seawater desalination process. *Journal of Membrane Science*, **183**, 259–267.

Tay, J.-H., Liu, J., and Delai Sun, D. (2002) Effect of solution physico-chemistry on the charge property of nanofiltration membranes. *Water Research*, **36**, 585–598.

Teixeira, M.R., Rosa, M.J., and Nyström, M. (2005) The role of membrane charge on nanofiltration performance. *Journal of Membrane Science*, **265**, 160–166.

Turan, M. (2004) Influence of filtration conditions on the performance of nanofiltration and reverse osmosis membranes in dairy wastewater treatment. *Desalination*, **170**, 83–90.

van de Lisdonk, C.A.C., Rietman, B.M., Heijman, S.G.J., et al. (2001) Prediction of supersaturation and monitoring of scaling in reverse osmosis and nanofiltration membrane systems. *Desalination*, **138**, 259–270.

Van der Bruggen, B., Koninckx, A., and Vandecasteele, C. (2004) Separation of monovalent and divalent ions from aqueous solution by electrodialysis and nanofiltration. *Water Research*, **38**, 1347–1353.

Van der Bruggen, B., Daems, B., Wilms, D., and Vandecasteele, C. (2001) Mechanism of retention and flux decline for the nanofiltration of dye baths from the textile industry. *Separation and Purification Technology*, **22–23**, 519–528.

van der Horst, H.C. (1995) Fouling of organic membranes during processing of dairy liquids. *International Dairy Federation [*Special Issue*] S.I.*, **9504**, 36–52.

van Kemenade, M.J.J.M. and de Bruyn, P.L. (1987) A kinetic study of precipitation from supersaturated calcium phosphate solutions. *Journal of Colloid and Interface Science*, **118**, 564–85.

Vembu, R. and Rathinam, V. (1997) Separation of minerals from whey permeate. United States patent 5,639,501.

Vourch, M., Frappart, M., Balannec, B., et al. (2003) Treatment of dairy industry wastewater by membrane cascade operation for production of reusable water. *Recents Progress Genie Procedes*, **90**, 699–706.

Vourch, M., Balannec, B., Chaufer, B. and Dorange, G. (2005) Nanofiltration and reverse osmosis of model process waters from the dairy industry to produce water for reuse. *Desalination*, **172**, 245–256.

Vrijenhoek, E.M., Hong, S., and Elimelech, M. (2001) Influence of membrane surface properties on initial rate of colloidal fouling of reverse osmosis and nanofiltration membranes. *Journal of Membrane Science*, **188**, 115–128.

Walstra, P. and Jenness, R. (1984) *Dairy Chemistry and Physics*, John Wiley & Sons, Inc., Canada.

Watkinson, W.J. (2008) Chemistry of detergents and disinfectants, in *Cleaning-in-Place: Dairy, Food and Beverage Operations* (ed. A.Y. Tamime), Blackwell Publishing, Oxford.

Zeman, L.J. and Zydney, A.L. (1996) *Microfiltration and Ultrafiltration: Principles and Applications*, Marcel Dekker, New York.

Zhang, Y., Yoshihito, T., Wang, H., et al. (2008) Patent CN101195652A.

Zhang, Y., Bai, J., Zhang, J., et al. (2012) Investigation of ultra- and nanofiltration for separation of proteins and lactose from dairy wastewater. *Asian Journal of Chemistry*, **24**, 2593–2597.

6 Development and application of reverse osmosis for separation

Masoumeh Zargar, Bo Jin, and Sheng Dai

School of Chemical Engineering, The University of Adelaide, Adelaide, South Australia, Australia

6.1 INTRODUCTION

Reverse osmosis (RO) is a pressure-driven separation technology in which the solution is pressed into the membrane module by applying a pressure higher than the osmotic pressure in the process, which leads to the separation of most solutes from solvent (e.g., more than 99% in the case of brackish water). RO is considered as the most efficient and affordable technique for desalination. It has broad applications in various industrial processes, in particular for food processing industries and water or wastewater treatment. A reverse osmosis membrane (ROM), the centre of the RO process, is expected to have desirable characteristics, including high selectivity for the solvent over the solute, low fouling, high chemical and mechanical stability, and durability for a long-term operation.

The first work on the ROMs was proposed in 1949 by Gerald Hassler from the University of California, Los Angeles (UCLA), who reported the osmotic properties of cellophane membranes. Although his proposed system never reached a high yield, he could establish the scientific and technical fundamentals of ROMs for future studies. Around 1955, Reid and Breton from the University of Florida introduced a high-pressure RO system with 98% salt rejection but low water flux, even with very thin membranes. Regardless of the inefficient system, their research findings did shape further studies on the development of current ROMs (Glater, 1998). Professor Samuel Yuster and his co-workers at the UCLA developed the modified cellulose acetate ultrafiltration (UF) membranes with a high water flux and acceptable salt rejection. They synthesized a very thin-film membrane supported by a thicker porous substrate with an asymmetric structure and only one active layer (Glater, 1998). The

Membrane Processing for Dairy Ingredient Separation, First Edition.
Edited by Kang Hu and James M. Dickson.

hard work of this group resulted in the fabrication of the first practical ROM with one order of magnitude higher performance compared with the available commercial membranes at that time (Loeb and Sourirajan, 1963). Studies at UCLA and other research centres on the development of economic ROMs was continued after this discovery, leading to the invention of the first practical spiral wound membrane module in 1963 and the first commercially successful hollow fibre module in 1967 (Kucera, 2010). Establishment of the first commercial membrane desalination plant at Coalinga, California, USA, by Loeb and Selover in 1967 might be the most important achievements in the development of the ROM desalination process (Loeb and Selover, 1967). As a result of several studies during the last decades on the development of the RO process, significant improvements have been made on the materials and performance of ROMs, such as polymeric ROM and thin-film composite (TFC) membranes, for industrial applications. A major concern for the industrial application of the ROM technology remains to be the high costs dictated by both energy consumption and membrane replacement. Recent R&D efforts have been focused on the modification of the membrane surface and structure to improve the functionality (permeability and selectivity) and applicability (mechanical, chemical, and biological stability). The current ROM market is dominated by the thin-film polymer membranes consisting of three layers: a polyester web acting as structural support, a microporous interlayer, and an ultra-thin barrier layer on the upper surface. The ultra-thin barrier layer and the porous support can be independently optimized with respect to structure, stability, and performance. Alongside the advancements in the ROM technology, the development of membrane materials has undeniably made the RO process more economic. The improvement of permeate flux and fouling resistance is deemed as the foremost priority for the development of new generation ROMs for a low cost process. To date, there is an ongoing need to develop suitable membranes that offer enhanced performance and low costs for industry.

This chapter provides an overview on the membrane materials and module configurations, and transport mechanisms of the ROMs. We summarize the key parameters in terms of membrane characteristics and operation conditions that affect the performance of ROMs, and technical and economic challenges in the application of RO processes. Special attention is given to discuss the process parameters and challenges using ROMs in the dairy industry. Finally, the prospects for the development of ROM materials and technologies are briefed.

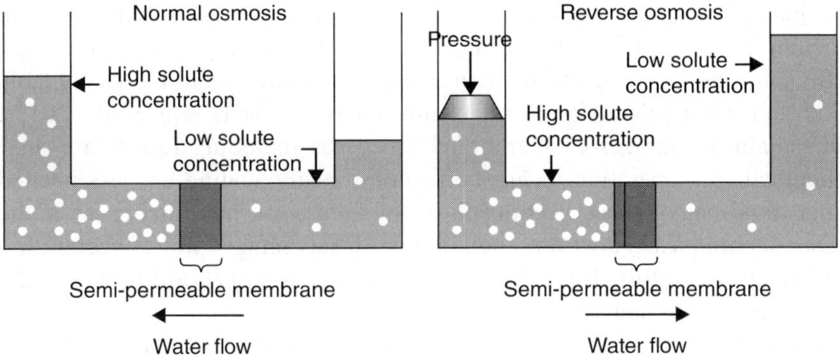

Figure 6.1 Schematic view of normal osmosis and reverse osmosis processes.

6.2 REVERSE OSMOSIS AND ITS WORKING MECHANISM

A selective/semi-permeable membrane is the centre of a reverse osmosis process. The membrane allows a part of solution (mostly solvent) to pass through it, while rejecting other components in the solution (solutes). When this membrane is placed between two sections, one only containing pure solvent and the other one comprising solvent plus some solutes, a normal flow for equilibration occurs from the solvent side to the solution one. This flow is the result of a chemical potential driving force due to the presence of solutes. An exact pressure applied for the purpose of ceasing the flow is called osmotic pressure. By increasing the applied pressure higher than the osmotic pressure, the normal osmosis flow will be reversed and the solvent will move from the high solute concentration part to the low solute concentration one. Whereas this process is exactly the reverse of the osmosis process, it is called reverse osmosis (RO) (Chen et al., 2011). Figure 6.1 presents a schematic view of the normal osmosis and reverse osmosis processes.

6.3 REVERSE OSMOSIS MEMBRANES

A membrane is defined as a semi-permeable barrier between two phases. In the case of reverse osmosis membranes (ROMs), the membrane allows water molecules to pass through it, while rejecting the majority of dissolved salts, bacteria, and organic materials. However, external pressure is required to dominate osmotic pressure resulted from the different solutes. A good membrane should be chemically and mechanically stable and energy efficient, and be designed in such a way as to have high and sustainable flux and good selectivity to reject unwanted materials (Ng et al., 2013). The main advantages of membrane processes including ROMs are their chemical free operation, energy efficiency, easily scaling-up, less needed area, automated process controlling, and ease of hybridizing with the other techniques. However, their major disadvantages are concentration polarization, fouling, limited chemical selectivity, and short lifetime of the membranes (Chen et al., 2011).

Membranes are usually classified by different parameters, including process, materials, morphology, preparation method, geometry, and separation regime. One of the most popular classification methods is based on membrane materials. Commercial ROMs are mainly fabricated from polymers. They also could be made using other materials such as metal, carbon, glass, ceramic, etc. (Ren and Wang, 2011). The membrane materials are generally classified into two main groups as inorganic and polymeric membranes, which are described as follows.

6.3.1 Inorganic reverse osmosis membranes

Ceramic membranes, the main commonly used inorganic membranes, are mostly made from titania, silica, alumina, zirconia, or a mixture of these materials. Whereas the cost of synthesis for the inorganic membranes is relatively high, application of these membranes is currently limited to special cases where polymeric membranes cannot be utilized (e.g., high temperatures and reactivity of the environment, highly contaminated feed). In the inorganic category, zeolite ROMs have gained significant interest in recent years. However, polymeric membranes need less membrane area toward an equivalent performance, which makes them more favorable than zeolite membranes. The zeolite membranes have at least three times greater thickness compared with the current polymeric ROMs, which leads to a lower water flux in the zeolite membranes compared with polymeric ones (Lee et al., 2011). Generally, the inorganic membranes have advantages such as chemical and mechanical resistance and high selectivity and applicability under harsh operational conditions. However, they are not applied extensively due to less control over pore size, higher thickness of the selective barrier layer, higher water flux, and most importantly higher operational cost compared with the polymeric membranes (Ng et al., 2013).

6.3.2 Polymeric reverse osmosis membranes

Recent research is highly focused on polymeric ROMs due to their several advantages over inorganic ones. These include controllable pore size, higher flexibility and selectivity to different chemicals, less installation area requirement, cost effectiveness, easy up-scaling and down-scaling, and good cooperation with the other processes. Due to these characteristics, polymers are considered as better materials for membrane fabrication. However, they face some issues such as sensitivity to biofouling, low mechanical stability, and high hydrophobicity, which need to be resolved for the industrial applications (Ng et al., 2013). Cellulose acetates (CA) and aromatic polyamides (PA) have been widely used as the main materials for fabrication of ROMs from the time of the first RO plant operation (Lee, Arnot, and Mattia, 2011). The first CA membrane was made in the late 1950s from cellulose diacetate (CDA) polymer and was used commercially in the 1960s. CA was widely recognized as the best material for fabrication of ROMs until 1969 due to its stable performance at the presence of influents containing fouling agents and cost effectiveness (El-Saied et al., 2003; Younos and Tulou, 2005). However, the industrial application of CA membranes became limited due to its narrow pH and temperature stability range, sensitivity to compaction under high pressures, susceptibility to hydrolysis, and the high chance of being attacked by biological agents (Lee, Arnot, and Mattia, 2011). Therefore, finding another material with fewer drawbacks became the focus of many studies in the following years (Li and Wang, 2010b; Lee, Arnot, and Mattia, 2011). The first commercial polymeric membrane without cellulose was an aromatic PA asymmetric hollow fibre membrane synthesized by Richer and Hoehn (1980). Although their membrane suffered from low salt retention and water flux, its functionality, stability, and lifetime were higher than the CA membranes. Compared to the CA spiral wound membrane, the PA membrane has a higher flux per unit module volume due to the good packing of the hollow fibres (Lee, Arnot, and Mattia, 2011). The schematic diagram of classification of polymeric membranes is shown in Figure 6.2.

6.3.3 Thin-film composite reverse osmosis membranes

Thin-film composite (TFC) membrane is a new generation of composite polymeric membranes that consist of an ultra-thin PA film (10–1000 nm) and a support layer. TFC membranes were firstly fabricated by in situ polycondensation on porous polysulfone (PSF) in the 1970s. It was a replacement for asymmetric ROMs synthesized via phase inversion of CA (Jeong et al., 2007; Humplik et al., 2011). Two step casting methods are used for the fabrication of the support film and the ultra-thin barrier layer. This makes it possible to individually optimize these layers and develop better characteristics in the TFC membranes. In the composite structure, the porous support layer acts as the membrane mechanical strength enhancer, while the ultra-thin barrier layer is mainly used for making the membrane selective. There is usually another support layer in the TFC membrane matrix

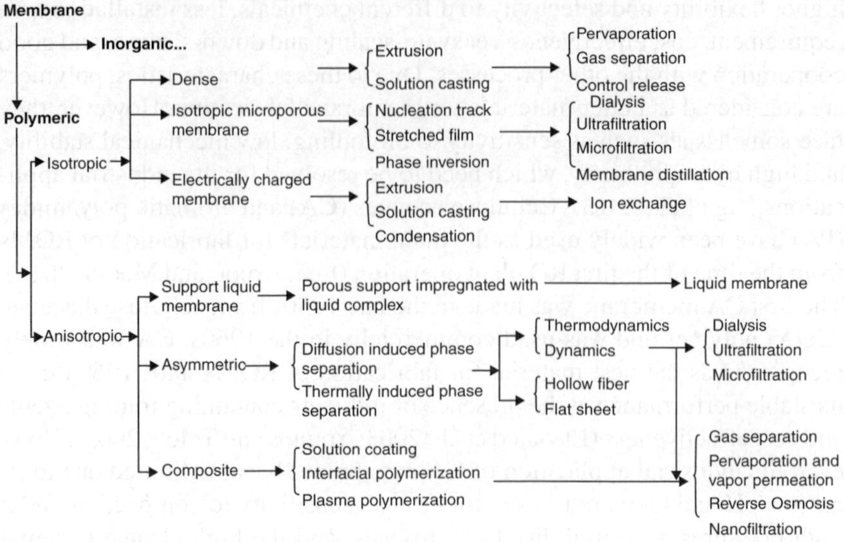

Figure 6.2 Classification of polymeric membranes (Ren and Wang, 2011). Reproduced with permission of Springer.

that is mostly made of a nonwoven fabric and applies to support the whole structure under a high-pressure RO process. These anisotropic composite membranes are referred as TFC membranes due to their peculiar structure (Li and Wang, 2010b; Lee, Arnot, and Mattia, 2011). TFC membranes are very popular in different industrial applications including the dairy industry due to the resistance of their materials to high temperatures and different chemicals (Ng et al., 2013).

Different polymers have been tested for the fabrication of both layers. For the barrier layer, promising permeability was obtained by vinyl acetate/acrylonitrile, vinylene carbonate/acrylonitirile, allylamine, acetylene/water/carbon monoxide combinations, and acetylene/water/nitrogen (Lee et al., 2012). Many polymers have also been successfully applied for synthesis of the porous support layer. Among the tested polymers, polysulfone has been widely used as the best commercial composite membrane and the most favorable material for the support layer of TFC/TFN-ROMs due to its compaction resistance and good flux, and, most significantly, its resistance in an acidic environment (Lee, Arnot, and Mattia, 2011; Lau et al., 2012). However, its application as a commercial membrane has been limited by the physical and chemical characteristics of hydrophobicity, sensitivity to different solvents, and low thermal stability (<50 °C). Therefore, recent studies have been focused on the development of new substrate membranes with more thermal and chemical stability, and higher water flux and solute

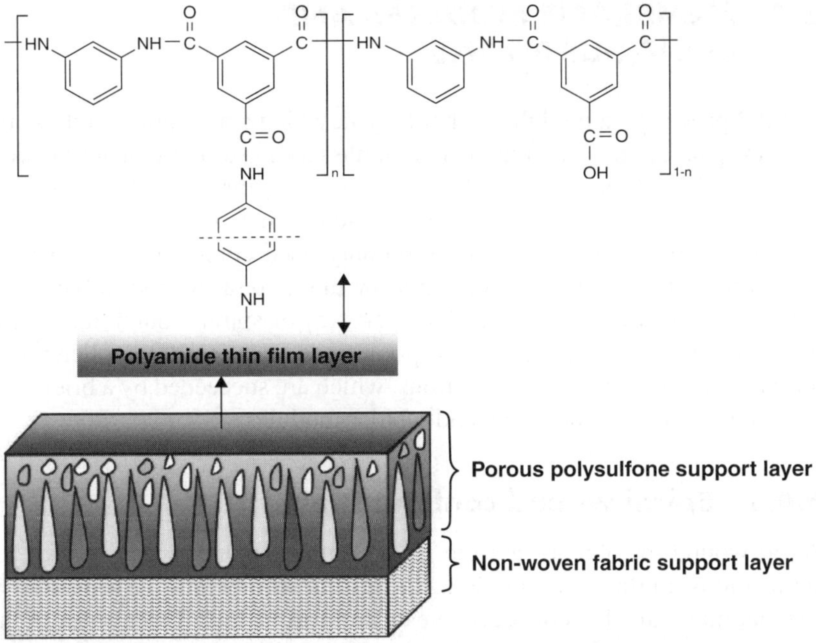

Figure 6.3 Schematic of the TFC-ROM and the chemical structure of the aromatic polyamide.

rejection (Lau et al., 2012). Figure 6.3 shows a schematic description of the TFC-ROMs and the chemical formula for the selective PA layer.

The main advantages of TFC-ROMs for the special application in the dairy industry are their high thermal stability and wider pH ranges. Higher chemical stability and bigger rejection of low molecular weight solutes are other interesting advantages of these membranes in dairy processes (Cheryan, Veeranjaneyulu, and Schlicher, 1990). Several studies have been performed on the application of TFC-ROMs in the dairy industry and the results indicated that composite membranes had a higher flux compared with CA membranes. As a result, energy savings could be even greater in them, especially when we apply them as a supplement to the evaporation systems (Hiddink, de Boer, and Nooy, 1980; Spangler and Amundson, 1986; Kulozik and Kessler, 1988; Morales, Amundson, and Hill, 1990).

Although TFC membranes have presented promising performances compared with the conventional CA membranes, they face a few key technical challenges such as sensitivity to fouling and chlorine attacks (Misdan, Lau, and Ismail, 2012). Addressing these problems can further enhance performance of these membranes and make them highly efficient in a wide range of industrial applications, including the dairy and food industries.

6.4 MEMBRANE MODULES AND CONFIGURATIONS

The RO process is a modular technology in which a membrane and its support are packed in a module. This module can make the contact between pressurized feed and the membrane surface possible and allows the collection of permeate flow from a side without being contaminated with the concentrate one. There are four commonly used commercial ROM module configurations, which include hollow fiber, spiral wound, tubular, and flat sheet. An overview on the characteristics of spiral wound and hollow fibre modules, which are the most popular designs for RO application, is presented in the following subsections, which are succeeded by a brief comparison of them with flat sheet and tubular modules.

6.4.1 Spiral wound configuration

Spiral wound membranes are the most common membranes used for RO applications. In this configuration, various flat membranes are separated by channel materials that collect permeate and form a leaf. Three sides of this arrangement are sealed and the fourth one remains open to provide an exit way for permeate flow. Some of these assemblies are wound around a central plastic tube, which has a number of holes to collect permeate from the leaves (Chen et al., 2011).

There are usually three to six spiral wound membranes within a pressure tube, which are attached together in a way that the feed flow of the first element becomes the feed of the next ones. Figure 6.4 presents a schematic view of a spiral wound ROM. Permeate of each element goes to the central

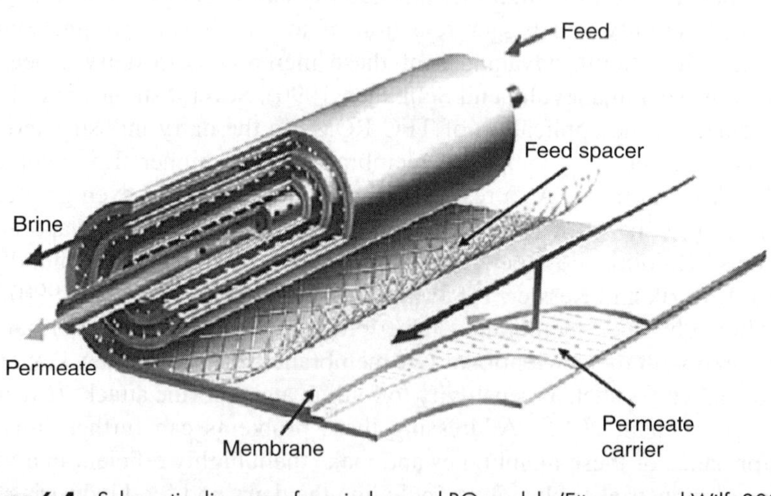

Figure 6.4 Schematic diagram of a spiral wound RO module (Ettouney and Wilf, 2009). Reproduced with permission of Springer.

permeate collector tube and leaves as a single permeate flow. The recovery of a typical spiral wound membrane with four to six elements is about 50% under the normal design situation. Using less space compared with flat sheet or tubular membranes, having a high specific membrane surface area, having less fabrication and replacement costs, and being easily cleaned by either chemical or hydraulic methods are the advantages of these membranes. However, due to the small flow passages, which might be blocked by the highly turbid influents, an extensive pretreatment is required before importing the feed flow to these membrane modules (Polasek and Talosharif, 2003; Chen et al., 2011).

6.4.2 Hollow fiber, flat sheet, and tubular configurations

Hollow membranes can be fabricated as thin as a human hair and are assembled into a cylindrical bundle with the same distance from each other. Figure 6.5 describes the schematic structure of the hollow fiber membranes. The ends of the cylindrical bundle are epoxy sealed and have formed a sheet-like permeate tube. The whole bundle is in a cylindrical shell and is called a permeator. The pressurized feed water enters the assembly, flows in a central distribution tube, and passes in a radial way around the fiber bundle to reach the shell. The permeate exits from the permeate tube, which is the end of the bundle. This configuration is usually highly compacted and thus requires less area. On the other hand, due to this high packing density

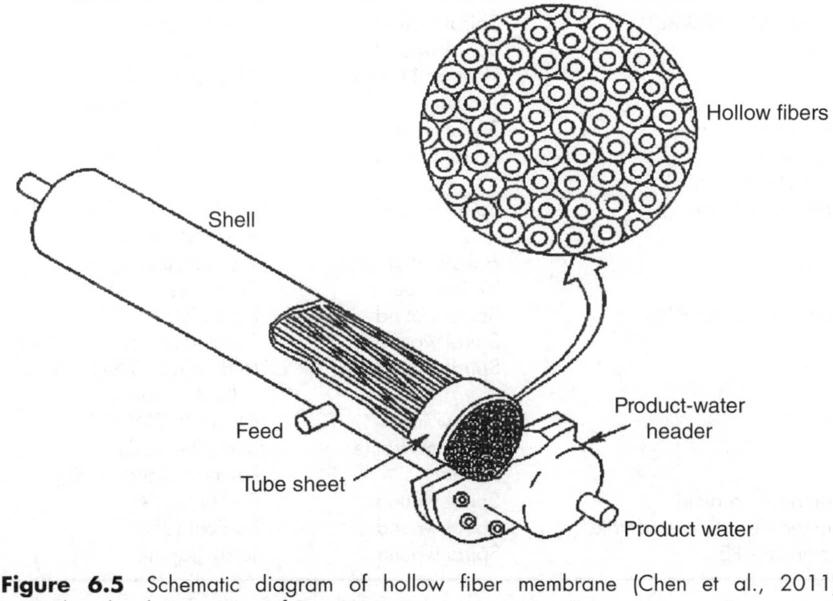

Figure 6.5 Schematic diagram of hollow fiber membrane (Chen et al., 2011). Reproduced with permission of Springer.

and small space between fibers, they are very prone to be fouled and could not be cleaned easily. Therefore, influent should be well pretreated before importing to the membrane. A typical single hollow fiber permeator could have 50% recovery and the minimum required reject flow.

Two other module types rarely used for RO are tubular and flat sheet configurations. Tubular systems have the benefit of operation on highly turbid feed and the ease of cleaning via mechanical or hydraulic methods. However, their capital cost is very high compared with the spiral wound and hollow fiber membranes. A flat configuration is also applicable for chemical processes but it is highly costly to operate on a large scale due to its complexity (Chen et al., 2011).

Currently, PA spiral wound membranes are the most commercially used membranes for RO and nanofilteration (NF) applications, which has taken approximate 91% of the marketing proportion, while asymmetric CA hollow fiber membranes are far behind. The PA spiral wound membranes have a higher solute retention and net pressure driving force, while the CA hollow fiber membranes have a higher chlorine resistance and, therefore, more stability against biofouling (Lee, Arnot, and Mattia, 2011). A number of commercial ROMs and their modules are presented in Table 6.1.

Table 6.1 Commercially available ROMs and modules (Chian et al., 2007). Reproduced with permission of Humana Press.

Membrane materials	Module types	Companies
Asymmetric cellulose acetate	Spiral wound	Toray (Japan)
	Spiral wound	Desalination System (USA)
	Plate and frame	DDS (Denmark)
	Tubular	Paterson Candy (UK)
Asymmetric aramid	Spiral wound	Du Pont (USA)
Asymmetric poly(acrylonitrile)	Spiral wound	Sumitomo (Japan)
Cellulose acetate	Spiral wound	Hydranautics/Nitto Denko (USA/Japan)
	Hollow fiber	Toyobo (Japan)
	Hollow fiber	Osmonics (USA)
Composite polyamide	Spiral wound	Fluid system/UOP
	Spiral wound	Desalination system (USA)
	Spiral wound	Hydranautics/Nitto Denko (USA/ Japan)
	Hollow fiber	Filmtec/DOW (USA)
	Plate and frame	DDS (Denmark)
	Tubular	Paterson Candy (UK)
Composite aramid	Spiral wound	Du Pont (USA)
Composite cellulose acetate	Spiral wound	Du Pont (USA)
Composite PEC	Spiral wound	Toray (Japan)

6.5 TRANSPORT MECHANISMS AND MODELS IN REVERSE OSMOSIS MEMBRANES

Several mathematical models have been developed and used to describe the fluid transportation mechanisms in the ROM process. The most important ones are diffusion-based and pore flow-based models. Although both of these models have been used and accepted by many researchers and engineers, there is a fundamental difference from one model to the others with respect to their assumptions. The diffusion models consider the ROM active layer as a nonporous material while the pore flow models assume the membrane active layer as a porous structure. Therefore, each of them explains the transport mechanisms using a different approach. Diffusion models consider only the diffusion while the pore ones consider both diffusion and convection (Williams, 2003).

In this section, the authors do not intend to present derivations and details of all of these models. Instead, a brief introduction on the fundamentals and final equations are presented. For a detailed description we refer to the exceptional published reviews on these transport models (Dickson, 1988; Mehdizadeh, 1990; Bhattacharyya and Williams, 1992; Moonkhum et al., 2010; Crittenden et al., 2012).

6.5.1 Diffusion-based models

6.5.1.1 Solution diffusion model (SDM)

Diffusion models are the most popular transport models in RO, dialysis, pervaporation, and gas permeation processes. The diffusion models are based on the material solubility in the membrane structure. The separation is obtained by dissolving different permeants in the membrane and various diffusion rates are achieved due to the difference in the concentration of permeants and diffusion rates. Three main models have been proposed based on this mechanism including: the solution diffusion model (SDM), the solution diffusion imperfection model (SDIM), and the extended solution diffusion model (ESDM).

In the SDM, the most important assumptions are:

1. The active surface layer of the membrane, which is responsible for the membrane selectivity, is a dense, homogeneous, and, most importantly, nonporous film.
2. Both solvent and solutes should first dissolve in the membrane and then diffuse through it due to their independent chemical potential.
3. Salt diffusion is solely governed by the concentration gradient and is not affected by the pressure gradient across the membrane.
4. Water concentration and diffusion coefficients are assumed constant.

Considering the assumptions of the SDM, permeation in the membrane is governed by concentration and pressure gradient across it. However, the effect of pressure gradient on the solute diffusion at low solute concentrations is negligible. On the other hand, because of consideration of the constant concentration of water on both sides of the membrane, the only driving force for water permeation is pressure difference (Williams, 2003).

Based on the SDM assumptions, water flux (J_w) can be obtained by this model via

$$J_w = \frac{D_{wm} C_{wm} \overline{V}_w}{R_g T \delta}(\Delta P - \Delta \pi) \xrightarrow{\text{yields}} J_w = A(\Delta P - \Delta \pi) \qquad (6.1)$$

where C_{wm} and D_{wm} are the water concentration and diffusivity in the membrane respectively, $\overline{V}_w$ is the partial molar volume of water, R_g is the universal gas constant, T is absolute temperature, δ is the membrane thickness, P is the applied pressure, π is the osmotic pressure, and, finally, A is the water permeability coefficient across the membrane (Figure 6.6).

For the solute flux (J_s), by considering the third assumption of the SDM, only the concentration difference is the driving force for the solutes and therefore the solute flux can be obtained as follows:

$$J_s = B(C_F - C_P) \equiv B(C_{sfm} - C_{spm}) \qquad (6.2)$$

where B is the solute permeability coefficient across the membrane, C_F and C_P are the bulk feed and bulk permeate concentrations, and C_{sfm} and C_{spm} are the solute concentrations at the feed side and permeate side of the membrane surface, respectively.

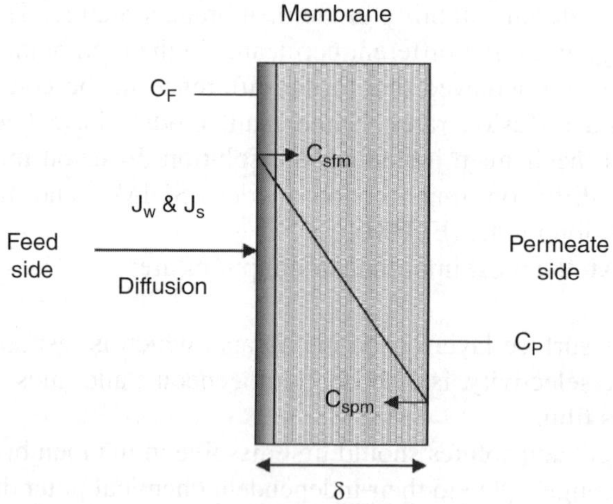

Figure 6.6 Schematic diagram of membranes for the solution diffusion mechanism.

It should be pointed out that the mass transfer coefficients for water and solutes across the membrane (*A* and *B*) should be achieved through experiments.

By applying the mentioned equations, the solute rejection (*R*) for the SDM can be calculated as follows:

$$\frac{1}{R} = 1 + \left(\frac{B}{A}\right)\left(\frac{1}{\Delta P - \Delta\pi}\right) \tag{6.3}$$

6.5.1.2 Solution diffusion imperfection model (SDIM)

The main assumption in SDM is the nonporous structure of the active membrane layer. Due to the possibility of existence of pore flow in the membrane transport mechanism, the solution diffusion imperfection model (SDIM) has been developed to modify the SDM. This model also considers the probability of existence of imperfections (pores) on the active layer surface causing the transport through them (Fritzmann et al., 2007). The water and solute flux across the membrane assuming the SDIM model can be obtained through the following equations:

$$J_{w_{(SDIM)}} = A(\Delta P - \Delta\pi) + C\Delta P \equiv J_{w_{(SDM)}} + C\Delta P \tag{6.4}$$

$$J_{s_{(SDIM)}} = B(C_F - C_P) + C\Delta P \equiv J_{s_{(SDM)}} + C\Delta P \tag{6.5}$$

where $J_{w_{(SDM)}}$ and $J_{s_{(SDM)}}$ are equivalent to the water and solute fluxes calculated based on the SDM from Equations (6.1) and (6.2), respectively, *C* is the mass transfer coefficient across the pores and $C\Delta P$ is the pore flow contribution to the water and solute fluxes across the membrane (Williams, 2003).

The rejection for the SDIM can be defined as follows:

$$\frac{1}{R} = 1 + \left(\frac{B}{A}\right)\left(\frac{1}{\Delta P - \Delta\pi}\right) + \left(\frac{C}{A}\right)\left(\frac{\Delta P}{\Delta P - \Delta\pi}\right) \tag{6.6}$$

6.5.1.3 Extended solution diffusion model (ESDM)

Another issue that SDM assumptions do not cover is negative solute rejection in the RO process. The extended solution diffusion model (ESDM) was developed by Burghoff, Lee, and Pusch (1980) to revise the SDM in this regard. They identified that, although the pressure gradient influence is negligible for inorganic salt solutions, it can be effective for organic salt solutions and therefore cannot be disregarded (Burghoff, Lee, and Pusch, 1980; Dickson, 1988). By considering the solute transport resulting from the pressure gradient across the membrane, the solute flux for this model would be different from that of the SDM and can be calculated from the following equation, in which *D* is the solute transfer coefficient related to the

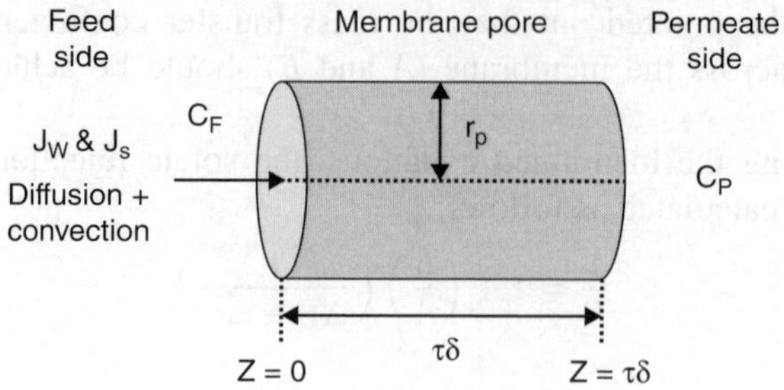

Figure 6.7 Schematic view of the membrane pore for pore flow models.

pressure gradient:

$$J_s = \frac{D_{sm}K_{sm}}{\delta}(C_F - C_P) + D\Delta P \tag{6.7}$$

The rejection by considering the ESDM assumption can also be defined by

$$\frac{1}{R} = \frac{1 + \left(\frac{B}{A}\right)\left(\frac{1}{\Delta P - \Delta\pi}\right)}{1 - \frac{D}{AC_F}\left(1 - \frac{\Delta\pi}{\Delta P}\right)} \tag{6.8}$$

6.5.2 Pore models

The second category of mathematical models for transport phenomena in ROMs is the pore flow model, in which the membrane is considered porous and the pore plays the most important part in the separation mechanism. In this model, some permeants are filtered from the flow while the others pass through the membrane pores (Williams, 2003). A schematic view of the membrane pore for the modelling based on pore models is presented in Figure 6.7.

6.5.2.1 Preferential sorption capillary flow (PSCF) model

The preferential sorption capillary flow (PSCF) model was an early pore model developed by Sourirajan in 1970. In the PSCF model it is assumed that the separation mechanism in ROMs is governed by both surface phenomena and fluid transport through the pores. In this model, the surface layer of ROMs is defined as a selective layer with the affinity to attract solvent and to transport it under pressure through its pores, while repelling the solutes. Therefore, pure solvent is adsorbed on the surface and transfers through the pores, leading to the successful separation process (Sourirajan and Matsuura, 1985).

Considering the assumptions of this model, the water flux can be obtained via

$$J_w = A(\Delta P - [\pi(X_F) - \pi(X_P)]) \tag{6.9}$$

where A is the water permeability coefficient across the membrane, π is the osmotic pressure, and X is the solute mole fraction either in the feed or permeate (Williams, 2003).

The solute flux can also be achieved by the following equation in which D_{sp} is the diffusivity of solute into the membrane pores, k_D is the distribution coefficient of solute from the feed solution to the pores of the membrane, and C_T is the total molar concentration:

$$J_s = \frac{D_{sp}k_D C_T}{\delta}(X_F - X_P) \tag{6.10}$$

The solute rejection can also be obtained via

$$\frac{1}{R} = 1 + \frac{D_{sp}k_D C_T}{\delta}\frac{1}{A(\Delta P - [\pi(X_F) - \pi(X_P)])} \tag{6.11}$$

6.5.2.2 Finely porous model

Another pore model, which was developed in 1966 by Merten and was modified by Jonnson and Boesen in 1975, is called the finely porous model. This model assumes that the solvent flow is through pores while solute transfer is governed by both diffusion and convection in pores (Merten, 1966; Jonsson and Boesen, 1975).

In this model, the water flux can be obtained by

$$J_w = \varepsilon u = \frac{\varepsilon r_p^2 \Delta P}{8\eta\tau\delta}\left[\frac{1}{1 + \frac{r_p^2 X_{sm} C_P}{8\eta}}\right] \tag{6.12}$$

where ε is the membrane porosity, u is the linear velocity of pore fluid, τ is the tortuosity, $\tau\delta$ is the membrane pore length, r_p is the membrane pore radius, η is the viscosity, and X_{sm} is the friction coefficient between the solute and the membrane pore (Williams, 2003).

Rejection through this model can be obtained from

$$\frac{1}{1-R} = \frac{a}{k_D} + \left(1 - \frac{a}{k_D}\right)\exp\left[-u\tau\delta\frac{X_{sw}}{R_g T}\right] \tag{6.13}$$

where a and X_{sw} are defined as follows, in which D_{sm} and D_{sw} are the solute diffusion coefficient in the membrane and water, respectively:

$$a = \frac{D_{sw}}{D_{sm}} \tag{6.14}$$

$$X_{sw} = \frac{R_g T}{D_{sw}} \tag{6.15}$$

Sourirajan and Matsuura modified the finely porous model by introducing the surface force pore flow (SFPF) model in 1985 through considering the radial solute concentration gradient in addition to just the axial one. The details and derivations of this model are presented elsewhere (Sourirajan and Matsuura, 1985; Bhattacharyya and Williams, 1992; Williams, 2003).

Both diffusion and pore models were developed and used in the nineteenth century. There has been always a debate on the precision of each of them. However, as of now the most popular model accepted for transport phenomena in ROMs is the solution diffusion model and only a few modellers still use pore flow models to explain the separation mechanism in ROMs (Wijmans and Baker, 1995).

6.6 REVERSE OSMOSIS PROCESS

6.6.1 Overview of the process and performance evaluation in reverse osmosis membranes

Performance of the ROMs is usually evaluated by the measurement of permeate flux and solute rejection (Pendergast, Ghosh, and Hoek, 2013). The permeate flux is the rate (weight or volume) of permeate collected per unit membrane area. The solute rejection is defined as the relative change in solute concentration after passing through the membrane. They are both greatly influenced by different operational and process variables such as pressure, temperature, pH, velocity, etc., which are discussed in detail in Section 6.6.2 of this chapter.

A schematic view of a regular RO process for considering the performance parameters is presented in Figure 6.8.

To calculate the permeate flux, the following equation is usually used:

$$\text{Flux} = \frac{V_3}{At} \tag{6.16}$$

where V_3 is the collected permeate volume, A is the active membrane surface area, and t is the length of time taken for the process.

Solute rejection can get complicated due to the concentration polarization effect, which can result in different solute concentrations in bulk solution and near the membrane surface. Considering the concentration polarization, two different methods can be applied to calculate separation performance in the RO membranes. The first approach, which is more regularly used and is specially for moderately dilute solutions, considers the solute concentration

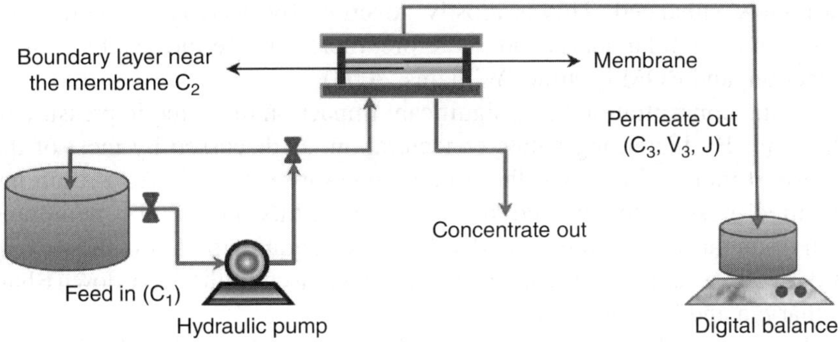

Figure 6.8 Schematic view of the ROM process.

in the feed (bulk) solution as the initial concentration and is calculated using

$$\%\text{Rejection} = \frac{C_1 - C_3}{C_1} \times 100 \tag{6.17}$$

The second way, which calculates the rejection by assuming a very good mixing on the high-pressure side of the membrane, considers the boundary-layer concentration as the initial concentration and is obtained via

$$\%\text{Rejection} = \frac{C_1 - C_2}{C_1} \times 100 \tag{6.18}$$

where C_1, C_2, and C_3 are the solute concentrations in feed, membrane boundary layer, and permeate solution respectively (Parekh, 1988).

6.6.2 Parameters affecting performance of reverse osmosis membranes

Permeate flux and solute rejection, which are two main criteria for evaluation of ROM performance, are both influenced by several factors. The main influential parameters are categorized as feed, membrane, and process variables (Bhattacharyya and Williams, 1992).

6.6.2.1 Feed variables

The most significant feed variables are the pretreatment operation, the concentration of solutes, and temperature and pH of the feed.

Pretreatment operations can define the quality of feedwater by considering the concentration of the particulates, precipitates, colloids, or microorganisms in the fluid. Through reducing different types of fouling and scaling agents during the pretreatment process, the efficiency of the ROMs can be extremely enhanced. This is mostly governed by decreasing fouling and concentration polarization and, therefore, increasing permeates flux, solute rejection, and ROM lifetime (Williams, 2003).

Solute concentration has a significant impact on the osmotic pressure of the fluid. By increasing solute concentration, as discussed by most of the transport models in ROMs, the osmotic pressure increases. As a result the required pressure to generate the same solvent flux through the membrane will be enhanced. Therefore, under the same operating pressure, the solvent flux declines as the solute concentration increases in the feed flow (Bhattacharyya and Williams, 1992).

Temperature has a direct impact on the permeate flux in ROMs. Based on the Arrhenius model for viscosity, there is a relationship between temperature and viscosity. Therefore, by increasing the temperature, the viscosity of the solvent decreases, which leads to higher solvent permeation through the membrane (Sourirajan and Matsuura, 1985; Dickson, 1988). On the other hand, temperature does not have a significant effect on the solute rejection in the RO process due to the relative enhancement of solute and solvent diffusion resulting from a temperature increment (Williams, 2003).

Feed pH does not have a considerable influence on the solvent flux in the RO process. However, it plays an important role in solute rejection, especially when the solute is an ionisable organic material. Previous studies have shown that salt rejection generally increases with pH increment and decreases at very high and very low pH values (<5 and >11). Van Wagner et al. (2010) studied the effect of pH on the performance of different commercial PA-ROMs and reported that salt rejection was linearly decreased by reducing pH from 9 to 5 while the water flux was stable and independent of the feed pH. The enhancement in negative charge by pH increment has been proved by increasing the zeta potential of the membrane surface, and this is probably the reason behind salt rejection increment by enhancing pH. The studies have also shown that the pH effect on the ROM performance is reversible in short-term exposure and the membranes exposed to low pH solutions for around one day have reached higher salt rejection again after increasing feed pH (Sourirajan and Matsuura, 1985; Dickson, 1988; Elimelech, Chen, and Waypa, 1994; Van Wagner et al., 2009).

6.6.2.2 Membrane variables

Membrane variables such as material, module configuration, and module arrangement should also be considered as influential parameters on the performance of ROMs. There have been enormous research efforts given to the development of novel ROMs and modification of them so as to enhance their performance and selectivity for different applications. Regarding the effect of materials and their physiochemical characteristics, different behavior and properties of organic and inorganic (polymeric) membranes in the ROM processes could be considered, depending on the specific requirements for different applications. For instance, polymeric membranes such as CA and PA could demonstrate different separation performance in terms of their sensitivity to fouling and chemical and biological attack. As discussed in Section 6.3, membrane material is one of the most predominating criteria for selection of the ROM process. Regarding the module geometry and arrangement, the specifications of each module type and presence and characteristic of different nonmembrane components in each (e.g., feed and permeate spacers in spiral wound modules) (see Section 6.4) should be taken into account as effective parameters on the final performance of the ROM.

6.6.2.3 Process variables

The feed flow rate and operating pressure and time, which are briefly described in this section, are the most influential process variables in ROM processes (Bhattacharyya and Williams, 1992).

The feed flow rate is a key operation parameter to determine not only the separation rate but importantly the efficiency of the ROM process. By increasing the feed flow rate and therefore enhancing the velocity, concentration polarization (which will be explained in Section 6.7.1) decreases. As a result, the feed flow rate increment results in a higher permeate flux through the ROMs. Increasing the feed flow rate can also enhance the solute rejection to some extent. However, as the increase in flow rate can also enhance the osmotic pressure near the membrane, it can moderate both permeate flux and solute rejection enhancements (Valentas, Rotstein, and Singh, 1997).

Operating pressure usually has a positive linear relationship with the solvent flux through the ROMs but it does not have a notable influence on the solute flux. For most of the chemicals, desirably pressure and solute retention also have a positive linear relationship. However, for some organics with intense interaction with the membrane, lower rejection is observed by operating pressure enhancement (Bhattacharyya and Williams, 1992).

Operating time is used to determine the life cycle of the ROMs and their economic feasibility in the RO process. ROMs usually have a certain lifetime and operational efficiency based on their materials and other process and operational variables. What happens in a normal RO process is deterioration of the ROM primal properties by passing the time in operation. For instance, due to chemicals and microorganisms involved and operation under high pressure, chemical degradation, fouling, and compaction might occur in most membranes after a long-term operation. Therefore, operating time has an inverse relationship with the performance of the ROMs (Bhattacharyya and Williams, 1992).

6.7 TECHNICAL AND ECONOMIC CHALLENGES

6.7.1 Concentration polarization

In the RO process, the feed moves toward the membrane in a convective flow and a major part of the solutes is rejected by the membrane. In order to equalize the solute concentration, the retained solutes should diffuse back to the bulk solution. Whereas the diffusion is far slower than the convection, the concentration of solutes gradually enhances near the membrane by the passage of time, making a boundary layer near the membrane wall. This causes concentration polarization, which is an undesirable phenomenon. The concentration polarization has a detrimental effect on the membrane performance and durability (Goulas and Grandison, 2008). Probable negative impacts of concentration polarization include:

- Increment of the osmotic pressure near the membrane wall, resulting in decreasing the water flux in the membrane.
- Scaling or particle fouling in the case of accumulation of solutes on the membrane with a higher concentration than the solubility of the solutes, leading to a lower flux.
- Increasing concentration gradient of the solutes across the membrane, leading to an enhanced solute flux and, therefore, a declined solute rejection.
- Membrane deterioration due to the fouling resulting from concentration polarization.

By increasing the back-diffusion rate of the solutes, the concentration polarization can be partially controlled, which can be achieved by increasing the temperature or mixing of the bulk feed with the solution near the boundary layer. Mixing can be done via increasing the cross-flow velocity across the membrane or addition of turbulence promoters, spacers, etc., to the membrane module (Williams, 2003). A schematic view of the concentration polymerization profile in a cross-flow filtration unit is presented in Figure 6.9.

6.7.2 Fouling

Fouling, which refers to the deposition of the solutes on the membrane surface or within its pores, is the most common and significant problem associated with the ROM processes. Foulants form an extra barrier layer on top of the membrane and/or block its pores, resulting in the decrement of solvent permeability and flux and enhancing the required transmembrane pressure. Therefore, control of membrane fouling is critical to control the economic

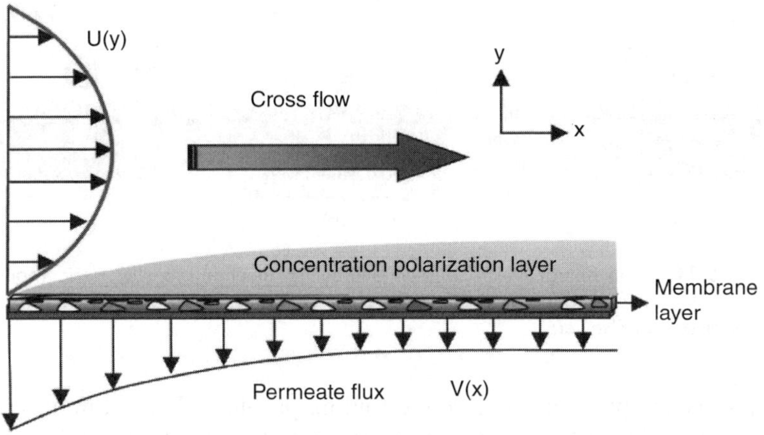

Figure 6.9 Schematic view of the concentration polarization profile in a cross-flow membrane system.

aspects of the process as the direct impact of fouling is to diminish membrane performance and increase operational costs (Ng et al., 2013; Misdan, Lau, and Ismail, 2012).

There are four different fouling categories taking place in ROMs, which encompass inorganic fouling, organic fouling, colloidal fouling, and biofouling (Ng et al., 2013; Misdan, Lau, and Ismail, 2012). Several interactions between the solutes and the ROMs can generate or exacerbate the fouling in ROMs. The most important factors affecting fouling are the physicochemical properties of the membrane, such as hydrophilicity, surface charge, roughness, and zeta potential (Louie et al., 2006). Prevention or minimization of fouling can be achieved by controlling operational conditions of the ROM process, including temperature, pressure, and cross-flow velocity, and modification of the membrane materials in terms of physicochemical properties (Louie et al., 2006; Shirazi, Lin, and Chen, 2010).

Minimizing fouling in polymeric membranes has been an on-going R&D subject for decades. The main approach is to hinder foulants from adhering to the membrane surface and sticking inside its pores. Generally, attachments of the foulants on the membrane can be minimized by pretreatment processes, ultrasonic fortification, physical and chemical membrane surface modification, and cleaning methods. Physical cleaning approaches such as relaxation using permeate to flush the membrane backwards are recognized as simple and economical methods, which have been used in most industrial ROM processes. To date, membrane surface modification has become one of the main research subjects in this regard (Ng et al., 2013).

Physical coating of the membrane surface with fouling-resistant and water-permeable polymers or surfactants has been utilized in some

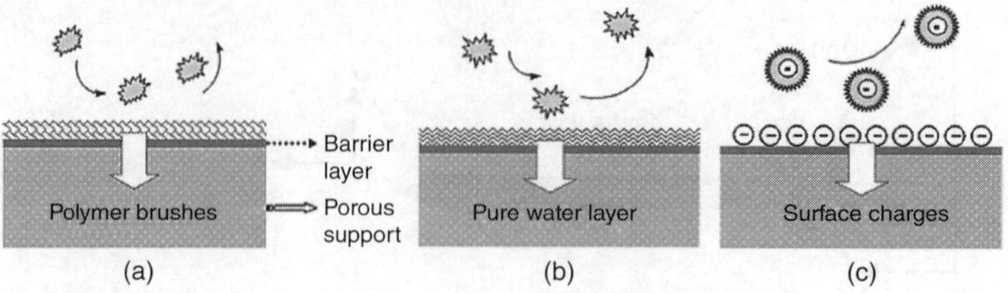

Figure 6.10 Conceptual illustrations of antifouling mechanisms: (a) steric repulsion, (b) formation of a water layer, (c) electrostatic repulsion (Kang and Cao, 2012). Reproduced with permission of Elsevier.

studies, as an approach to solve the fouling problem. The commonly used water-permeable polymers or surfactants include polyethyleneimine (PEI), polyvinyl alcohol (PVA), and polyetherpolyamide block copolymer (PEBA 1657). Chemical surface modification is also used to attach monomers covalently on the membrane surface through free-radical, radiation-, photochemical-, redox-, and plasma-induced grafting methods (Li and Wang, 2010a).

The fouling is more serious in hydrophobic membranes rather than hydrophilic ones (Ng et al., 2013). Thus, enhancement of the membrane hydrophilicity can be a promising approach to improve membrane fouling resistance (Kang and Cao, 2012). For example, grafting or creation of long-chain hydrophilic polymers like PEG on the membrane surface has shown a strong repulsion of macromolecules like proteins from the membrane surface as a result of steric repulsion (Figure 6.10a) (Kang and Cao, 2012). By having a highly hydrophilic membrane surface, a film of water is formed on the membrane surface, which does not allow hydrophobic foulants to attach on the membrane surface. However, it cannot prevent adhesion of hydrophilic foulants on the surface, as shown in Figure 6.10b (Kang and Cao, 2012). Enormous studies have focused on chemical modification of the ROMs to enhance their hydrophobicity (Lau et al., 2012). Incorporation of nanoparticles to the membrane structure has been the most promising method and has been studied intensely in recent years (Ng et al., 2013).

The formation of biofilm on the membrane surface is one of the key mechanisms of membrane fouling in the ROM process. For instance, the presence of microorganisms on the membrane surface could lead to an increase in the amount of dissolved ions near the membrane surface and therefore a significant enhancement in concentration polarization. Furthermore, accumulation of different microorganisms on the membrane surface enhances the formation of acidic by-products and therefore results in membrane deterioration. It is believed that the resistance of the membrane is enhanced after forming biofilm, hence resulting in lowering separation efficiency of the ROM process (Misdan, Lau, and Ismail, 2012). To control biofouling in ROMs, to date four main methods have been used: (1) applying biocides, such as ozone or chlorine, which make a subsequent problem for the membrane, (2) UV excitation for deactivating the bacteria, (3) developing fouling-resistant membranes (e.g., incorporating antibiofouling nanoparticles like silver, TiO_2, etc., to the membrane barrier layer), and (4) installation of pretreatment stages before the RO membrane process, such as microfiltration (MF) or ultrafiltration (UF) membranes (Vrouwenvelder, Van Loosdrecht, and Kruithof, 2011).

The roughness and electrostatic charge of the membrane surface are very important properties related to membrane fouling. Previous studies have shown that surface morphology is effective against membrane fouling in the way that the more surface smoothness results in less fouling on the membrane surface. Therefore, a reduction of surface roughness can improve the antifouling property of the ROMs. However, it should be pointed out that low roughness may result in a lower membrane flux. Regarding the membrane charge, although electrostatic repulsion between the charged membrane surface and the same charge foulants can be beneficial for fouling reduction, these membranes are sensitive to the foulants with the opposite charge and could be easily fouled by them. Therefore, their charged surface is dependent on the feed characteristics and cannot be used for various feeds (Kang and Cao, 2012).

Several solvent flux and solute rejection models that are developed by considering the effect of both fouling and concentration polarization on the flow mechanism can be found in the literature, including Pellegrino (2000), Goulas and Grandison (2008), and Shirazi, Lin, and Chen (2010).

6.7.3 Energy

RO is considered as a simple and cheap separation process compared with thermal evaporation processes. However, it is still a very energy-intensive technology. The high energy and capital costs due to the low water permeability and fouling resistance associated with the ROM process has been an ongoing challenge for its industrial applications. The consumed energy in this process is due to the osmotic pressure resulting from differences in solute concentration, which should be dominated by high-pressure pumps to press water into the membrane module (Mirza, 2008).

The energy consumption and performance of ROMs can be affected by these process variables in terms of physiochemical properties of the membrane and feed and operational factors, as discussed earlier in Section 6.6.2. The key variables might be related to concentration polarization, fouling, and costs associated with enhancing the recovery ratio and performing pretreatment. Pretreatment is considered expensive because of the low generated permeate volume compared with the large amount of feed that should be pretreated before the RO process. Enhancing the recovery ratio, which is performed sometimes to decrease pretreatment costs, increases the costs at the membrane due to the increase in the solute concentration. Therefore, a process engineering strategy is needed to optimize the process so as to minimize the energy costs (Gude, 2011).

Several studies have been performed on the development of the RO process and ROMs to reduce specific energy consumption (SEC) in industrial applications, which has led to the establishment of some general strategies to decrease the SEC. Some of these are:

- Increasing membrane permeability through advanced materials and modification methods (Hoek and Ghosh, 2009; Lee, Arnot, and Mattia, 2011; Kang and Cao, 2012).
- Applying energy recovery device (ERD), which decrease the workload of the main pump (Geisler, Krumm, and Peters, 1999; Avlonitis, Kouroumbas, and Vlachakis, 2003; Peñate and García-Rodríguez, 2011; Li, 2011).
- Using renewable energy resources such as wind, solar, and hydro energy as a substitution or supplement to the electricity driven pumps (Tzen and Morris, 2003).
- Intermediate chemical demineralization (ICD) by removing minerals from the concentrate flow of a primary RO to decrease solute concentration in the feed solution of the secondary RO system (concentrate of the primary RO) (Li, 2011).
- Optimization of the RO system in both operational parameters and membrane properties by applying mathematical modelling (Oh, Hwang, and Lee, 2009).

6.8 REVERSE OSMOSIS PROCESS IN THE DAIRY INDUSTRY

6.8.1 Application of reverse osmosis membranes in the dairy industry

In the food industry, a large amount of energy is consumed for concentration of the liquid products or wastes, which is mostly governed by thermal operations. Many liquids and fluids in the food industry are sensitive to high temperatures and their properties (flavour, components, etc.) may be changed due to the thermal process. Thus, the membrane technology can be a suitable separation process as a substitution for thermal operations for food, especially in the dairy industry.

In the dairy industry, membrane technology plays a key role in the processing, concentration, fractionation, demineralization, and bacterial reduction. The most widely used membrane in the dairy industry is the ultrafiltration (UF) membrane, which is applied in different stages in dairy processing such as processing of fermented milks and various cheeses and whey protein concentrates (Carić, 1994). Nanofiltration is usually used for demineralization of whey and UF concentrate and permeate flows. Microfiltration (MF) is also applied mostly for removal of bacteria in whey and skim milk and fat reduction for production of whey protein concentrates (WPCs) (Carić et al., 2009).

RO in the dairy industry is mostly used for concentration of milk and whey. The RO process has been receiving high attention in recent years due to the several advantages of it over the other regular processes, especially in the concentration and effluent treatment stages. The principal applications of the ROMs in the dairy industry are summarized below:

1. Concentration of milk and whey
   - Supplement to evaporation stage (removing water before evaporation)
   - Pre-concentration of whey before spray drying operations
   - On-farm concentration of milk to reduce transportation costs
   - Lactose concentration after UF of whey
2. Treatment of wastewater generated during different stages of dairy processing.

6.8.1.1 Concentration of milk and whey and recovery of milk solids

RO is a very flexible and energy-efficient approach as a supplementation for the evaporation process used in milk and whey concentration and milk powder production. For instance, the consumed energy of the RO process for whey preconcentration is around 9 kW h/t (of the removed water), while it is around 9–150 kW h/t for vacuum evaporation (Daufin et al., 2001). Cheryan, Veeranjaneyulu, and Schlicher (1990) reported that even if only 25% of the condensed evaporated or dried milk could be processed by ROMs, it meant a huge energy saving of at least 3×10^{11} BTU per annum (Cheryan, Veeranjaneyulu, and Schlicher, 1990).

Concentration of full cream or skim milk can be performed by ROMs up to 2–3 times at operating pressure of around 300–1000 psi with the concentrated milk as a concentrate flow and almost pure water as the permeate (only small inorganic molecules, nicotinic acid, and vitamin B6 have been reported that had trace elements in the permeate flow of the dairy process) (Glover, 1971). On the other hand, a very much lower concentration can be achieved by the RO process compared with the evaporation because of the limiting effect of osmotic pressure and membrane fouling. However, as mentioned, ROM is usually used as a preceding stage before evaporation due to its lower energy consumption and effect on thecapacity enhancement of evaporation plants (Goulas and Grandison, 2008). Therefore, in the case of the requirement to a new evaporation line or extension of a previously established one, the use of RO as the supplement can save a lot of energy. As a result, integration of an RO plant with evaporators can be a cost-effective engineering strategy to increase the capacity of evaporators and make the process more economical.

The dairy concentrates from an RO process can be applied for other dairy processes as well. For instance, skim milk concentrated from ROMs can be used as a supplement for milk powders for consolidation of yoghurt and ice cream. Using this concentrate is very cost-effective compared with using milk powders and increases the concentration rate by approximately 1.5 times. In addition, the milk concentrates from ROMs can decrease the quantity of requisite starters in a cheese production process. Cheese whey can also be concentrated by up to 4 times using ROMs, with the main goal of reducing the transformation costs by decreasing the volume and, as mentioned, preconcentration prior to the evaporation stages (Bundgaard, 1974; Barbano and Bynum, 1984; Goulas and Grandison, 2008).

ROMs can also be applied for permeate fractionation in UF membranes used for whey concentration. The retentate from UF concentration of whey contains the concentrated protein while the permeate is a flow of lactose and salt. The mineral part can be recovered by different approaches such as filtration, drying, or precipitation of the minerals, while the remainder lactose can be separated by ROM and evaporation to finally generate crystalline lactose through crystallization, centrifugation, or drying (Suárez et al., 1992).

6.8.1.2 Dairy effluent treatment via reverse osmosis membranes

The dairy industry is the biggest polluting food industry in terms of the large amount of effluents (0.2–10 L per L of produced milk). The produced effluent mainly consists of diluted milk. It is generated from starting and rinsing stages in different dairy processes, such as milk concentration, cheese manufacturing, and standardization of the proteins. Various studies have been performed on the dairy wastewater treatment via different individual membrane processes or different combinations of them (Delbeke, 1981; Koyuncu et al., 2000; Akoum et al., 2004; Vourch et al., 2008).

Around 1–3% of the milk components (mainly lactose and proteins) will find their way in the dairy effluent and most of the lactose and ionic contents must be removed from this wastewater to make it recyclable. The final stage of dairy effluent treatment should be either NF or RO (Daufin et al., 2001; Akoum et al., 2004). The results of previous studies have suggested the RO process as a better choice for this application. Balannec et al. (2005)

compared the performance of NF and RO membranes for dairy wastewater treatment and concluded that the RO process generated higher treated water flux with better water quality compared with the NF process. The same group also investigated the efficiency of effluent treatment with the RO process in 11 dairy plants in France. They reported that the RO process could essentially decrease total organic carbon (TOC) and mineral contents in the effluent and therefore provide acceptable water for reuse in the plants (Vourch et al., 2008).

An important advantage in using the ROM process is the recovery of valuable chemicals and recycled water. Treated water and separated elements can also be used in other applications. Lactose and proteins recovered by the RO process from the effluent can be used for animal consumption. The treated water can be used in the plant itself as the heating/cooling water, for boilers, and for cleaning stages. This can significantly reduce the water consumption and the volume of disposed effluent. However, the quality of this water should be at least equal to the produced water from drying/evaporation stages to be suitable for reusing in these processes (Cheryan, Veeranjaneyulu, and Schlicher, 1990; Balannec et al., 2002, 2005).

6.8.2 Factors affecting the reverse osmosis membrane performance in dairy processing

Similar to the general ROM process, the physiochemical properties of the feed and membrane and processing variables are also effective on the ROM performance in dairy processing. In this section, a summary of effective parameters and the effect of each on the flux and solute rejection of some ROM examples in the dairy industry are presented.

6.8.2.1 Effect of feed composition

Comparing the concentration of four different dairy streams including acid whey, sweet whey, whole milk. and skim milk has shown that the composition of feed has a significant influence on the flux and solute rejection of the ROMs. Under designed operating conditions, the flux obtained with these streams has the following trend:

$$\text{Sweet whey} > \text{acid whey} > \text{skim milk} > \text{whole milk}$$

The reason behind this is the effect of concentration polarization and fouling arising from the composition of these materials. For example, acid whey has a pH of around 4.35, which is close to the isoelectric point of whey proteins (4.5–4.6), while sweet whey has a higher pH (around 5.5). Therefore, acid whey has higher protein fouling and thus less flux compared with sweet whey (Blatt et al., 1970). The lower flux of milk compared with whey is attributed to the presence of casein protein in milk, which has been proved to enhance concentration polarization effects, resulting in increasing the resistance of the membrane against permeation (Morales, Amundson, and Hill, 1990).

The effect of feed composition on the rejection properties of ROMs in the dairy industry is also attributed to the concentration polarization due to different compositions of the dairy streams, which are concentrated through the RO processes. As stated previously, the concentration polarization is higher in milk compared with whey due to the presence of casein and other macrosolutes in milk, which form a denser gel layer compared with whey on the membrane surface. However, the experiments already performed on different feed solutions revealed that skim and whole milk both have similar solute rejections, suggesting that the fat globules do not have too much influence on the concentration polarization. Similar to the effect on flux, acid whey has been reported to have a high solute rejection, which is governed by its high concentration polarization sensitivity (Morales, Amundson, and Hill, 1990).

6.8.2.2 Effect of membrane structure

The studies on the membrane module type and materials on ROM efficiency in dairy processes concentrated on the composite and cellulose acetate membranes and tubular and spiral wound modules. These reports revealed that the tubular membrane modules had a higher flux compared with the spiral wound ones. However, the tubular membranes consumed more energy per unit volume of generated permeate than the spiral wound ones. The results also suggested that the composite membranes had higher flux and solute rejection compared with CA membranes and were therefore considered to be more energy efficient (Cheryan, Veeranjaneyulu, and Schlicher, 1990; Morales, Amundson, and Hill, 1990).

6.8.3 Concentration polarization and fouling of reverse osmosis membranes during milk concentration

Concentration polarization is one of the biggest problems in all membrane processes including the RO process in the dairy industry. Concentration polarization leads to the formation of a gel layer with a higher concentration of rejected solutes near the membrane surface. When water is not the carrier solvent in the solution passing from ROMs, the permeate flux dramatically decreases due to the effect of concentration polarization. For the case of milk concentration, the flux with similar membranes and operating parameters is almost 10 times smaller than that for water as the solvent (Goulas and Grandison, 2008). Concentration polarization continuously makes a decrement in the permeate flux until the flux becomes stable. The point in which the flux levels off is different for each process and is dependent on the membrane, solution concentration, and operating variables (Glover and Brooker, 1974).

Some feed contents including proteins, salts (mainly calcium phosphates), and lipids are the most problematic materials in the dairy industry with regard to exacerbating the concentration polarization problem (Van Boxtel, Otten, and van der Linden, 1991). The most responsible content for the concentration polarization in dairy ROMs is protein, which forms a gel layer adhering to the membrane surface, making another filtration layer and as a result causing membrane flux decrement. Calcium phosphate has also been proved to have an important effect on the concentration polarization and fouling in the dairy industry (Schmidt, 1987; Hanemaaijer et al., 1989). Supersaturation of calcium phosphate in milk and whey has been reported as the main reason for its precipitation in the ROM process. Although protein inhibits the precipitation of calcium phosphate through forming a complex, the generated complex together with calcium phosphate can precipitate in the membrane module and enhance membrane fouling. Some methods have been reported to decrease fouling using calcium phosphate. For instance, by increasing the temperature for a short period of time (e.g. 60 °C for 10 min) before importing the feed solution to the membrane module, supersaturation of calcium phosphate can be decreased through precipitating it outside the membrane module (Van Boxtel, Otten, and van der Linden, 1991). Solubility of calcium phosphate can also be enhanced by the addition of some acid to the milk/whey and decreasing the feed pH to around 5.8–6 (Hiddink, de Boer, and Nooy, 1980).

Generally, membrane flux decline is the result of either hydrodynamic resistance to permeate flow or effective osmotic pressure enhancement. Hydrodynamic resistance is mainly due to the presence of fats and proteins in the feed flow. An increase in osmotic pressure results from an increase in the solute concentration (salt, lactose, etc.) during the process. When the polarized layer on the surface of the membrane forms a gel or precipitates, severe fouling can take place. The only way to enhance the flux is to decrease concentrated solutes near the membrane and enhance convective transfer of them to the bulk solution. The best approach to achieve this is to increase the turbulence of flow near the membrane or to enhance the feed flow rate. On the other hand, concentration polarization enhances as the operating pressure increases. Therefore, raising the flow rate is more influential in controlling concentration polarization at higher pressures (Cheryan, Veeranjaneyulu, and Schlicher, 1990; Goulas and Grandison, 2008).

Fouling is another big problem associated with all membrane processes including ROMs. In the dairy industry, fouling in ROMs is governed by (1) enhancing the thickness of the protein gel layer on the membrane, leading to a higher membrane resistance to the feed flow, and (2) deposition of calcium phosphate and protein complexes. Protein is the major substance that causes concentration polarization and fouling in membranes. Whereas whey has a very low protein concentration compared with milk (only 0.6–0.8 weight precent), permeate flux in whey concentrating ROMs is superior to milk ones.

Van Boxtel, Otten, and van der Linden (1991) investigated the effect of concentration polarization in cheese whey concentration via a tubular ROM pilot plant. They proposed some useful methods to control the fouling in ROMs during whey concentration, including (1) increasing flow velocity and decreasing temperature and (2) controlling pressure and temperature, leading to the achievement of a constant permeate flow rate.

6.9 CURRENT DEVELOPMENT IN REVERSE OSMOSIS MEMBRANES

6.9.1 Thin-film nanocomposite reverse osmosis membranes

Nanotechnology opens a new frontier in the development of advanced membranes with improved performance. Recent discoveries reported that the incorporation of hydrophilic inorganic nanoparticles in the thin layer of the ROM could increase the water flux by almost double compared to current commercially available ROMs (Lee, Arnot, and Mattia, 2011). Introduction of functional nanoparticles in the thin film generates thin-film nanocomposite ROM (TFN-ROM), which leads to the breakthrough in ROMs by further enhancing permeability with no sign of solute rejection loss. Nanoparticles are very promising to improve the properties of thin films due to their tunable characteristics with respect to sizes, solubilities, shapes, and hydrophilicity (Funk and Lloyd, 2008; Lee, Arnot, and Mattia, 2011). It is generally accepted that permeability correlates to hydrophilicity and surface roughness whilst solute rejection corresponds with the extent of cross-linking, chain stiffness, and chain packing density of the skin active layer (Kong et al., 2011). Nanoparticles with different sizes could create a "percolation threshold" through the polymeric matrix. These novel approaches have attracted considerable interest to produce a new class of ROMs with desired functions and properties.

Recent studies showed that TFC-ROMs composed of nanosized zeolite A (Fathizadeh, Aroujalian, and Raisi, 2011) and silica (Jadav and Singh, 2009) exhibited a dramatic improvement in both permeability and fouling resistance. The results from the studies published so far strongly imply that a high-performance ROM with desirable properties can be fabricated by incorporating nanoparticles. A few reports proposed that the hydrophilic functional groups within the pores or on the surface of nanoparticles could alter membrane functionality, such as water permeability, fouling resistance, and salt rejection (Kong et al., 2011). On the other hand, introducing nanoparticles leads to a change in the membrane microstructure. The resulting polymer–inorganic interfacial region and the interfacial reactions between these nanoparticles and polymers govern the membrane properties and performance (Jin et al., 2013). Furthermore, those interactions also control the stability and physical properties of the nanoparticles in the polymer membranes, and consequently affect the ROM life cycle. Although scattered published findings provide a promising insight into the development of novel TFN-ROMs, so far no detailed studies on describing the interfacial interactions between the nanoparticles and polymers are found. There is a lack of knowledge to understand how to govern the polymer–inorganic interfacial region and interfacial reactions with respect to nanoparticle characteristics and the polymerization process. Such information is extremely important for fabrication of ROMs with desirable properties and performance.

6.9.2 Membrane surface modification and process optimization

There have been also several studies in recent years to investigate how to enhance separation performance of ROMs, especially antifouling and permeability. These studies have been mostly focused on membrane surface modification and application of new interfacial polymerization monomers or improving the polymerization reaction.

An effective approach to improve antifouling characteristics and amending surface properties of ROMs is surface modification. Methods for surface modification are characterized into two major groups, namely physical and chemical treatments (Kang and Cao, 2012). Physical techniques that are commonly used are surface adsorption and surface coating. The chemical methods mainly include hydrophilization treatment, radical grafting, chemical coupling, plasma polymerization or plasma-induced polymerization, and initiated chemical vapor deposition. Another important approach to mitigate the fouling problem is the incorporation of nanoparticles via direct coating by depositing them on to the membrane surface or introducing them to the membrane matrix, for example, during the interfacial polymerization process, which is used for TFC/TFN-ROMs (Kang and Cao, 2012).

Optimization of the interfacial polymerization reaction process has also been used recently to enhance membrane surface properties such as hydrophilicity and surface roughness. Some of the optimized parameters include the coefficients of reactant diffusion, kinetics, nucleation rate, solvent solubility, reaction time, curing time, polymer molecular weight range, solution composition, and properties of the microporous support (Lee. Arnot, and Mattia, 2011).

6.10 CONCLUSIONS AND OUTLOOK

Novel ROMs with superior performance (namely higher permeate flux and solute rejection) are greatly demanded in order to further decrease capital and operational costs in the industrial processes that use ROMs. In this regard, membranes should be more tolerant to chlorine (the antibacterial agent usually used) and have higher antifouling capability and chemical and mechanical resistivity. The cleaning agents and methods also should be modified in order to reduce membrane replacement costs. Whereas fouling and the concentration polymerization rate in the dairy industry are high, the evolution of new modified ROMs can be very vital and cost effective in this industry. In this regard, both modification of the membrane material and optimization of the operating conditions should be taken into account. Considering the current state of research on developing novel modified ROMs, it can be concluded that there is a very promising future for the application of a new generation of composite ROMs with advanced properties in the dairy and other food industries.

Nomenclature

A = water permeability coefficient across the membrane (m^3/N s)
A = active membrane surface area(m^2)
B = solute permeability coefficient across the membrane (m^3 m/mol s)
$C\Delta P$ = pore flow contribution to the water and solute flux across the membranein SDIM (m^3/m^2 s)
C_1 = solute concentrations in feed (mol/m^3)
C_2 = solute concentrations in the membrane boundary layer (mol/m^3)
C_3 = solute concentrations in permeate solution (mol/m^3)
C_F = bulk feed concentration (mol/m^3)
C_P = bulk permeate concentration (mol/m^3)
C_{spm} = solute concentration at permeate side of the membrane surface (mol/m^3)
C_{sfm} = solute concentration at feed side of the membrane surface (mol/m^3)
C_T = total molar concentration (mol/m^3)
C_{wm} = water concentration in the membrane (mol/m^3)
D = solute transfer coefficient related to the pressure gradient (m^3/N s)

D_{sm} = solute diffusion coefficient in the membrane (m^2/s)
D_{sp} = diffusivity of solute into the membrane pores (m^2/s)
D_{sw} = solute diffusion coefficient in water (m^2/s)
D_{wm} = water diffusivity in the membrane (m^2/s)
J_s = solute flux (m^3/m^2 s)
$J_{s(SDM)}$ = solute flux calculated based on SDM (m^3/m^2 s)
$J_{w(SDM)}$ = water flux calculated based on SDM (m^3/m^2 s)
J_w = water flux (m^3/m^2 s)
k_D = distribution coefficient of solute from the feed solution to the poresof membrane (m^3/mol)
K_{sm} = partition coefficient relating solute concentration at the feed and permeate side of the membrane to the bulk feed and permeate solute concentration ($K_{sm} = \frac{C_{sfm}}{C_F} = \frac{C_{spm}}{C_P}$)
P = applied pressure (N/m^2 Pa)
r_p = membrane pore radius (m)
R_g = universal gas constant (8.314) (J/mol K)
R = solute rejection
T = absolute temperature (K)
t = length of time taken for the process (s)
u = linear velocity of pore fluid (m/s)
$\overline{V}_w$ = partial molar volume of water (m^3/mol)
V_3 = collected permeate volume (m^3)
X_{sm} = friction coefficient between solute and the membrane pore (kg mol/m^3 s)
X_F = solute mole fraction in feed
X_P = solute mole fraction in permeate

Greek letters

δ = membrane thickness (m)
π = osmotic pressure (N/m^2 Pa)
ε = membrane porosity
τ = tortuosity
$\tau\delta$ = membrane pore length (m)
η = viscosity (Pa s)

References

Akoum, O., Jaffrin, M.Y., Ding, L.H., and Frappart, M. (2004) Treatment of dairy process waters using a vibrating filtration system and NF and RO membranes. *Journal of Membrane Science*, **235** (1–2), 111–122.

Avlonitis, S.A., Kouroumbas, K., and Vlachakis, N. (2003) Energy consumption and membrane replacement cost for seawater RO desalination plants. *Desalination*, **157** (1–3), 151–158.

Balannec, B., Gésan-Guiziou, G., Chaufer, B., Rabiller-Baudry, M., and Daufin, G. (2002) Treatment of dairy process waters by membrane operations for water reuse and milk constituents concentration. *Desalination*, **147** (1–3), 89–94.

Balannec, B., Vourch, M., Rabiller-Baudry, M., and Chaufer, B. (2005) Comparative study of different nanofiltration and reverse osmosis membranes for dairy effluent treatment by dead-end filtration. *Separation and Purification Technology*, **42** (2), 195–200.

Barbano, D.M. and Bynum, D.G. (1984) Whole milk reverse osmosis retentates for cheddar cheese manufacture: cheese composition and yield. *Journal of Dairy Science*, **67**, 2839–2849.

Bhattacharyya, D. and Williams, M. (1992) Theory – reverse osmosis, in Membrane Handbook (eds. W. Ho and K. Sirkar), Van Nostrand Reinhold, New York, pp. 269–280.

Bhattacharyya, D., Williams, M., Ray, R., and McCray, S. (1992) Reverse osmosis, in Membrane Handbook (eds. W. Ho and K. Sirkar), Van Nostrand Reinhold, New York, pp. 263–390.

Blatt, W., Dravid, A., Michaels, A., and Nelsen, L. (1970) Solute polarization and cake formation in membrane ultrafiltration: causes, consequences, and control techniques, in *Membrane Science and Technology* (ed. J. Flinn), Springer, US, pp. 47–97.

Burghoff, H.G., Lee, K.L., and Pusch, W. (1980) Characterization of transport across cellulose acetate membranes in the presence of strong solute–membrane interactions. *Journal of Applied Polymer Science*, **25** (3), 323–347.

Bundgaard, A.G. (1974) Hyperfiltration of skim milk for ice cream manufacture. *Dairy Industry*, **39**, 119–122.

Carić, M. (1994) *Concentrated and Dried Dairy Products*, VCH Publishers, Inc., New York, pp. 7–55.

Carić, M., Akkerman, J.C., Milanović, S., Kentish, S.E., and Tamime, A.Y. (2009) Technology of evaporators, membrane processing and dryers, in *Dairy Powders and Concentrated Products*, Wiley-Blackwell, pp. 99–148.

Chen, J.P., Chian, E.S.K., Sheng, P.X., Nanayakkara, K.G.N., Wang, L.K., and Ting, E.P. (2011) Desalination of seawater by reverse osmosis, in *Membrane and Desalination Technologies* (eds. L. Wang, J. Chen, Y.-T. Hung, and N. Shammas), Humana Press, pp. 559–601.

Cheryan, M., Veeranjaneyulu, B., and Schlicher, L.R. (1990) Reverse osmosis of milk with thin-film composite membranes. *Journal of Membrane Science*, **48** (1), 103–114.

Chian, E.K., Chen, J.P., Sheng, P.-X., Ting, Y.-P., and Wang, L. (2007) Reverse osmosis technology for desalination, in *Advanced Physicochemical Treatment Technologies*, Humana Press, pp. 329–366.

Crittenden, J.C., Rhodes Trussell, R., Hand, D.W., Howe, K.J., and Tchobanoglous, G. (2012) Reverse osmosis, in *MWHs Water Treatment: Principles and Design*, third edition, John Wiley & Sons, Inc., Hoboken, NJ, USA, p. 1921.

Daufin, G., Escudier, J.P., Carrère, H., Bérot, S., Fillaudeau, L., and Decloux, M. (2001) Recent and emerging applications of membrane processes in the food and dairy industry. *Food and Bioproducts Processing*, **79** (2), 89–102.

Delbeke, R. (1981) Recovery of milk solids by hyperfiltration. *Milchwissenschaft*, **36** (11), 669–672.

Dickson, J. (1988) Fundamental aspects of reverse osmosis, in *Reverse Osmosis Technology* (ed. B. Parekh), Marcel Dekker, Inc., New York, pp. 1–51.

Elimelech, M., Chen, W.H., and Waypa, J.J. (1994) Measuring the zeta (electrokinetic) potential of reverse osmosis membranes by a streaming potential analyzer. *Desalination*, **95** (3), 269–286.

El-Saied, H., Basta, A.H., Barsoum, B.N., and Elberry, M.M. (2003) Cellulose membranes for reverse osmosis. Part I. RO cellulose acetate membranes including a composite with polypropylene. *Desalination*, **159** (2), 171–181.

Ettouney, H. and Wilf, M. (2009) Commercial desalination technologies, in *Seawater Desalination* (eds. G. Micale, L. Rizzuti, and A. Cipollina), Springer, Berlin and Heidelberg, pp. 77–107.

Fathizadeh, M., Aroujalian, A., and Raisi, A. (2011) Effect of added NaX nano-zeolite into polyamide as a top thin layer of membrane on water flux and salt rejection in a reverse osmosis process. *Journal of Membrane Science*, **375** (1–2), 88–95.

Fritzmann, C., Löwenberg, J., Wintgens, T., and Melin, T. (2007) State-of-the-art of reverse osmosis desalination. *Desalination*, **216**(1–3), 1–76.

Funk, C.V. and Lloyd, D.R. (2008) Zeolite-filled microporous mixed matrix (ZeoTIPS) membranes: prediction of gas separation performance. *Journal of Membrane Science*, **313** (1–2), 224–231.

Geisler, P., Krumm, W., and Peters, T. (1999) Optimization of the energy demand of reverse osmosis with a pressure-exchange system. *Desalination*, **125** (1–3), 167–172.

Glater, J. (1998) The early history of reverse osmosis membrane development. *Desalination*, **117** (1–3), 297–309.

Glover, F.A. (1971) Concentration of milk by ultrafiltration and reverse osmosis. *Journal of Dairy Research*, **38**, 373–379.

Glover, F.A. and Brooker, B.E. (1974) The structure of the deposit formed on the membrane during the concentration of milk by reverse osmosis. *Journal of Dairy Research*, **41** (01), 89–93.

Goulas, A. and Grandison, A.S. (2008) Applications of membrane separation, in *Advanced Dairy Science and Technology*, Blackwell Publishing Ltd, pp. 35–74.

Gude, V.G. (2011) Energy consumption and recovery in reverse osmosis. *Desalination and Water Treatment*, **36** (1–3), 239–260.

Hanemaaijer, J.H., Robbertsen, T., van den Boomgaard, T., and Gunnink, J.W. (1989) Fouling of ultrafiltration membranes. The role of protein adsorption and salt precipitation. *Journal of Membrane Science*, **40** (2), 199–217.

Hiddink, J., de Boer, R. and Nooy, P.F.C. (1980) Reverse osmosis of dairy liquids. *Journal of Dairy Science*, **63**(2), 204–214.

Hoehn, H.H. and Richter, J.W. (1980) Aromatic polyamide, polyester and polyamide separation membranes. US Patent Application RE30351.

Hoek, E.M.V. and Ghosh, A.K. (2009) Nanotechnology-based membranes for water purification, in *Nanotechnology Applications for Clean Water* (eds. N. Savage, M. Diallo, J. Duncan, A. Street, and R. Sustich), William Andrew Publishing, Boston, pp. 47–58.

Humplik, T., Lee, J., OHern, S.C., Fellman, B.A., Baig, M.A., Hassan, S.F., Atieh, M.A., Rahman, F., Laoui, T., Karnik, R., and Wang, E.N. (2011) Nanostructured materials for water desalination. *Nanotechnology*, **22** (29), 1–19.

Jadav, G.L. and Singh, P.S. (2009) Synthesis of novel silica-polyamide nanocomposite membrane with enhanced properties. *Journal of Membrane Science*, **328** (1–2), 257–267.

Jeong, B.-H., Hoek, E.M.V., Yan, Y., Subramani, A., Huang, X., Hurwitz, G., Ghosh, A.K., and Jawor, A. (2007) Interfacial polymerization of thin film nanocomposites: a new concept for reverse osmosis membranes. *Journal of Membrane Science*, **294** (1–2), 1–7.

Jin, F., Zhang, W., Lv, C., Li, Z., Su, R., Qi, W., Yang, Q., and He, Z. (2013) RSC Adv. DOI: 10.1039/C3RA42908C.

Jonsson, G. and Boesen, C. (1975) Water and solute transport through cellulose acetate reverse osmosis membranes. *Desalination*, **17**, 145–165.

Kang, G. and Cao, Y. (2012) Development of antifouling reverse osmosis membranes for water treatment: a review. *Water Research*, **46** (3), 584–600.

Kong, C., Shintani, T., Kamada, T., Freger, V., and Tsuru, T. (2011) Co-solvent-mediated synthesis of thin polyamide membranes. *Journal of Membrane Science*, **384** (1–2), 10–16.

Koyuncu, I., Turan, M., Topacik, D., and Ates, A. (2000) Application of low pressure nanofiltration membranes for the recovery and reuse of dairy industry effluents. *Water Science and Technology*, **41** (1), 213–221.

Kucera, J. (2010) *Reverse Osmosis: Industrial Applications for Processes*, Scrivener Publishing, Salem, MA.

Kulozik, U. and Kessler, H.-G. (1988) Permeation rate during reverse osmosis of milk influenced by osmotic pressure and deposit formation. *Journal of Food Science*, **53** (5), 1377–1383.

Lau, W.J., Ismail, A.F., Misdan, N., and Kassim, M.A. (2012) A recent progress in thin film composite membrane: a review. *Desalination*, **287**, 190–199.

Lee, K.P., Arnot, T.C., and Mattia, D. (2011) A review of reverse osmosis membrane materials for desalination – Development to date and future potential. *Journal of Membrane Science*, **370** (1–2), 1–22.

Lee, Y.-F., Chang, K.-H., Hu, C.-C., and Chu, Y.-H. (2012) Designing tunable microstructures of Mn_3O_4 nanoparticles by using surfactant-assisted dispersion. *Journal of Power Sources*, **206**, 469–475.

Li, M. (2011) Reducing specific energy consumption in reverse osmosis (RO) water desalination: an analysis from first principles. *Desalination*, **276** (1–3), 128–135.

Li, D. and Wang, H. (2010a) Development of reverse osmosis membranes, in Desalination: A Review, Chemeca (2010): *Engineering at the Edge*, Hilton Adelaide, South Australia, Barton, ACT, Engineers Australia.

Li, D. and Wang, H. (2010b) Recent developments in reverse osmosis desalination membranes, *Journal of Materials Chemistry*, **20** (22), 4551–4566.

Loeb, S. and Selover, E. (1967) Sixteen months of field experience on the Coalinga pilot plant. *Desalination*, **2**(1), 75–80.

Loeb, S. and Sourirajan, S. (1963) Sea water demineralization by means of an osmotic membrane, in *Saline Water Conversion-II*, American Chemical Society, pp. 117–132.

Louie, J.S., Pinnau, I., Ciobanu, I., Ishida, K.P., Ng, A., and Reinhard, M. (2006) Effects of polyether–polyamide block copolymer coating on performance and fouling of reverse osmosis membranes. *Journal of Membrane Science*, **280** (1–2), 762–770.

Mehdizadeh, H. (1990) Modelling of transport phenomena in reverse osmosis membranes. Dissertation, J. Dickson, Director, Department of Chemical Engineering, McMaster University, Hamilton, Ontario, Canada.

Merten, U. (1966) Transport properties of osmotic membranes, in *Desalination by Reverse Osmosis* (ed. U. Merten), The MIT Press, Cambridge, MA, pp. 15–54.

Mirza, S. (2008) Reduction of energy consumption in process plants using nanofiltration and reverse osmosis. *Desalination*, **224** (1–3), 132–142.

Misdan, N., Lau, W.J., and Ismail, A.F. (2012) Seawater reverse osmosis (SWRO) desalination by thin-film composite membrane – Current development, challenges and future prospects. *Desalination*, **287**, 228–237.

Moonkhum, M., Lee, Y.G., Lee, Y.S., and Kim, J.H. (2010) Review of seawater natural organic matter fouling and reverse osmosis transport modeling for seawater reverse osmosis desalination. *Desalination and Water Treatment*, **15** (1–3), 92–107.

Morales, A., Amundson, C.H., and Hill, C.G. (1990) Comparative study of different reverse osmosis membranes for processing dairy fluids: I. Permeate flux and solute rejection studies1. *Journal of Food Processing and Preservation*, **14** (1), 39–58.

Ng, L.Y., Mohammad, A.W., Leo, C.P., and Hilal, N. (2013) Polymeric membranes incorporated with metal/metal oxide nanoparticles: a comprehensive review. *Desalination*, **308**, 15–33.

Oh, H.-J., Hwang, T.-M., and Lee, S. (2009) A simplified simulation model of RO systems for seawater desalination. *Desalination*, **238** (1–3), 128–139.

Parekh, B.S. (1988) Reverse Osmosis Technology: Application for High-Purity-Water Production, Marcel Dekker Inc., New York.

Pellegrino, J. (2000) Filtration and ultrafiltration equipment and techniques. *Separation and Purification Methods*, **29**, 91–118.

Peñate, B. and García-Rodríguez, L. (2011) Energy optimisation of existing SWRO (seawater reverse osmosis) plants with ERT (energy recovery turbines): technical and thermoeconomic assessment. *Energy*, **36** (1), 613–626.

Pendergast, M.M., Ghosh, A.K., and Hoek, E.M.V. (2013) Separation performance and interfacial properties of nanocomposite reverse osmosis membranes. *Desalination*, **308**, 180–185.

Polasek, V., Talo, S., and Sharif, T. (2003) Conversion from hollow fiber to spiral technology in large seawater RO systems – process design and economics. *Desalination*, **156** (1–3), 239–247.

Ren, J. and Wang, R. (2011) Preparation of polymeric membranes, in *Membrane and Desalination Technologies* (eds. L.K. Wang, J.P. Chen, Y.-T. Hung, and N.K. Shammas), Humana Press, pp. 47–100.

Schmidt, D.G. (1987) *Studies on the Precipitation of Calcium Phosphate. I. Experiments in the PH Range 5.3 to 6.8 at 25 °C and 50 °Cin the Absence of Additives*, Nederlands Instituut voor Zuivelonderzoek.

Shirazi, S., Lin, C.-J., and Chen, D. (2010) Inorganic fouling of pressure-driven membrane processes – A critical review. *Desalination*, **250** (1), 236–248.

Sourirajan, S. and Matsuura, T. (1985) *Reverse Osmosis/Ultrafiltration Principles*, National Research Council of Canada, Ottawa, Canada.

Spangler, P.L. and Amundson, C.H. (1986) Concentration of milk and whey using composite, spiral wound, reverse osmosis membranes. *Journal of Dairy Science*, **69** (6), 1498–1509.

Suárez, E., San Martín, F., Alvarez, R., and Coca, J. (1992) Reverse osmosis of whey. Determination of mass transfer coefficients. *Journal of Membrane Science*, **68** (3), 301–305.

Tzen, E. and Morris, R. (2003) Renewable energy sources for desalination. *Solar Energy*, **75** (5), 375–379.

Valentas, K.J., Rotstein, E., Singh, R.P. (1997) *Handbook of Food Engineering Practice*, CRC Press LLC, New York.

Van Boxtel, A.J.B., Otten, Z.E.H., and van der Linden, H.J.L.J. (1991) Evaluation of process models for fouling control of reverse osmosis of cheese whey. *Journal of Membrane Science*, **58** (1), 89–111.

Van Wagner, E.M., Sagle, A.C., Sharma, M.M., and Freeman, B.D. (2009) Effect of crossflow testing conditions, including feed pH and continuous feed filtration, on commercial reverse osmosis membrane performance. *Journal of Membrane Science*, **345** (1–2), 97–109.

Van Wagner, E.M., Freeman, B.D., Sharma, M.M., Hickner, M.A., and Altman, S.J. (2010) *Polyamide Desalination Membrane Characterization and Surface Modification to Enhance Fouling Resistance*, Technical Information Center Oak Ridge Tennessee, pp. 1–154.

Vourch, M., Balannec, B., Chaufer, B., and Dorange, G. (2008) Treatment of dairy industry wastewater by reverse osmosis for water reuse. *Desalination*, **219** (1–3), 190–202.

Vrouwenvelder, J.S., Van Loosdrecht, M.C.M., and Kruithof, J.C. (2011) A novel scenario for biofouling control of spiral wound membrane systems. *Water Research*, **45** (13), 3890–3898.

Wijmans, J.G. and Baker, R.W. (1995) The solution-diffusion model: a review. *Journal of Membrane Science*, **107** (1–2), 1–21.

Williams, M.E. (2003) *A Review of Reverse Osmosis Theory*, EET Corporation and William Engineering Services Company, Inc., Harriman, TN.

Younos, T. and Tulou, K.E. (2005) Overview of desalination techniques. *Journal of Contemporary Water Research and Education*, **132** (1), 3–10.

7 Pervaporative extraction of dairy aroma compounds

Boya Zhang[1], Panida Sampranpiboon[2], and Xianshe Feng[1]

[1] Department of Chemical Engineering, University of Waterloo, Waterloo, Ontario, Canada
[2] Department of Chemical and Material Engineering, Rangsit University, Pathumthani, Thailand

7.1 INTRODUCTION

Over the past few decades, a considerably large number of flavored compounds in dairy products have been identified. Parliament and McGorrin (2000) and McGorrin (2001) provided a comprehensive review on the most potent flavor components in milk, cream, butter, cheese, and other cultured dairy products. Recently, the recovery of these flavor compounds from dairy products has attracted significant attention. During processing of dairy products, some volatile aroma compounds may be lost due to evaporation or thermal degradation. Even a small loss of the aroma compounds may significantly affect the sensory quality of the products. Nowadays, artificial flavors can no longer satisfy the needs of consumers who are gradually shifting away from artificial flavors, and naturally flavored products are becoming increasingly popular. This is especially true for dairy flavors because it is very difficult to produce authentic flavors artificially due to the complex profile of the compounds involved. Thus, extraction and recovery of aroma compounds from natural sources as favoring ingredients are of particular interest to the food, pharmaceutical, and cosmetic industries.

Traditionally, aroma compounds are concentrated and recovered by solvent extraction, distillation, partial condensation, and gas stripping (Karlsson and Tragardh, 1998; Baudot and Marin, 1997). Thermal processes that involve phase changes of evaporation and condensation are often unsatisfactory for recovering heat-sensitive aroma compounds, even at moderately high temperatures. Solvent extraction, on the other hand, requires the use of an extracting agent, which needs to be removed from the extract subsequently. Flavor extraction with liquid solvents is uncommon for food processing and supercritical fluids may be used, but the process

Membrane Processing for Dairy Ingredient Separation, First Edition.
Edited by Kang Hu and James M. Dickson.

is costly and other undesired components (e.g., fat) are often extracted along with the aroma compounds. Gas stripping is generally ineffective for extraction of aromas with low volatilities. Pervaporation, as a relatively new separation process, has attracted attention as an alternative to the conventional flavor recovery technologies. Pervaporation separation is influenced but not determined by the volatility or the molecular size of the species to be separated. Rather, it is based on selective solubility and diffusivity of the permeant in the membrane. Aroma compounds with higher boiling points and bigger molecular sizes than water can still permeate through appropriate organophilic membranes to be enriched in the permeate stream.

Pervaporation has the following major advantages:

1. No additives are needed, and thus there is no secondary contamination to the products recovered (which can be regarded as natural products).
2. It can operate at ambient or moderate operating temperatures, a feature that is particularly important to prevent thermal degradation of flavor compounds.
3. Only the permeate, which is a small fraction of the feed, undergoes phase change, and thus the energy consumption of the process is relatively low.
4. With current organophilic membranes, many hydrophobic flavor compounds can be enriched in the permeate beyond their solubility limits, and consequently a much higher purity may be achieved in the organic phase upon phase separation of the permeate.

Pervaporation separation is a membrane process where the components in a liquid mixture pass through a membrane selectively to produce a vaporous stream on the downstream side of the membrane, and the permeate vapor can be condensed and collected as liquid. For aroma extraction from aqueous solutions, organophilic membranes should be used, and the aroma compounds will permeate through the membrane preferentially over water, resulting in an aroma-enriched permeate. In this chapter, the basic principles of pervaporative extraction and enrichment of aroma compounds and the current status of pervaporation membranes for concentration of dairy aromas are reviewed. The characteristics of pervaporation separation and some potential technical issues related to pervaporative extraction of dairy aromas are also discussed.

7.2 PERVAPORATION – FUNDAMENTALS

Pervaporation is a relatively new membrane separation process that has some elements in common with reverse osmosis and membrane gas separation. It is known to be effective for the separation of close-boiling and

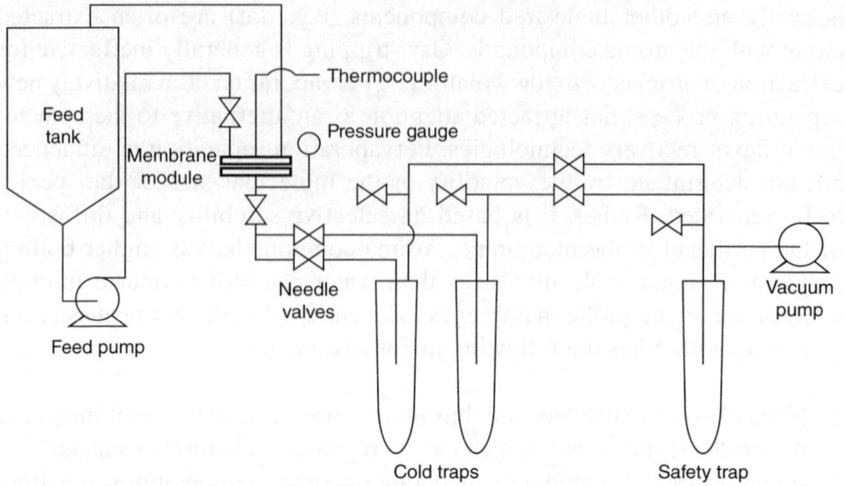

Figure 7.1 Schematic diagram of a typical laboratory pervaporation set-up (Overington et al., 2008). Reproduced with permission of Elsevier.

azeotropic mixtures, isomers, and heat-sensitive biochemicals (Shao and Huang, 2007). In pervaporation, the liquid mixture to be separated (feed) is placed in contact with a membrane, and the permeated product (permeate) is removed as a low-pressure vapor from the other side (Figure 7.1). The permeate vapor can be condensed and collected as liquid. The driving force for the mass transport in pervaporation is normally provided by applying a vacuum pump on the permeate side to maintain a permeate vapor pressure lower than the saturated vapor pressure of the feed (Borjesson, Karlsson, and Tradardh, 1996; Trifunovic and Tragardh, 2005; Ortiz, 2008; She and Hwang, 2006a; Overington et al., 2008). In industrial applications, the vacuum on the permeate side may also be partially generated by condensation of the permeate vapor. Alternatively, the driving force may be created by using a purge gas on the permeate side, and the permeate stream is subjected to an additional processing step (e.g., partial condensation) in order to separate the membrane permeated species from the sweeping gas. Vacuum pervaporation is the most widely utilized mode of operation, while the purge gas pervaporation is of interest when the permeate can be discharged without condensation (e.g., solvent dehydration with highly water-selective membranes). For pervporative extraction of aroma compounds, vacuum pervaporation is more appropriate.

The usefulness of pervaporation for separation of liquid mixtures was recognized in the mid 1930s (Farber, 1935). However, it was not until the 1980s was the first industrial scale pervaporation plant commercialized for dehydration of ethanol when a composite membrane comprising of a thin layer of cross-linked poly(vinyl alcohol) supported on a porous polyacrylonitrile substrate was developed. This remains the primary industrial application of pervaporation today. In spite of a great deal of research on pervaporation for aroma recovery, industrial-scale units for this application are still lacking, partially because the sensory profiles of the permeate have not been well studied and little is known about the long-term membrane performance. The first reported work on pervaporation for extraction and concentration of volatile aroma compounds may be attributed to Voilley et al. (1988) who investigated recovery of volatile aroma compounds (1-octene-3-ol and 2,5-dimethylpyrazine) from water, although strictly speaking it is not a pervaporation process because a microporous polypropylene membrane (pore diameter 0.2 μm, porosity 50%) was used and the aroma extraction is based on air stripping via the microporous membrane. They later used vacuum pervaporation through nonporous poly(dimethyl siloxane) (PDMS) membranes and zeolite-filled PDMS membranes, and the model aromas tested were expanded to include ethyl ethanoate, ethyl butanoate, and ethyl hexanoate (Voilley et al., 1989; Voilley, Charbit, and Gobert, 1990; Lamer and Voilley, 1991). The vacuum pervaporation with nonporous membranes was shown to be more efficient than the membrane-based air stripping process. In the following years, a great deal of laboratory studies has been carried out on pervaporative extraction of aroma compounds with different sensory attributes, as shown in Figure 7.2, which shows the number of publications each year that are indexed within the "Web of Science Core Collection" (as of June 2014).

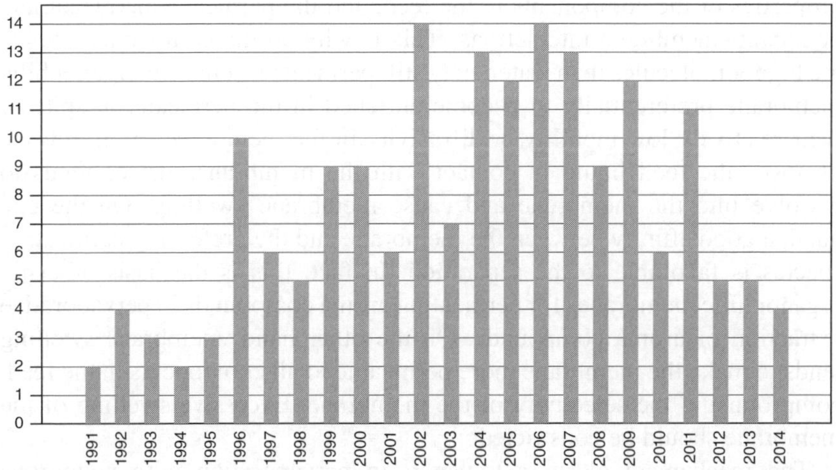

Figure 7.2 Number of publications each year indexed within the "Web of Science Core Collection" (as of June 2014).

Pervaporation is different from other membrane processes in that there is a phase change involved. The permeate leaving the membrane is in a vaporous state. Because of the nonporous structure of the membrane, the permeation flux is generally low. This makes pervaporation useful for certain niche applications where conventional separation processes are ineffective or the permeate products are highly value added. It is not particularly suitable for circumstances where a high permeate throughput is required. As only a small fraction of the feed that has permeated through the membrane undergoes a phase change from liquid to vapor, the energy needed for the vaporization of the permeate is generally not a significant issue. From an energy consumption point of view, pervaporation is especially advantageous when the concentration of the preferentially permeating species in the feed is low, which is the case for recovery of aroma compounds from dilute feed solutions. In practice, the heat required for the vaporization of permeate during the course of permeation can simply be provided by the feed liquid in the form of sensible heat, and no heat supply directly to the membrane is needed.

Unlike reverse osmosis, mass transport in pervaporation is not limited by the osmotic pressure of the feed, and the driving force for mass transfer through the membrane is provided by lowering the chemical potential of the permeate stream on the downstream side. Consequently, the feed pressure is not critical; pervaporation is typically operated at atmospheric pressure (or slightly above, depending on the hydrodynamics of the feed flow). Unlike ultrafiltration or microfiltration, where the separation is primarily based on size sieving of the permeating species, pervaporation separation is governed by the chemical nature of the macromolecules that form the membrane, the physical structure of the membrane, the physicochemical properties of the components in the feed, and the permeant–permeant and permeant–membrane interactions. This is why aroma compounds, which are bigger molecules than water, may still permeate through an organophilic membrane preferentially to become enriched in the permeate, in spite of their relatively low mobility or diffusivity in the membrane. As in reverse osmosis, the feed liquid in contact with the membrane surface tends to dissolve into the membrane and cause membrane swelling. On the one hand, a good affinity between the membrane and the preferably permeating species is favorable to the separation. In fact, this is the basis of using organophilic membranes for permeating aroma compounds in pervaporative extraction of aroma compounds. On the other hand, membrane swelling tends to make the membrane more permeable to all components in the feed, compromising the selectivity of the membrane. Excessive swelling of the membrane should be constricted.

The mechanism of mass transport in pervaporation with nonporous membranes can be described by the solution-diffusion model. It was originally proposed by Graham (1866) to describe gas permeation through polymer membranes, and then adopted by Binning et al. (1961) to describe pervaporation transport. According to this mechanism, pervaporative transport through a nonporous membrane consists of three sequential steps: (1) sorption of the permeant from the liquid feed to the membrane through the feed/membrane interface, (2) diffusion of the sorbed molecules in the membrane, and (3) desorption of the permeant molecules from the membrane to the vapor phase on the downstream side. This is shown schematically in Figure 7.3. Both the sorption and desorption steps are generally considered to be very fast and equilibria are established instantaneously on both sides of the membrane. Unlike the sorption step, where selective sorption may occur due to specific affinities of the membrane to certain components in the feed (i.e., preferential sorption), the desorption step is nonselective and all permeating molecules are removed from the membrane upon arrival at the downstream side (Fleming, 1990). As such, the diffusion of the permeant in the membrane is the rate-determining step in pervaporation.

In general, both solubility and diffusivity of the permeant in the membrane are concentration dependent. A number of mathematical equations for mass transport have been formulated on the basis of Fick's diffusion equation using different empirical correlations of concentration dependencies of the solubility and/or diffusivity. However, caution should be exercised in using the equations as they are valid only within the established range for which the relationships for diffusion and thermodynamic equilibria are applicable. As an approximation, when the solubility and diffusivity coefficients can be treated as constant, the permeability coefficient of the

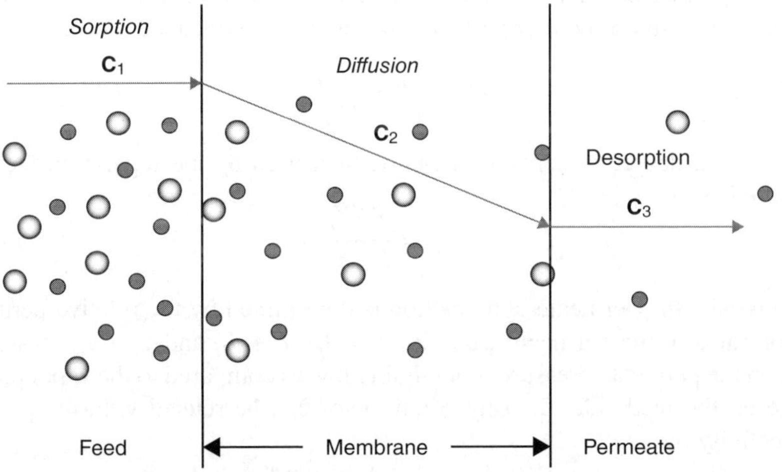

Figure 7.3 Illustration of solution-diffusion model for mass transport in pervaporation.

membrane can be related to the solubility and diffusivity coefficients by (Shao and Huang, 2007; Feng and Huang, 1997):

$$P_i = S_i D_i \tag{7.1}$$

where P_i, S_i, and D_i are the permeability, solubility, and diffusivity coefficients of permeating species i, respectively. The permeability is an intrinsic property of the membrane material and is related to the permeation flux by

$$J_i = \left[\frac{P_i}{l}\right](X_i \gamma_i p_i^{sat} - Y_i p^p) \tag{7.2}$$

where J is the permeation flux, p^{sat} and γ are the saturated vapor pressure and activity coefficient of the permeating component in the feed liquid, respectively, p^p is the permeate pressure, l is the effective thickness of the membrane, and X and Y are the concentrations of the permeant (in mole fractions) in the feed and permeate, respectively. This equation is derived for pervaporation in analog to gas permeation through a membrane, assuming an equivalent partial pressure difference $(X_i \gamma_i p_i^{sat} - Y_i p^p)$ across the membrane as the driving force for permeation. The ratio (P_i/l), which is the membrane permeability normalized by the membrane thickness, is called the permeance of the membrane. It is equal to the permeation flux normalized by the transmembrane driving force. Unlike the permeability coefficient, which is a property of the membrane material, the permeance is a property of the membrane as the membrane thickness comes into play. Thus, the membrane permeance is more relevant for practical applications, especially when asymmetric or composite membranes, whose effective thicknesses are often difficult to determine accurately, are used for increased permeation fluxes.

The permselectivity of a membrane to a pair of permeating species i and j can be measured by their permeability ratio (or permeance ratio):

$$\alpha_{ij}^0 = \frac{P_i}{P_j} = \frac{[P_i/l]}{[P_j/l]} \tag{7.3}$$

The actual degree of separation can be measured by the separation factor, defined as

$$\alpha_{ij} = \frac{Y_i/Y_j}{X_i/X_j} \tag{7.4}$$

Obviously, the permeate composition is determined by the relative permeation rates of the permeants, that is, $Y_i = J_i/(J_i + J_j)$ and $Y_j = J_j/(J_i + J_j)$. When the permeate pressure is negligibly low as compared to the vapor pressure on the feed side, the separation factor can be related with the permselectivity by

$$\alpha_{ij} = \alpha_{ij}^0 \times \left(\frac{\gamma_i}{\gamma_j}\right) \times \left(\frac{p_i^{sat}}{p_j^{sat}}\right) \tag{7.5}$$

Equation (7.5) reveals that the separation factor is determined by three parameters. The first parameter (α_{ij}^{0}) is the membrane permselectivity, an intrinsic permeability property of the membrane material. The second parameter ($\gamma_{\mathrm{i}}/\gamma_{\mathrm{j}}$) is the ratio of their activity coefficient in the feed liquid, which is a thermodynamic property of the feed solution determined by the excess Gibbs energy of the feed liquid. The activity coefficient ratio is affected by the liquid composition and temperature; however, if the feed liquid behaves as an ideal solution (which is uncommon for most pervaporation applications), then the second parameter will become unity. The third parameter (p_i^{sat}/p_j^{sat}) reflects the effect of saturated vapor pressure of pure permeant on the separation performance. Generally speaking, all three parameters are affected by the operating temperature. Equation (7.5) demonstrates how the membrane, the nature of the permeants, and more explicitly the operating conditions (composition and temperature) affect the separation.

For pervaporative extraction of aroma compounds from aqueous solutions, the concentration and enrichment of the aroma in the permeate is also commonly characterized by an enrichment factor (β). It is simply defined as a ratio of the aroma concentration in the permeate (Y_i) to the aroma concentration in the feed (X_i):

$$\beta = \frac{Y_i}{X_i} \tag{7.6}$$

When the aroma concentration in the feed liquid is considerably low ($X_i \ll 1$) and a high vacuum is applied on the permeate side ($p = 0$), the enrichment factor can be related to the separation factor by

$$\beta = \frac{\alpha}{1 + \alpha X} \tag{7.7}$$

where X is the aroma mole fraction in the feed and α is the separation factor for aroma/water; for convenience their subscripts are removed. Equation (7.7) shows the following features: (1) when the feed aroma concentration X is sufficiently low such that $\alpha X \ll 1$, then α and β will be equal numerically and (2) when the separation factor is sufficiently high such that $\alpha X \gg 1$, the enrichment factor will approach an upper limit that is equal to ($1/X$).

In pervaporation, the effect of the operating temperature on the separation performance is often represented by temperature dependence of the permeation flux, which can usually be described by an Arrhenius type of correlation:

$$J_i = J_{0i} \exp\left(-\frac{E_{Ji}}{RT}\right) \tag{7.8}$$

where E_J is the *apparent* activation energy for permeation of component i, J_0 is a preexponential factor, R is the universal gas constant, and T is the temperature. It should be pointed out that the apparent activation energy measures the *overall* effects of temperature on the permeation flux, which has accounted for the effect of temperature on the driving force. As mentioned above, the activity coefficient and the saturated vapor pressure are affected by temperature as well. The apparent activation energy is not a true representation of the activation energy for permeation in the membrane. The activation energy (E_P) that characterizes the temperature effect on the intrinsic permeability of the membrane should be distinguished from the apparent activation energy. As mentioned earlier, the permeability coefficient (P) is a product of the diffusivity (D) and solubility (S) coefficients (Equation (7.1)). Both D and S are normally dependent on temperature, and their temperature dependencies can be expressed as

$$D_i = D_{0i} \exp\left(-\frac{E_{Di}}{RT}\right) \tag{7.9}$$

$$S_i = S_{0i} \exp\left(-\frac{\Delta H_i}{RT}\right) \tag{7.10}$$

where E_D and ΔH are the activation energy of *diffusion* and the enthalpy change of *dissolution* of the permeant in the membrane, respectively, and D_0 and S_0 are their preexponential factors. As such, the following relation results:

$$P_i = P_{0i} \exp\left(-\frac{E_{Pi}}{RT}\right) \tag{7.11}$$

where E_P is the activation energy of *permeation*, which is a combination of the activation energy of diffusion and the enthalpy change of dissolution of the permeant in the membrane (i.e., $E_P = E_D + \Delta H$), and P_0 is the preexponential factor for the permeability coefficient ($P_0 = D_0 S_0$). Rearranging Equations (7.2) and (7.11),

$$\left[\frac{P_i}{l}\right] = \frac{J_i}{(X_i \gamma_i p_i^{sat} - Y_i p^p)} = \left[\frac{P_{0i}}{l}\right] \exp\left(-\frac{E_{Pi}}{RT}\right) \tag{7.12}$$

Thus, the activation energy E_P can be evaluated from the temperature dependence of membrane permeance $[P_i/l]$ (which is equal to the permeation flux normalized by the driving force for permeation). Plotting ln $[P_i/l]$ versus $1/T$ will yield a straight line with a slope from which E_P can be determined. On the other hand, since the permeate pressure in pervaporation is generally low, the driving force for permeation is largely determined by the equilibrium vapor pressure. As a first approximation, if the saturated vapor pressure (p^{sat}) of a liquid follows the Clausius–Clapeyron equation and the temperature dependence of the activity coefficient of the permeant is insignificant, then the activation energy E_P will be equal to the apparent activation energy minus the heat of evaporation of the permeant (Feng and Huang, 1997):

$$E_P = E_J - \Delta H_V \tag{7.13}$$

where ΔH_V is the heat of evaporation of the permeant. Because evaluating the apparent activation energy E_J from the ln J versus $1/T$ data is much simpler than evaluating the activation emery for permeation E_P from the ln $[P_i/l]$ versus $1/T$ data, especially when the permeate pressure is not accurately known, a simple yet useful approach of estimating E_P is to subtract the heat of evaporation from the apparent activation energy. This also explicitly shows how the enthalpy change due to the phase change in pervaporation influences the permeation behavior.

For pervaporative extraction of aroma compounds, the separation performance is often measured in terms of permeation flux and separation factor (or aroma enrichment factor), which have accounted for the effects of operating conditions (feed composition and temperature) on the separation. Alternatively, the membrane performance can also be characterized in terms of the aroma permeability (or permeance) and aroma/water permeability ratio, given the operating conditions.

7.3 PERVAPORATION FOR RECOVERY OF AROMA COMPOUNDS USING ORGANOPHILIC MEMBRANES

The separation performance of pervaporation for aroma recovery depends on the following factors: the nature (chemical structure and physicochemical properties) of the aroma compounds, the properties (material and morphology) of the membrane, the feed composition, and the operating conditions (temperature and permeate pressure), which will be discussed later. Most of the work reported in the literature deals with dilute aqueous solutions containing model aroma compounds, mainly binary water–organic mixtures and, to a lesser extent, dilute aqueous solutions of multiple dairy aroma components. When multiple aroma compounds are present in the feed, the permeation of one aroma compound may be affected by the presence of other aromas due to the coupling effect among the permeating species caused by the permeant–permeant interactions. However, this effect is often negligible in highly diluted solutions (Olsson and Tragardh, 1999a; Peng and Liu, 2003; She and Hwang, 2006b; Shepherd, Habert, and Borges 2002). Almost all of the aroma compounds have bigger molecular sizes than water, and the diffusivity aspect involved in pervaporation is unfavorable to the enrichment of aroma on the permeate side. Thus, the membrane needs to have a good affinity to the aroma compounds in order to compensate for the unfavorable diffusivity selectivity. In other words, the solubility aspect involving pervaporation should be exploited to achieve the desired separation. As such, organophilic membranes are appropriate for pervaporative extraction of the aroma compounds from aqueous solutions.

7.3.1 Dairy aroma compounds

A large number of aroma compounds have been identified in dairy products that produce different sensory attributes, including esters, ketones, acids, alcohols, aldehydes, lactones, and sulfur compounds (Toso, Procida, and Stefanon, 2002; Routray and Mishra, 2011; Cheng, 2010; Urbach, 1997; Nursten, 1997). Among them, there are over 30 flavor compounds typically present in dairy products that have been used for research on pervaporative extraction. They are listed in Table 7.1, along with their sensory attributes and solubilities in water. In general, all aroma compounds have very low concentrations in dairy products and the flavor of a product is given by a combination of the odors from all the aroma compounds. It may be pointed out that the aroma composition can be different, depending on the feedstock of the animal, grazing or silage, and the processing and storage conditions of the products as well. Real dairy solutions often contain more than one flavor compound and their concentration profile determines their overall smells and tastes. In addition, proteins, fats, and sugars are present in dairy products, and these substances may potentially affect the separation performance due to fouling of the membrane surface. Thus, model feed solutions containing single or multiple aroma compounds are extensively used in pervaporation research.

Based on a literature survey, the top 10 most widely used aroma compounds in pervaporation research are shown in Figure 7.4. Three ester compounds (i.e., ethyl acetate, ethyl butanoate, and ethyl hexanoate) are on the top of the list, which is not surprising because they are also common aromas in fruit and vegetable juices. This does not mean that the esters are more important flavorants than other aromas in dairy products. To get a rough idea about the magnitude of enrichment factor that can be achieved by pervaporation, the range of selectivity for pervaporative extraction of some dairy aroma compounds with current membranes are presented in Figure 7.5. A general rule of thumb is that aroma compounds with higher hydrophobicities tend to be separated more effectively by pervaporation using organophilic membrane (Baudot and Marin, 1997).

Ester aromas are a typical group of hydrophobic aroma molecules with relatively small molecular sizes. They are normally fruity flavored. As shown in Figure 7.5, the membranes generally have a higher selectivity to the ester molecules than other aroma compounds. Among all the ester aromas shown, ethyl hexanoate is best concentrated by pervaporation

Table 7.1 Dairy aroma compounds studied for pervaporative extraction.

Aroma compounds	Formula	Odor	Molecular weight (g/mol)	Boiling point (°C)	Solubility in water at 20 °C (g/L)
Esters					
Methyl acetate	$C_3H_6O_2$	Fragrant	74	57	250
Ethyl acetate	$C_4H_8O_2$	Sweet	88	77	83
Ethyl propionate	$C_5H_{10}O_2$	Pineapple-like	102	99	24
Ethyl butanoate	$C_6H_{12}O_2$	Pineapple-like	116	120	6
Ethyl hexanoate	$C_8H_{16}O_2$	Pineapple-like	144	228	Hardly soluble
Ethyl octanoate	$C_{10}H_{20}O_2$	Fruity, apricot-like	172	207	Hardly soluble
Propyl acetate	$C_5H_{10}O_2$	Fruity, pear-like	102	102	19
Butyl acetate	$C_6H_{12}O_2$	Banana/apple-like	116	127	7
Ketones					
2-Heptanone	$C_7H_{14}O$	Banana-like	114	151	4
2-Nonanone	$C_9H_{18}O$	Fruity	142	192	0.5
Diacetyl	$C_4H_6O_2$	Buttery	86	88	20
2-Butanone	C_4H_8O	Butterscotch-like	72	80	280
3-Octanone	$C_8H_{16}O$	Grassy	128	167	Hardly soluble
Acids					
Acetic acid	$C_2H_4O_2$	Sour, pungent	60	118	Miscible
Butanoic acid	$C_4H_8O_2$	Obnoxious	88	164	Miscible
Hexanoic acid	$C_6H_{12}O_2$	Goaty	116	206	11
Octanoic acid	$C_8H_{16}O_2$	Irritating	144	240	0.7
Propionic acid	$C_3H_6O_2$	Slightly rancid	74	141	Miscible
Heptanoic acid	$C_7H_{14}O_2$	Rancid	130	223	2.4
Alcohols					
1-Octen-3-ol	$C_8H_{16}O$	Mushroom-like	128	174	Hardly soluble
1-hexanol	$C_6H_{14}O$	Acoholic	102	155	5.9
Aldehydes					
Hexanal	$C_6H_{12}O$	Green	100	130	5
Phenylacetaldehyde	C_8H_8O	Sweet, rose	120	195	2
Acetaldehyde	C_2H_4O	Etherial	44	20	Miscible
3-Methylbutanal	$C_5H_{10}O$	Green, malty	86	91	1.4
Pentanal	$C_5H_{10}O$	Pungent	86	102	Slightly soluble
Heptanal	$C_7H_{14}O$	Obnoxious	114	153	Slightly soluble
Lactones					
γ-Decalactone	$C_{10}H_{18}O_2$	Coconut-like	170	281	Slightly soluble
Sulfur compounds					
Dimethyl trisulfide	$C_2H_6S_3$	Sulfury	126	365	Hardly soluble
Methyl thiobutanoate	$C_5H_{10}OS$	Cheese-like	118	301	Slightly soluble
Aromatic compounds					
Vanillin	$C_8H_8O_3$	Vanilla	152	285	10

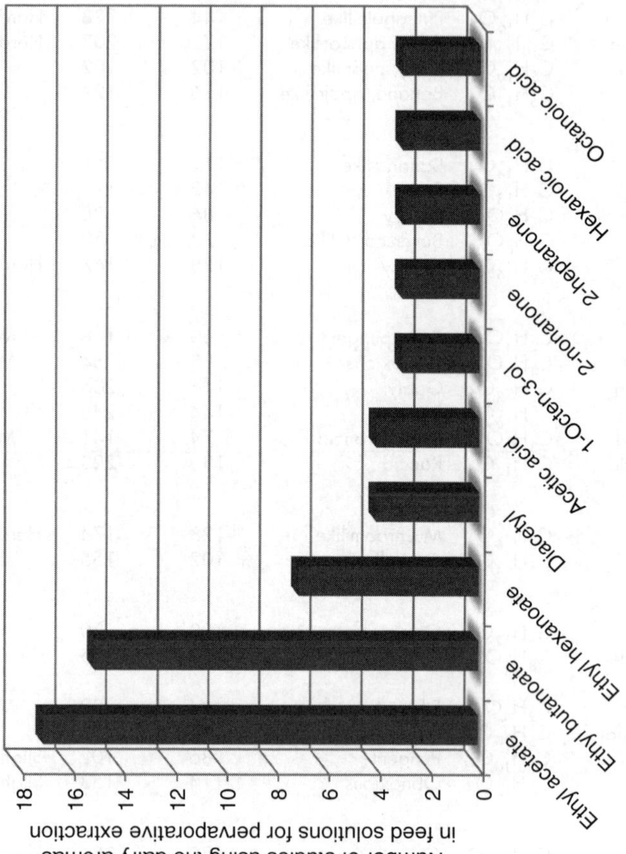

Figure 7.4 Dairy aromas most widely used as model compounds in pervaporation extraction.

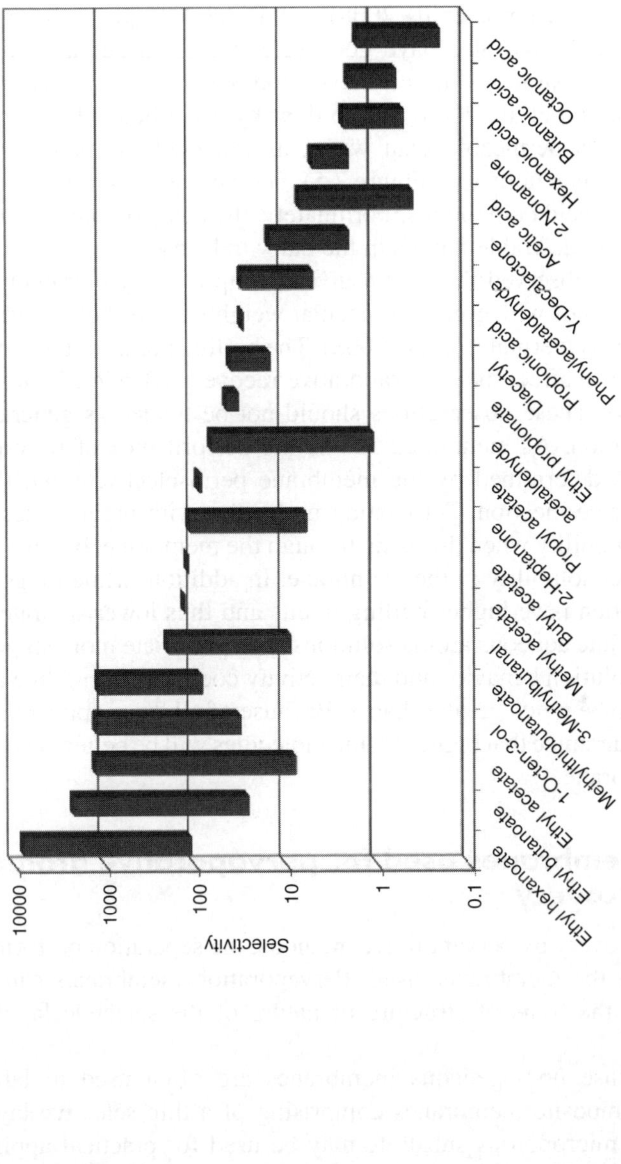

Figure 7.5 Range of membrane selectivities for pervaporative extraction of dairy aroma compounds.

with an enrichment factor in the range of 112–8200, depending on the membranes used (Pereira et al., 2005; Sampranpiboon et al., 2000a). The broad range of selectivity is a result of the different membrane types and structures and different operating conditions used in the pervaporation processes. Hydrophobic alcohols (e.g., 1-octen-3-ol (Pereira et al., 2005)), aldehydes (e.g., acetaldehyde (Wu et al., 2012) and 3-methylbutanal (Kanani et al., 2003; Raisi and Aroujalian, 2011)), and ketones (e.g., 2-heptanone (Overington et al., 2008) and diacetyl (Rajagopalan, Cheryan, and Matsuura, 1994)) also have good selectivities in pervaporation. The pervaporative separations of acid flavor compounds from water are rather poor (Overington et al., 2008) due to their hydrophilicity. They are sometimes even less permeable than water, as reflected by their enrichment factors of less than unity (see Figure 7.5). Pervaporation is thus not effective to concentrate acid compounds. Fortunately, the acid components are often not particularly desirable flavors in the dairy industry.

It has been observed that, for a given group of aroma compounds, the aroma compounds with greater molecular weights can be better enriched by pervaporation (Overington et al., 2008). The boiling points of the aromas are also reported to affect their pervaporative recovery (Baudot, Souchon, and Marin, 1999). These observations should not be treated as general trends in pervaporation. As mentioned above, the performance of pervaporative separation is determined by the membrane permselectivity and the driving force for permeation. Permeating molecules with bigger sizes tend to have lower mobility when diffusing through the membrane, but they usually exhibit higher solubility in the membrane. In addition, while bigger aroma molecules often have higher boiling points and thus lower saturated vapor pressures, dilute aqueous aroma solutions tend to deviate more significantly from ideal solution behavior and their activity coefficients in dilute aqueous solutions can be much greater than 1. Because of all these opposing effects, there is no guarantee that bigger aroma molecules will be better pervaporated by the membrane.

7.3.2 Membranes used for pervaporative aroma recovery

In aroma recovery by pervaporative methods, the separation performance is governed by the membranes used. Pervaporation membranes can be categorized on the basis of structure or nature of the selective layer of the membrane.

While dense homogeneous membranes are often used in laboratory research, composite membranes comprising of a thin selective layer supported on a microporous substrate may be used for practical applications in order to enhance the permeation flux. The membranes may be in the form of hollow fibers or flat sheets, and therefore appropriate module designs should be used. Hollow fiber modules with either shell-side feed or tube-side feed are commonly used configurations. Flat sheet membranes are usually packed as plate-and-frame or spiral wound modules. Generally hollow fiber modules have a larger membrane packing density than flat membranes. Hollow fiber membranes are self-supporting and no external support is needed. During manufacturing of flat membranes, the substrate membrane is often cast on to a nonwoven backing material, which provides additional support to the resulting composite membrane (Figure 7.6). Flat composite membranes with a poly(dimethyl siloxane) (PDMS) selective layer are available from GKSS, Sulzer, and Dow Corning. The microporous substrates are primarily based on poly(vinylidene fluoride) (PVDF), polyacrylonitrile (PAN), polysulfone (PS), or polyetherimide (PEI), with pore sizes in the range of ultrafiltration membranes. Huang, Moon, and Pal (2002) have reported that the pore size and porosity of the substrate, rather than the substrate material itself, also influence the separation performance, although the top layer is more dominant. When the permeate vapor passes through a microporous substrate, the mass transport resistance can still be significant due to Knudsen diffusion (Rautenbach and Helmus, 1994).

Most of the selective surface layers in pervaporation membranes are homogeneous. To improve the separation performance, hydrophobic fillers may be incorporated into the polymer matrix, thereby enhancing the sorption selectivity and restricting membrane swelling. Zeolites (Bowen, Noble, and Falconer, 2003; Vankelecom, De Beukelaer and Uytterhoeven, 1997), silica (Shirazi, Ghadimi, and Mohammadi, 2012), activated carbon (Ji, Sikdar, and Hwang, 1995), and carbon black (Panek and Konieczny, 2007) have been used as the filler materials. Both glassy and rubbery polymers can be used to form the selective layer of the pervaporation membranes. Glassy polymers tend to favor the permeation of water and may be used for dehydration of organic solvents, whereas elastomeric

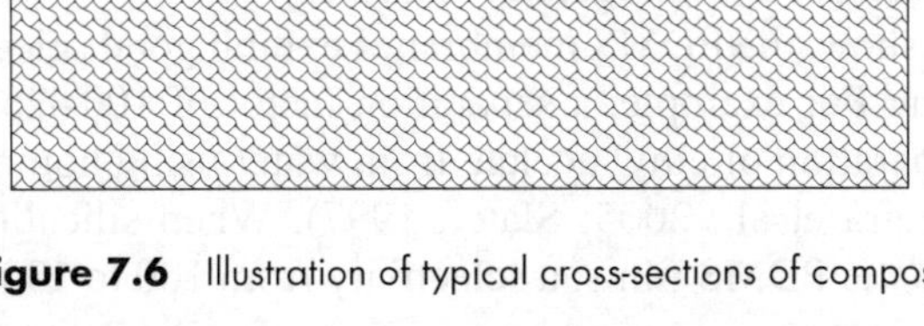

Figure 7.6 Illustration of typical cross-sections of composite membranes.

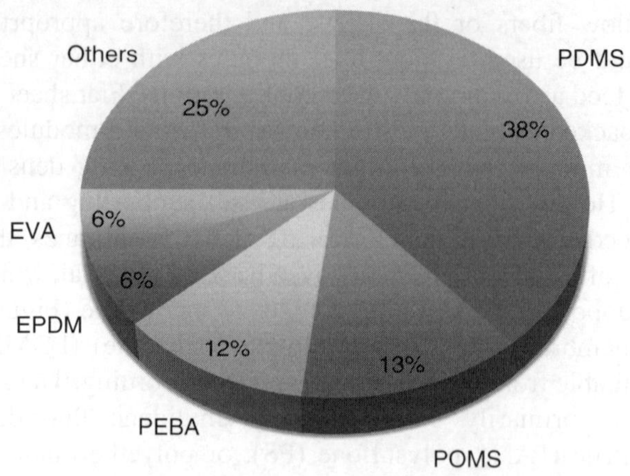

Figure 7.7 Organophilic membranes used for pervaporative extraction of dairy aroma compounds.

polymers are suitable for preferential permeation of organic compounds. A survey of current literature shows that PDMS is by far the most widely used rubbery material for pervaporative extraction of aroma compounds, followed by poly(octhylmethyl siloxane) (POMS) and poly(ether block amide) (PEBA). These three materials account for about two-thirds of all membranes used for aroma recovery by pervaporation. This is illustrated in Figure 7.7. Ethylene propylene diene monomer (EPDM) and ethylene vinyl acetate copolymer (EVAC) are also used to a lesser extent. The chemical structures of these elastomers are shown in Figure 7.8.

7.3.2.1 Poly(dimethyl siloxane) (PDMS)

Poly(dimethyl siloxane) (PDMS), often referred to as "silicone rubber," is considered to be the current benchmark material for organophilic pervaporation membranes. PDMS has good mechanical and chemical properties and can be used to fabricate supported or unsupported flat sheets. It can also be used to produce hollow fibers by coating on suitable hollow fiber substrates. There is a great deal of research on dairy aroma recovery by pervaporation using PDMS membranes, as shown in Table 7.2. Flat sheet PDMS membranes, either standing alone or supported on a substrate, with or without hydrophobic nanofillers, have been investigated extensively.

As the data in Table 7.2 show, the PDMS membranes exhibit good selectivity for aroma enrichment. For example, a separation factor of 100–250 can be achieved for the separation of ethyl acetate from water using homogeneous PDMS films (Pereira et al., 2005; Slater, 1997). When silicalite fillers are incorporated into the PDMS films, a selectivity from 100 to 1300

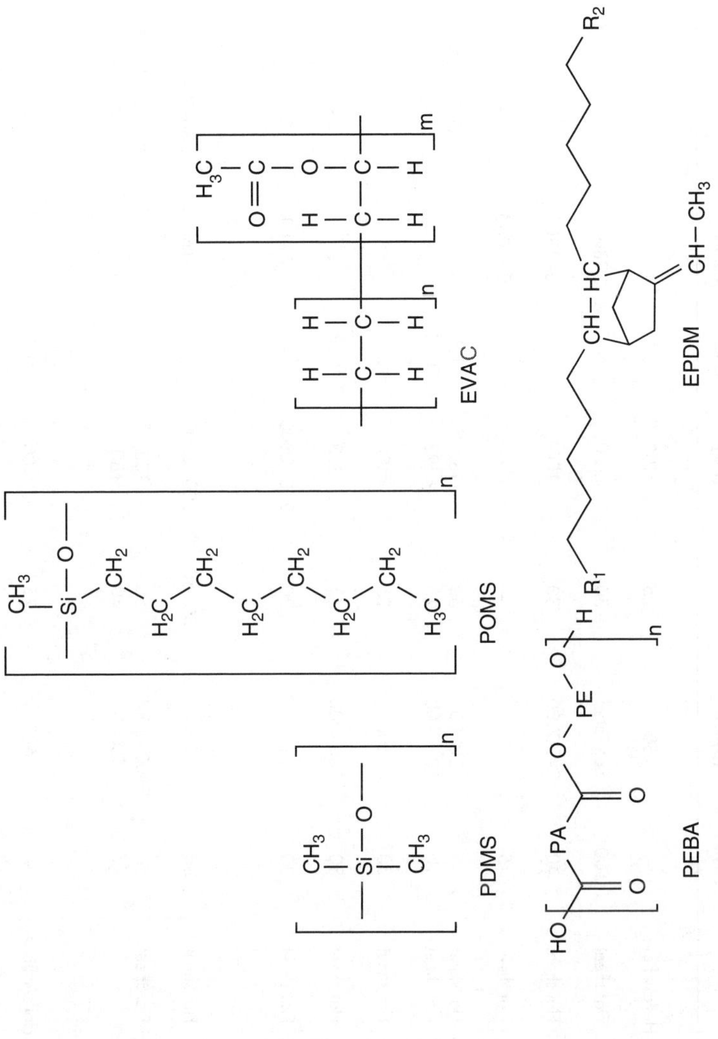

Figure 7.8 Structure of PDMS, POMS, PEBA, EPDM, and EVAC polymers.

Table 7.2 Pervaporative extraction of dairy aromas by PDMS-based membranes.

Membranes	Membrane geometry	Effective thickness (μm)	Feed conc. (ppm)	Temperature (°C)	Permeate pressure (Pa)	Feed flow rate (L/min)	Aroma Flux (g/m^2 h)	Separation factor	Enrichment factor	Reference
Ethyl acetate										
PDMS*	Hollow fiber	–	600	25	<400	0.5	–	–	96.8	Pereira et al., 2005
PDMS	Flat sheet	500	Saturated	30	<100	–	2980	–	11.6	Djebbar et al., 1998
PDMS	Flat sheet	40	0–70000	30	100	–	0–1300	–	–	Nguyen et al., 2000
PDMS	Flat sheet	75	–	–	–	–	0–3000	–	–	Nguyen et al., 2000
PDMS 1060	Flat sheet	–	20300	50	133	1.5	–	237	–	Slater, 1997
PDMS 1060	Flat sheet	10	100–900	25	<293	1.53	–	141	–	Pereira, et al., 2005
PDMS 1070*	Flat sheet	30	600	25	<400	0.5	–	–	124.4	Pereira, et al., 2005
PDMS 1070	Flat sheet	30	100–900	25	<293	1.53	–	393	–	Pereira, et al., 2005
PDMS 1070	Flat sheet	30	100	30	250–2000	–	1.7–1.1	300	–	Baudot, Souchon, and Marin, 1999
PDMS 1070	Flat sheet	30	500	30–50	300–2500	–	4–16	230–280	–	Baudot and Marin, 1996
PDMS 1070	Flat sheet	–	9500–80000	50	133	1.5	–	309–1281	–	Slater, 1997
PDMS/PMHS–alumina	–	3.5	50000	40	180	–	–	140	–	Tanaka et al., 2010
Ethyl butanoate										
PDMS*	Hollow fiber	–	600	25	<400	0.5	–	–	516	Pereira et al., 2005
PDMS*	Flat sheet	0.5	100	20–40	900–2400	–	0.56–3.59 at 2 kPa	–	6–24	Overington et al., 2008

PDMS*	Flat sheet	1.5	100	20–40	2000	–	0.64–0.94	–	6 at 40 °C	Overington et al., 2008
PDMS	Flat sheet	130	88	25	50	$Re=10^5$	2.61	–	1944	Lamer et al., 1994
PDMS	Flat sheet	500	Saturated	30	<100	–	5000	–	158	Djebbar et al., 1998
PDMS	Flat sheet	10	300	30–50	400–2000	0.49–0.65	0.010–0.019	77–234	–	Song and Lee, 2005
PDMS	Hollow fiber	–	50–1,000	25–45	–	0.33–5	5.7–7	–	400–1000	Shepherd, Habert, and Borges, 2002
PDMS DS-7	Flat sheet	–	148	23	2000	0.285	–	–	30	She and Hwang, 2006
PDMS 1060	Flat sheet	10	100–900	25	<293	1.53	10–33	510	–	Pereira et al., 2005
PDMS–PVDF–PP	Flat sheet	4	148	23	2,000	0.285	–	–	19	She and Hwang, 2006
PDMS–PVDF–PP	Flat sheet	6	148	23	2,000	0.285	–	–	27	She and Hwang, 2006
PDMS–PVDF	Flat sheet	2.5	148	23	2,000	0.285	–	–	26	She and Hwang, 2006
PDMS 1070	Flat sheet	30	100–900	25	<293	1.53	5–22	875	–	Pereira et al., 2005
PDMS 1070	Flat sheet	80	100	50	400	$Re=33000$	0.59	–	41	Isci, Sahin, and Sumnu, 2006
PDMS 1070*	Flat sheet	30	600	25	<400	0.5	–	–	410	Pereira et al., 2005
Ethyl hexanoate										
PDMS*	Hollow fiber	–	100	25	<400	0.5	–	–	194	Pereira et al., 2005
PDMS*	Flat sheet	0.5	100	20–40	900–2400	–	0.13–2.4 at 2 kPa	–	5.2–7.2	Overington et al., 2008
PDMS*	Flat sheet	1.5	100	20–40	2000	–	0.44–0.93	–	6 at 40 °C	Overington et al., 2008

(*continued overleaf*)

Table 7.2 *(continued)*

Membranes	Membrane geometry	Effective thickness (μm)	Feed conc. (ppm)	Temperature (°C)	Permeate pressure (Pa)	Feed flow rate (L/min)	Aroma Flux (g/m² h)	Separation factor	Enrichment factor	Reference
PDMS	Flat sheet	130	88	25	50	$Re=10^5$	8.13	–	3150	Lamer et al., 1994
PDMS	Flat sheet	10	300	30–50	2000	0.49–0.65	0.014–0.024	112–252	–	Song and Lee, 2005
PDMS 1060	Flat sheet	10	100–900	25	<293	1.53	–	599	–	Pereira et al., 2005
PDMS 1070	Flat sheet	30	100–900	25	<293	1.53	–	750	–	Pereira et al., 2005
PDMS 1070*	Flat sheet	30	100	25	<400	0.5	–	–	213.2	Pereira et al., 2005
Ethyl octanoate										
PDMS*	Flat sheet	0.5	100	20–40	900–2400	–	0.003–0.1 at 2 kPa	–	1.1–2.3	Overington et al., 2008
PDMS*	Flat sheet	1.5	100	20–40	2,000	–	0.017–0.045	–	2.5 at 40 °C	Overington et al., 2008
Methylthiobutanoate										
PDMS 1060	Flat sheet	10	15	30	250–2000	–	0.6–0.4	300–400	–	Baudot, Souchon, and Marin, 1999
PDMS 1070	Flat sheet	30	15	30	285–2000	–	0.17–0.03	84–291	–	Baudot and Marin, 1996

Ethyl propionate										
PDMS	Flat sheet	500	Saturated	30	<100	–	2810	–	43.5	Djebbar et al., 1998
2–Heptanone										
PDMS*	Flat sheet	0.5	100	20–40	900–2400	–	0.0045–0.34 at 2 kPa	–	6–20	Overington et al., 2008
PDMS*	Flat sheet	1.5	100	20–40	2000	–	0.11–0.17	–	10 at 40 °C	Overington et al., 2008
2–Nonanone										
PDMS*	Flat sheet	0.5	100	20–40	900–2400	–	0.0054–0.13 at 2 kPa	–	2–4	Overington et al., 2008
PDMS*	Flat sheet	1.5	100	20–40	2000	–	0.029–0.066	–	4 at 40 °C	Overington et al., 2008
Diacetyl										
PDMS–PC	Flat sheet	25	20–2000	25–43	650–3000	–	0.044–63	33–40	–	Rajagopalan, Cheryan, and Matsuura, 1994
PDMS 1070	Flat sheet	30	80	30	250–2000	–	0.2–0.1	15	–	Baudot, Souchon, and Marin, 1999
PDMS 1070	Flat sheet	30	80	30, 50	240–2500	–	0.078–0.408	15, 19	–	Baudot and Marin, 1996
γ-Decalactone										
PDMS 1060	Flat sheet	10	500	30	250–2000	–	1–0.08	15–2	–	Baudot, Souchon, and Marin, 1999

(continued overleaf)

Table 7.2 *(continued)*

Membranes	Membrane geometry	Effective thickness (μm)	Feed conc. (ppm)	Temperature (°C)	Permeate pressure (Pa)	Feed flow rate (L/min)	Aroma Flux (g/m² h)	Separation factor	Enrichment factor	Reference
Acetic acid										
PDMS*	Flat sheet	0.5	100	20–40	900–2400	–	0.011–0.31 at 2 kPa	–	0.4–0.7	Overington et al., 2008
PDMS*	Flat sheet	1.5	100	20–40	2000	–	0.054–0.089	–	0.5 at 40 °C	Overington et al., 2008
Butanoic acid										
PDMS*	Flat sheet	0.5	100	20–40	900–2400	–	0.015–0.58 at 2 kPa	–	0.6–1.3	Overington, et al., 2008
PDMS*	Flat sheet	1.5	100	20–40	2000	–	0.092–0.19	–	1 at 40 °C	Overington, et al., 2008
Hexanoic acid										
PDMS*	Flat sheet	0.5	100	20–40	900–2400	–	0.013–1.07 at 2 kPa	–	0.5–2.3	Overington et al., 2008
PDMS*	Flat sheet	1.5	100	20–40	2000	–	0.092–0.25	–	1.5 at 40 °C	Overington et al., 2008
Octanoic acid										
PDMS*	Flat sheet	0.5	100	20–40	900–2400	–	0.0067–0.67 at 2 kPa	–	0.2–1.6	Overington et al., 2008
PDMS*	Flat sheet	1.5	100	20–40	2000	–	0.024–0.15	–	1 at 40 °C	Overington et al., 2008
1–Octen–3–ol										
PDMS	Flat sheet	130	87	25	50	Re=10^5	1.77	–	1272	Lamer et al., 1994
PDMS 1060	Flat sheet	10	100–900	25	<293	1.53	–	154	–	Pereira et al., 2005
PDMS 1070	Flat sheet	30	100–900	25	<293		252	–	1.53	Pereira et al., 2005

Phenylacetaldehyde									
PDMS	Flat sheet	160	180–12000	30	667	–	5–30	–	Kanani et al., 2003
Acetaldehyde									
ZSM–5–PDMS–Nylon	Flat sheet		30000–130000	15–45	–	0.83	6–65	–	Wu et al., 2012
3–Methylbutanal									
PDMS	Flat sheet	160	180–12000	30	667	–	120–200	–	Kanani et al., 2003
PDMS–PVDF–PP	Flat sheet	1	100	30	133	5	14	–	Raisi and Aroujalian, 2011
PDMS–PVDF–PP	Flat sheet	10	100	30	133	5	19	–	Raisi and Aroujalian, 2011
PDMS–PVDF–PP	Flat sheet	18	100	30	133	5	24	–	Raisi and Aroujalian, 2011.

* Multicomponent feed solution

can be obtained (Pereira et al., 2005; Baudot, Souchon, and Marin, 1999; Baudot and Marin, 1999). Dotremont et al. (1995) reported that the silicalites have a "reservoir" effect towards hydrophobic molecules, which enhances the membrane performance. Furthermore, when the feed solution contains strong nonpolar molecules, it will be more difficult for water molecules to penetrate silicalites. Similar results were observed by Te Hennepe et al. (1991), who showed that the exclusion and tortuosity effects arising from zeolite fillers in PDMS membranes resulted in a decrease in water permeability. The rationale of incorporating such hydrophobic fillers as silicalites in the membrane is to hinder water transport in the membrane due to organic sieving of the hydrophobic fillers. Baudot, Souchon, and Marin (1999) investigated the separation of ethyl acetate from water using a PDMS-1070 membrane, which is a composite membrane composed of 30-μm thick PDMS filled with 60% of silicalites; they observed a decrease in the transmembrane mass transfer coefficient of water. This is also supported by the work of Slater (1997), who compared the performance of unfilled PDMS 1060 membrane and silicalite-filled PDMS 1070 membrane for ethyl acetate separation from water under the same operating conditions. At a feed concentration of 2.03% ethyl acetate and at a temperature of 50 °C, the separation factor achieved with PDMS 1070 is about 2.5-fold higher than that with PDMS 1060. However, the enhanced selectivity is at the expense of significantly reduced flux. The experimental data of Slater (1997) showed that the permeation flux with PDMS 1070 membrane was only one-third that of the flux produced with the unfilled PDMS 1060 membrane. While a simple comparison of the fluxes between the two membranes is insufficient to conclude how the membrane permeability is affected by the silicalite fillers because of their different thicknesses, an examination of the differences in the partial fluxes of water and aroma between the two membranes revealed that the presence of the hydrophobic zeolites in PDMS 1170 not only significantly reduced the water flux but the aroma flux was also reduced.

However, the permselectivity of PDMS membranes are not always increased by incorporating silicate fillers in the membrane. Because of the molecular sieving effect of zeolites, bulky molecules can in principle be partly excluded from entering the pores, and consequently incorporation of the zeolite into the membrane may entail an important disadvantage of reduced aroma flux. Baudot and Marin (1996) and Baudot, Souchon, and Marin (1999) investigated the separation of methyl thiobutanoate from water under similar conditions by using both unfilled PDMS 1060 and silicalite-filled PDMS 1070 membranes. Both the selectivity and the partial permeation flux of the aroma obtained with the PDMS 1070 membrane were found to be 50% lower than those obtained with the PDMS 1060 membrane. Obviously, the use of silicalite fillers is not advantageous if the aroma molecules are big enough for the zeolite to induce a significant exclusion effect to the aroma compounds.

It should be pointed out that for silicalite-filled PDMS membranes, in addition to the molecular sieving effects that will decrease the membrane permeability to bulky permeating molecules, the strong binding between the aroma molecules and the silicalite can also decrease the aroma flux. Pereira et al. (2005) investigated the separation of ethyl butanoate from dilute aqueous solutions (100–900 ppm) by pervaporation using PDMS 1060 and 1070 membranes, and observed that the ethyl butanoate flux was 30–40% less with the PDMS 1070 membrane as compared to the silicalite-free PDMS 1060 membrane.

Additional efforts have been made to improve the pervaporation performance of PDMS membranes. One approach is to cross-link PDMS with other polymers to improve the membrane stability. Tanaka et al. (2010) cross-linked PDMS with poly(methyl hydrogen siloxane) (PMHS) in the pores of a tubular alumina substrate with an average pore size of 5 nm, followed by thermal treatment. Figure 7.9 illustrates the cross-linking reaction between PDMS and PMHS, where the Si–H bonds of the PMHS react with the hydroxyl groups of PDMS, forming a cross-linked structure while liberating hydrogen. An advantage with the pore-filled membranes is that the polymers are confined in the pores of the substrate, thereby making the membrane more stable. Compared to a PDMS/PMHS cross-linked membrane formed on a silicon substrate, the pore-filled PDMS/PMHS cross-linked membrane showed a reduced swelling degree (by 40%) and an enhanced selectivity. At a feed concentration of 5 wt% ethyl acetate, a separation factor of 140 and an overall permeation flux of 1.5 kg/m^2 h were obtained for pervaporative separation of ethyl acetate from water at 40 °C. It may be noted that an earlier study with a PDMS/PMHS cross-linked membrane supported on an asymmetric polyetherimide substrate showed much lower performance (separation factor 24, total flux 0.26 kg/m^2 h) for ethyl acetate separation from water under the same operating conditions (Osorio-Galindo et al., 2001). One may wonder whether the good performance obtained with Tanaka's PDMS/PMHS cross-linked membrane was indeed attributed to the pore-filling of the hydrophobic siloxane polymer into the substrate pores.

7.3.2.2 Poly(octhylmethyl siloxane) (POMS)

Poly(octhylmethyl siloxane) (POMS) is another promising organosiloxane polymer for pervaporative extraction of dairy aromas. Compared to PDMS membranes, POMS membranes usually exhibit a better selectivity, with a similar or a lower aroma permeability. The studies on dairy aroma recovery by pervaporation using POMS membranes are summarized in Table 7.3. Sampranpiboon et al. (2000a) studied the separation of ethyl

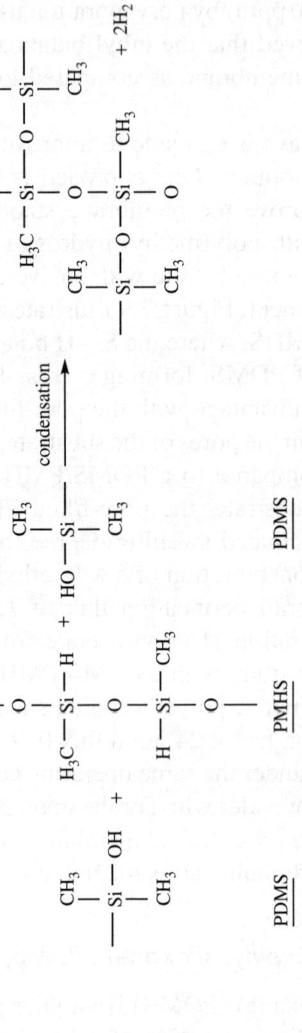

Figure 7.9 Possible mechanism of cross-linking reaction between PDMS and PMHS (Tanaka et al., 2010).

Table 7.3 Pervaporative extraction of dairy aromas by POMS-based membranes.

Membrane	Membrane geometry	Effective thickness (μm)	Feed conc. (ppm)	Temperature (°C)	Permeate pressure (Pa)	Feed flow rate (L/min)	Aroma Flux (g/m² h)	Separation factor	Enrichment factor	Reference
Ethyl acetate										
POMS	Single channel	10	1–50	–	–	–	0.01–0.7	–	–	Schafer, 2004
Ethyl butanoate										
POMS*	Flat sheet	5–6	100	20–40	2000	–	0.42–0.88	–	27 at 40 °C	Overington et al., 2008
POMS	Flat sheet	10	300	30–50	400–2000	0.49–0.65	0.011–0.021	118–281	–	Sampranpiboon et al., 2000a
POMS–PVDF–PP	Flat sheet	10	148	23	2000	0.285	–	–	46	She and Hwang, 2006b
Ethyl hexanoate										
POMS	Flat sheet	10	300	30–50	400–2000	0.49–0.65	0.013–0.043	244–307	–	Sampranpiboon et al., 2000a
POMS *	Flat sheet	5–6	100	20–40	2000	–	0.17–0.2	–	7.5 at 40 °C	Overington et al., 2008
Ethyl octanoate										
POMS *	Flat sheet	5–6	100	20–40	2000	–	0.004–0.004	–	1.5 at 40 °C	Overington et al., 2008
3–Heptanone										
POMS *	Flat sheet	5–6	100	20–40	2000	–	0.031–0.070	–	22.5 at 40 °C	Overington et al., 2008
3–Nonanone										
POMS*	Flat sheet	5–6	100	20–40	2000	–	0.0092–0.013	–	4.5 at 40 °C	Overington et al., 2008
1–Octen–3–ol										
POMS–PEI 03/011	Flat sheet	–	0.1–10	26	400	0.42–1.53	0.001–0.002	32–42	–	Martinez, Sanz, and Beltran, 2011

(continued overleaf)

Table 7.3 *(continued)*

Membrane	Membrane geometry	Effective thickness (μm)	Feed conc. (ppm)	Temperature (°C)	Permeate pressure (Pa)	Feed flow rate (L/min)	Aroma Flux (g/m^2 h)	Separation factor	Enrichment factor	Reference
Phenylacetaldehyde										
POMS–PEI	Flat sheet	5	180–12000	30	667	–	–	5–30	–	Kanani et al., 2003
Acetic acid										
POMS *	Flat sheet	5–6	100	20–40	2000	–	0.0062–0.024	–	0.7 at 40 °C	Overington et al., 2008
Butanoic acid										
POMS *	Flat sheet	5–6	100	20–40	2000	–	0.012–0.067	–	2 at 40 °C	Overington et al., 2008
Hexanoic acid										
POMS*	Flat sheet	5–6	100	20–40	2000	–	0.0084–0.047	–	1.5 at 40 °C	Overington et al., 2008
Octanoic acid										
POMS*	Flat sheet	5–6	100	20–40	2000	–	0.0036–0.0093	–	0.3 at 40 °C	Overington et al., 2008
3–Methylbutanal										
POMS–PEI	Flat sheet	5	180–12000	30	667	–	–	50–120	–	
POMS–PPSU–PP	Flat sheet	0.16	100	30	133	5	0.24	9	–	Kanani et al., 2003
POMS–PVDF–PP	Flat sheet	3.3	100	30	133	5	0.4	35	–	Raisi and Aroujalian, 2011
POMS–PVDF–PP	Flat sheet	20*	100	30	133	5	0.3	45	–	Raisi and Aroujalian, 2011

* Multicomponents feed solution.

butanoate and ethyl hexanoate from aqueous solutions using PDMS and POMS membranes of the same thicknesses (10 μm). The POMS membrane showed better permselectivity to the aroma compounds than PDMS. However, their aroma fluxes appear to be similar. An aroma enrichment of 118–281 was obtained with the POMS membrane as compared to an enrichment of 77–234 with the PDMS membrane. In addition, both membranes showed better selectivity for separation of ethyl hexanoate from water than ethyl butanoate/water separation. Raisi and Aroujalian (2011), who studied the separation of 3-methylbutanal from water using POMS and PDMS membranes, showed that POMS membrane is more selective to the permeation of 3-methylbutanal and a few other fruit aromas than the PDMS membrane. They also reported that the coupling effects between aroma compounds are not negligible when multiple aroma compounds are present in the feed solution. In addition, the liquid boundary layer effect was shown to be important as well in their pervaporation experiments under the operating conditions tested, and thicker membranes reportedly yielded a higher selectivity. This is not unexpected because a thicker membrane provides a greater resistance to mass transfer, and thus the overall mass transfer will be controlled more significantly by the membrane. As a result, the liquid boundary-layer effect becomes less significant to the separation performance.

Interestingly, Kanani et al (2003), who also investigated pervaporation separation of 3-methylbutanal from water using PDMS and POMS membranes, reported an opposite order in the selectivities of the two membranes for enrichment of 3-methylbutanal. Under the same operating conditions, 3-methylbutanal was enriched in the permeate more significantly with the PDMS membrane than the POMS membrane, and PDMS was reported to give better separation than POMS for the aldehyde. This, however, cannot be attributed fully to a higher selectivity of PDMS membrane because the PDMS membrane (thickness 160 μm) used was much thicker than the POMS membrane (5 μm). The observed separation performance is determined by the membrane and the boundary-layer effect. The mass transfer resistance of the membrane is directly related to the membrane thickness. If the liquid boundary layer effect is not negligible, the use of thicker membranes will give a better separation, although the permeation flux will be compromised. For comparison of membrane performance, Kanani et al. (2003) normalized the total and partial (aroma) fluxes to a 5-μm thickness for the PDMS membrane. This treatment is valid when the liquid boundary-layer effect on the membrane surface is insignificant. Otherwise, the true membrane permeation behavior will be masked by the liquid phase mass transfer resistance. The boundary-layer effect should be minimized in practice in order for the membranes to work at their full potential.

7.3.2.3 Poly(ether block amide) (PEBA)

Poly(ether block amide) (PEBA) is a group of block copolymers comprising of soft polyether segments and rigid polyamide segments developed by Atofina (now Arkema). A broad range of physical properties can be acquired by changing the nature or lengths of the rigid and soft blocks (Flesher Jr, 1986). PEBA represents a group of nonsilicone organophilic membranes for pervaporation. The work on PEBA membranes for dairy aroma recovery by pervaporation is listed in Table 7.4. Djebbar et al. (1998) prepared and evaluated the performance of a series of PEBA membranes (thickness ca. 100 μm) with different contents of ether segments. Their applicability for pervaporation was tested with three ester aromas, including ethyl acetate, ethyl propionate, and ethyl butanoate. Both the aroma permeation flux and enrichment factor were found to increase with an increase in the content of polyether segments in the copolymer. However, among the PEBA membranes tested, the one with the best performance showed a similar enrichment factor as the PDMS membrane (thickness ca. 500 μm) when the feed was a saturated aqueous solution of ethyl acetate (8.4 wt%), propionate (2 wt%), or butyrate (0.6 wt%) at 30 °C. The permeability of the aromas in the PEBA membranes is lower than in the PDMS membrane, which is understandable in view of the generally high diffusivities in the PDMS matrix due to the extraordinary flexibility of polymer chains arising from their siloxane linkages.

PEBA membranes appear to be advantageous over PDMS for the recovery of high-boiling hydrophobic aroma compounds. An enrichment factor up to 1205 has been reported (Baudot and Marin, 1996) for extracting methyl thiobutanoate (normal boiling point 142 °C) from water using a PEBA 40 membrane (thickness 70 μm). Compared to a silicalite-filled PDMS 1070 membrane (thickness 30 μm), the partial flux of methyl thiobutanoate through PEBA 40 was two- to four-fold higher, while the water flux was still lower. At 30 °C and 15 ppm of methyl thiobutanoate in the feed, the permeability coefficient of the aroma in PEBA 40 is 4 times that in PDMS 1070, whereas the water permeability coefficient in PEBA 40 is only 40% of the water permeability in PDMS 1070. This clearly demonstrates the excellent pervaporation performance of the PEBA membranes. In a subsequent study, Baudot, Souchon, and Marin (1999) also compared the performance of the PEBA 40 membrane with the silicalite-free PDMS 1060 membrane. While both membranes had similar permeabilities to the thioester, the PEBA 40 membrane was much less permeable to water.

For rather hydrophilic and low-boiling aromas such as diacetyl (normal boiling point 88 °C), both PEBA 40 and PDMS 1070 membranes showed similar selectivities (Baudot and Marin, 1996). With the PEBA 40 membrane, the sorption is more dominating than diffusion as far as the mass transport is concerned. Consequently, this membrane favors

Table 7.4 Pervaporative extraction of dairy aromas by PEBA-based membranes

Membranes*	Membrane geometry	Effective thickness (μm)	Feed conc. (ppm)	Temperature (°C)	Permeate pressure (Pa)	Aroma Flux (g/m^2 h)	Separation factor	Enrichment factor	Reference
Ethyl acetate									
PEBA (PE/PA 57.8/42.2)	Flat sheet	100	Saturated	30	<100	301	–	10	Djebbar et al., 1998
PEBA (PE/PA 40/60)	Flat sheet	100	Saturated	30	<100	40	–	8.2	Djebbar et al., 1998
Ethyl butanoate									
PEBA (PE/PA 89.4/10.6)	Flat sheet	100	Saturated	30	<100	800	–	137	Djebbar et al., 1998
PEBA (PE/PA 84.5/15.5)	Flat sheet	100	Saturated	30	<100	660	–	142	Djebbar et al., 1998
PEBA (PE/PA 73.2/26.8)	Flat sheet	100	Saturated	30	<100	282	–	132	Djebbar et al., 1998
PEBA (PE/PA 57.8/42.2)	Flat sheet	100	Saturated	30	<100	86	–	128	Djebbar et al., 1998
PEBA (PE/PA 40/60)	Flat sheet	100	Saturated	30	<100	10	–	91	Djebbar et al., 1998
PEBA 3533	Flat sheet	100	100–900	30–60	667	1.5–25	60–175	–	Sampranpiboon et al., 2000b
PEBA(PE/PA 57/43)	Flat sheet	–	200–900	25	–	–	50–130	–	Mohammadi, Kikhavandi, and Moghbeli, 2008
PEBA(PE/PA 57/43)	Flat sheet	–	200–900	25	–	–	70–125	–	Mohammadi, Kikhavandi, and Moghbeli, 2008

(*continued overleaf*)

Table 7.4 *(continued)*

Membranes	Membrane geometry	Effective thickness (μm)	Feed conc. (ppm)	Temperature (°C)	Permeate pressure (Pa)	Aroma Flux (g/m^2 h)	Separation factor	Enrichment factor	Reference
Methylthiobutanoate									
PEBA 40	Flat sheet	70	15	30	250–2000	0.55–0.2	700 – 1200	–	Baudot, Souchon, and Marin, 1999
PEBA 40	Flat sheet	70	15	30	190–2000	0.3–0.14	700 – 1205	–	Baudot and Marin, 1996
Ethyl propionate									
PEBA (PE/PA 89.4/10.6)	Flat sheet	100	Saturated	30	<100	1180	–	40.5	Djebbar et al., 1998
PEBA (PE/PA 84.5/15.5)	Flat sheet	100	Saturated	30	<100	1000	–	42.1	Djebbar et al., 1998
PEBA (PE/PA 73.2/26.8)	Flat sheet	100	Saturated	30	<100	295	–	39	Djebbar et al., 1998
PEBA (PE/PA 57.8/42.2)	Flat sheet	100	Saturated	30	<100	88	–	35	Djebbar et al., 1998
PEBA (PE/PA 40/60)	Flat sheet	100	Saturated	30	<100	20	–	33.5	Djebbar et al., 1998
Diacetyl									
PEBA 40	Flat sheet	30	80	30, 50	220–5000	0.019–0.087	17, 18	–	Baudot and Marin, 1996
Vanillin									
PEBA 25	Flat sheet	100	5,000	15–80	–	2–40	–	–	Boddeker, 1997
PEBA 40	Flat sheet	25–140	0–7000	15–80	–	0–25	–	–	Boddeker, 1997

* The numbers refer to the contents of polyether (PE) and polyamide (PA) segments or the codes of the polymer grades.

permeation of nonpolar and bulky permeant (e.g., thioesters), but it is not very competitive to selective permeation of such smaller and more polar molecules as diacetyl.

Mujiburohman and Feng (2007) prepared a PEBA membrane using Pebax 2533 for pervaporative separation of propyl propionate from water. This copolymer has a high content of polyether segments (86 wt% of polytetramethyleneoxide), with polyamide-12 being the hard segments. It was found that the permselectivity of the membrane was mainly derived from its excellent sorption selectivity. The diffusivity of propyl propionate through the membrane from its dilute aqueous solutions was affected exponentially by the aroma concentration in the fee solution.

7.3.2.4 Ethylene propylene diene monomer (EPDM) and ethylene vinyl acetate copolymer (EVAC)

Ethylene propylene diene monomer (EPDM) and ethylene vinyl acetate copolymer (EVAC) are two elastomer materials that have also been exploited for extraction of dairy aromas by pervaporation. The performance of EPDM and EVAC membranes for aroma separation are summarized in Tables 7.5 and 7.6, respectively. EPDM is a terpolymer of ethylene and propylene with ethylidene norbornene as a diene comonomer inserted in the chains. It has a strong resistance to ozone and other chemicals due to the saturated backbones (Huang, Moon, and Pal, 2002). Several EPDM membranes with different propylene contents have been investigated to separate ethyl butanoate from water (Huang, Moon, and Pal, 2002). An increase in the propylene content in the EPDM membrane resulted in an improved permselectivity to ethyl butanoate, but the membrane permeability was lowered due to the greater rigidity of propylene that hinders mass transport in the membrane.

Huang, Moon, and Pal (2002) fabricated thin-film composite membranes with a top layer of EPDM of different propylene contents supported on microporous substrates, and the resistance-in-series model was used to analyze the transport phenomenon. This study showed that the liquid boundary-layer resistance was as important as the membrane resistance itself. When different substrates were used, it was the porosity of the substrates, rather than other properties, that affected the separation performance of the composite membranes.

Another commercially available elastomer with desirable properties for pervaporative aroma recovery is ethylene vinyl acetate copolymer (EVAC). It consists of nonpolar crystalline ethylene segments and polar noncrystalline vinyl acetate segments. The vinyl acetate content affects the properties of the copolymer, including the degree of crystallization, glass transition temperature, mechanical strength, and thermal stability.

Table 7.5 Pervaporative extraction of dairy aromas by EPDM-based membranes.

Membranes	Membrane geometry	Effective thickness (μm)	Feed conc. (ppm)	Temperature (°C)	Permeate pressure (Pa)	Feed flow rate (L/min)	Aroma Flux (g/m² h)	Separation factor	Enrichment factor	Reference
Ethyl acetate										
EPDM[a]	Hollow fiber	–	600	25	<400	0.5	–	–	125.3	Pereira et al., 2005
EPDM	Flat sheet	10	100–900	25	<400	1.53	–	174	–	Pereira et al., 2005
Ethyl butanoate										
EPDM	Flat sheet	10	100–900	25	<400	1.53	3–15	2271	–	Pereira et al., 2005
V86 EPDM–PAN	Flat sheet	76	100–400	30	400	–	0.2–0.65	25–50	–	Huang, Moon, and Pal, 2002
V86 EPDM–PMMA	Flat sheet	76	100–400	30	400	–	0.18–0.35	55–105	–	Huang, Moon, and Pal, 2002
V86 EPDM–PSf	Flat sheet	76	100–400	30	400	–	0.2–0.65	70–80	–	Huang, Moon, and Pal, 2002
V86 EPDM–PVDF	Flat sheet	76	100–400	30	400	–	0.2–0.65	40–55	–	Huang, Moon, and Pal, 2002
K5 EPDM–PVDF	Flat sheet	32	100–400	30	400	–	0.4–3.5	–	–	Huang, Moon, and Pal, 2002
K5 EPDM–PVDF	Flat sheet	81	100–400	30	400	–	0.3–1.3	–	–	Huang, Moon, and Pal, 2002
K5 EPDM–PVDF	Flat sheet	76	200	15–35	400	–	0.4–0.8	–	–	Huang, Moon, and Pal, 2002
K5 EPDM–PVDF	Flat sheet	96	100–400	30	400	–	0.2–0.7	–	–	Huang, Moon, and Pal, 2002
V86 EPDM–PVDF	Flat sheet	76	100–600	30	400	0.2–1.6	0.4–3.2	50–210	–	Huang, Moon, and Pal, 2002
V88 EPDM–PVDF	Flat sheet	76	200–600	30	400	0.2–1.6	0.6–3.4	25–145	–	Huang, Moon, and Pal, 2002
V86 EPDM–PVDF	Flat sheet	76	200	15–35	400	–	0.5–1.5	–	–	Huang, Moon, and Pal, 2002
Ethyl hexanoate										
EPDM[a]	Hollow fiber	–	100	25	<400	0.5	–	–	1161	Pereira et al., 2005
EPDM	Flat sheet	10	100–900	25	<400	1.53	–	8171	–	Pereira et al., 2005
EPDM	Flat sheet	–	–	–	–	–	–	8171	–	Pereira et al., 2002
1–Octen–3–ol										
EPDM	Flat sheet	10	100–900	25	<400	1.53	–	957	–	Pereira et al., 2005

Table 7.6 Pervaporative extraction of dairy aromas by EVA-based membranes

Membranes	Membrane geometry	Effective thickness (μm)	Feed conc. (ppm)	Temperature (°C)	Permeate pressure (Pa)	Feed flow rate (L/min)	Aroma Flux (g/m^2 h)	Separation factor	Enrichment factor	Reference
Ethyl acetate										
EVA	Flat sheet	7	100–900	25	<400	1.53	–	31	–	Pereira et al., 2005
EVA38–PAN	Flat sheet	20	10000–25000	30–50	180	Re=10^5	31-1235	40–136	–	Bai et al., 2007
Ethyl butanoate										
EVA	Flat sheet	7	100–900	25	<400	1.53	2–10	655	–	Pereira et al., 2005
Ethyl hexanoate										
EVA	Flat sheet	7	100–900	25	<400	1.53	–	4443	–	Pereira et al., 2005
EVA	Flat sheet	–	–	–	–	–	–	4443	–	Bai et al., 2007
1–Octen–3–ol										
EVA	Flat sheet	7	100–900	25	<400	1.53	–	636	–	Pereira et al., 2005

Bai et al. (2007) studied the separation performance of EVAC membranes with different vinyl acetate contents to recover ethyl acetate from aqueous solutions. The separation factor was shown to decrease continuously when the content of vinyl acetate in the membrane increased from 26 wt% to 100 wt% (i.e., poly(vinyl acetate)). However, the aroma flux showed a different trend; increasing the content of vinyl acetate in the membrane increased the permeation flux of ethyl acetate to 38% of vinyl acetate in the membrane, and then the aroma flux began to decrease when the membrane contained more vinyl acetate. These trends were reportedly due to an increase in the hydrophilicity and swelling degree of the membrane. A maximum aroma flux of 550 g/m^2 h was obtained for a feed of 2.5 wt% ethyl acetate solution at 30 °C, and the corresponding separation factor was 118. At 50 °C, the aroma flux and the separation factor were increased to 1250 g/m^2h and 135 g/m^2h, respectively. Pereira et al. also prepared a composite membrane using EVAC supplied by Poliolefinas SA with an undisclosed content of vinyl acetate for ethyl acetate separation and the membrane showed much lower selectivity.

To further investigate EPDM and EVAC membranes for aroma separation by pervaporation, Pereira et al. (2005) carried out a comparative pervaporation study using PDMS membranes (PDMS 1060, PDMS 1070) as a baseline. Table 7.7 summarizes the performance of the membranes for pervaporative separation of ethyl acetate, ethyl butanoate, ethyl hexanoate, and octenol. The EPDM membranes showed the best performance among the four membranes tested. Based on a process simulation, it was concluded that more permeable membranes (e.g., PDMS 1060) would be more advantageous if the feed concentration was high enough to cause phase separation in the permeate after its condensation.

7.3.2.5 Other materials

The investigation of membranes for pervaporative recovery of dairy aromas has been extended to other materials, primarily polymeric materials based on poly(vinylidene fluoride) (PVDF) and poly[1-(trimethylsilyl)-1-propyne] (PTMSP), as shown in Table 7.8. Zeolite-based inorganic membranes and zeolite-filled polymeric membranes have also been used for separation of acid and aldehyde aromas. Bai et al. (2008) used a cross-linked hydroxyl-terminated polybutadiene-based polyurethaneurea (HTPB-DVB-PU) membrane for pervaporative recovery of ethyl acetate from water. The thermal resistance of the membrane was significantly enhanced by the introduction of the cross-linker. The separation factor and the total flux increased with an increase in the divinyl benzene content, and at a feed concentration of 2.5 wt% ethyl acetate and at 30 °C, the membrane showed a separation factor of 655 and a total flux of 256 g/m^2 h. In addition,

Table 7.7 A comparison of membrane performance for diary aroma extraction by pervaporation

Membrane	Effective thickness (μm)	Feed conc. (ppm)	Permeate pressure (Pa)	Aroma flux (g/m^2 h)	Separation factor
Ethyl acetate					
PDMS 1060	10	100–900	<293	–	141
PDMS 1070	30	100–900	<293	–	393
EPDM	10	100–900	<400	–	174
EVA	7	100–900	<400	–	31
Ethyl butanoate					
PDMS 1060	10	100–900	<293	10–33	510
PDMS 1070	30	100–900	<293	5–22	875
EPDM	10	100–900	<400	3–15	2271
EVA	7	100–900	<400	2–10	655
Ethyl hexanoate					
PDMS 1060	10	100–900	<293	–	599
PDMS 1070	30	100–900	<293	–	750
EPDM	10	100–900	<400	–	8171
EVA	7	100–900	<400	–	4443
1-Octen-3-ol				–	
PDMS 1060	10	100–900	<293	–	154
PDMS 1070	30	100–900	<293	–	252
EPDM	10	100–900	<400	–	957
EVA	7	100–900	<400	–	636

Based on data reported in Pereira et al. (2005). All membranes are flat sheets. Operating temperature 25 °C; feed flow rate 1.53 L/min.

fluoropolymers, mainly poly(vinylidene fluoride) and its copolymers, have also been used for fabricating membranes to extract ethyl acetate by pervaporation (Zhu, Tian, and Xu, 2005; Tian and Jiang, 2008).

Most pervaporation membranes for aroma extraction are organophilic. Surprisingly, Yongquan et al. (2012) prepared a hybrid membrane by incorporating a hydrophilic ionic liquid into poly(vinylidene-fluoride-co-hexafluoropropene) for separation of ethyl acetate from water. The ionic liquid used, 3-butylimidazolium tetrafluoroborate ([bmim]BF_4), is hydrophilic and miscible with water in any proportion (Huddleston et al., 2001; Sheldon and van Rantwijk, 2008). As expected, the sorption uptake of the oleophilic ethyl acetate in the membrane decreased significantly as the [bmim]BF_4 content in the membrane increased (Yongquan et al., 2012). However, it is intriguing that incorporating the hydrophilic ionic liquid in the organophilic polymer matrix was reported to have enhanced the membrane permselectivity to ethyl acetate. It was hypothesized that water molecules in the feed

Table 7.8 Pervaporative extraction of dairy aromas by other membranes

Membranes	Membrane geometry	Effective thickness (μm)	Feed conc. (ppm)	Temperature (°C)	Permeate pressure (Pa)	Feed flow rate (L/min)	Aroma Flux (g/m^2 h)	Separation factor	Enrichment factor	Reference
Ethyl acetate										
PTMSP	Flat sheet	–	84–137	–	–	$Re=2\times10^4$	50	133	1.5	Slater, 1997
HTPB–DVB–PU	Flat sheet	100	10000–25000	30	180	$Re=10^5$	25–250	370–650	–	Bai et al., 2008
P(VDF–HFP)	Flat sheet	20	10000–50000	30–50	400–533	–	100–2000	70–180	–	Zhu, Tian, and Xu, 2005
P(VDF–HFP)	Flat sheet	10	15000–30000	30–50	300–400	–	200–1800	45–180	–	Tian and Jiang, 2008
P(VDF–HFP)/[bmim]BF_4	Flat sheet	10	10000–50000	25–65	200	–	–	60–180	–	Yongquan et al., 2012
Ge–ZSM–5 zeolite	Tubular	30	50000	30	<250	1.2	130	90	–	Bowen et al., 2003
Surface–modified alumina	Tubular	–	1500–6000	30–50	1333	–	32–380	56–95	49–63	Song and Lee, 2005; Song, Lee, and Rim, 2004
Ethyl butanoate										
PHDFNMA–filled PTMSP	Flat sheet	–	–	–	–	–	0.01–4.8	–	40–170	Mishima and Nakagawa, 2002
Surface–modified alumina	Tubular	–	1500–6000	30–50	1333	–	55–512	89–131	60–102	Song, Lee and Rim, 2004
Propyl acetate										
Surface–modified alumina	Tubular	–	1500–6000	40	1333	–	75–290	85–90	–	Song, Lee, and Rim, 2004

Butyl acetate										
Surface–modified alumina	Tubular	–	1500–6000	40	1333	–	104–318	118–119	98–102	Song, Lee, and Rim, 2004
Methyl acetate										
Ge–ZSM–5 zeolite	Tubular	30	50000	30	<250	20	280	130	–	Bowen et al., 2003
Ethyl propionate										
Surface–modified alumina	Tubular	–	1500–6000	30–50	1333	–	547–482	75–114	55–93	Song, Lee, and Rim, 2004
Diacetyl										
Oleyl alcohol liquid membrane	Hollow fiber	22	50–450	30	1333	0.015	0.032	40	–	Ishii et al., 1995
Acetic acid										
Ge–ZSM–5 zeolite	Tubular	30	50000	30	<250	20	4.3	7	–	Bowen et al., 2003
Propionic acid										
Ge–ZSM–5 zeolite	Tubular	30	50000	30	<250	20	9.4	30	–	Bowen et al., 2003
Acetaldehyde										
PDMS filled with ZSM–5	Flat sheet		30000–130000	15–45		0.83	50–350	6–65	–	Wu et al., 2012
PU filled with 3A zeolite	Flat sheet		20000–200000	25	<1000	–	–	1.1–7.4	–	Wu et al., 2011

$[bmim]BF_4$: 1-butyl-3-methyl-imidazolium-tetrafluoroborate, DVB: divinyl benzene, HFP: hexafluoropropene, HTPB: hydroxyl-terminated polybutadiene, PHDFNMA: poly(1*H*,1*H*,9*H*-hexadecafluorononyl methacrylate), PTMSP: poly[1-(trimethylsilyl)-1-propyne], PU: poly(urethane).

would form a hydrate with [bmim]BF_4 loaded in the membrane, which would slow down water diffusion while enhancing the diffusion of ethyl acetate in the membrane. This is, however, yet to be verified experimentally. It is very uncommon to use a hydrophilic "additive" in an organophilic membrane to enhance selective permeation of oleophilic permeant.

Bowen, Noble, and Falconer (2003) used Ge-ZSM-5 zeolite (germanium substituted, MFI structure) membranes to separate carboxylic acids (acetic acid, propionic acid), esters (methyl acetate, ethyl acetate), and a few other organic compounds from aqueous solutions. This membrane was reported to be more hydrophobic than silicalite-1 membranes prepared by similar procedures. It is well known that simultaneous incorporation of Ge and Al in an MFI framework (Ge–ZSM-5) can change its catalytic properties as compared to other frameworks incorporated with either Ge or Al alone. It is interesting to note that when a proper amount of Ge and Al is incorporated into a silicalite-1 framework, the resulting membrane can be superhydrophilic. Such membranes may be very effective for solvent dehydration when the feed contains a small content of water (Wang et al., 2014), but not particularly suitable for pervaporation extraction of aromas from dilute aqueous solutions.

Lee and co-workers (Song and Lee, 2005; Song, Lee, and Rim, 2004) prepared hydrophobic membranes by surface modification of a tubular alumina substrate (pore diameter of 0.1 mm) with a silane coupling agent (perfluoro-alkylsilane). The membranes were tested for pervaporative extraction of ethyl acetate, propyl acetate, and butyl acetate from dilute aqueous solutions. Compared to PDMS, the surface-modified form showed a much higher flux but a lower selectivity. At a feed ester concentration of 0.15–0.60 wt%, the enrichment factors were in the range of 49–63 for ethyl acetate and 98–102 for butyl acetate, respectively. The membrane selectivity was still high enough for the ester concentration in the permeate to exceed its solubility limits.

Immobilized liquid membranes have also been attempted for aroma recovery by pervaporation. Ishii et al. (1995) used an oleyl alcohol liquid membrane immobilized in the pores of microporous propylene hollow fibers to extract diacetyl from an aqueous fermentation solution. Oleyl alcohol has a low saturated vapor pressure (est. 0.0124 Pa at 25 °C). The separation factor for diacetyl/water separation was 40. Because pervaporation is operated at vacuum on the permeate side, the long-term stability of liquid membranes is a significant concern. Evaporative loss over the long run or breakthrough across the membrane substrate under transmembrane pressures is a potential technical problem.

7.3.3 Influence of feed composition on dairy aroma recovery pervaporation

In dilute feed solutions, the permeation flux of an aroma usually increases almost linearly with the aroma concentration in the feed, while water flux is not significantly affected by the feed concentration (Pereira et al., 2005; Isci, Sahin, and Sumnu, 2006; Song, Lee, and Rim, 2004; Jullok et al., 2013; Mishima and Nakagawa, 2000). The linear relationship can be explained using the solution-diffusion model (Kanani et al., 2003), where the diffusivity and solubility are approximately constant. Under such circumstances, the membrane selectivity measured by the permeability ratio will be independent of feed concentrations. However, if the membrane has a strong affinity to the aroma compounds (i.e., a high solubility coefficient), considerable membrane swelling by the penetrant may occur, which will increase the diffusivities of both water and aroma in the membrane. This normally happens when the aroma concentration in the feed is over a few thousands of ppm. Sometimes, the aroma diffusivity can be affected exponentially by its concentration in the feed, resulting in a nonlinear concentration dependence of the aroma flux (Gu et al., 2013; Mujiburohman and Feng, 2007). While membrane swelling tends to increase both the aroma and water fluxes (Feng and Huang, 1997; Peng and Liu, 2003), the selectivity can be increased (Wu et al., 2012) or lowered (Sampranpiboon et al., 2000b) at higher feed aroma concentrations. Generally speaking, excessive membrane swelling will lower the membrane selectivity.

The effect of feed concentration on the membrane selectivity is apparently related to the aroma compounds and to the membrane. For instance, Mohammadi, Kikhavandi, and Moghbeli (2008) found that with the same PEBA membrane (57 wt% poly(tetramethylene glycol) segment and 43 wt% polyamide 12 segments) the separation factor for ethyl butanoate enrichment increased with an increase in the ethyl butanoate concentration in the feed solution, whereas the opposite was true for the separation of isopropanol enrichment. Sampranpiboon et al. (2000b) also investigated ethyl butanoate enrichment and found that the separation factor for ethyl butanoae/water separation is roughly proportional to the feed aroma concentration when a PEBA 3533 membrane was used and the separation factor appeared to be inversely proportional to the feed concentration when PDMS and POMS membranes were used (Sampranpiboon et al., 2000a). In addition, when multiple aromas are present in the feed, the interactions among the aromas will also influence the permeation of individual aromas, which is reflected by the commonly observed coupling effects.

Compared to model feed solutions, the composition of real feed solutions in the dairy industry are much more complicated. Proteins, lactose, and fats are the major components of dairy products. They can interact with both volatile and nonvolatile aromas in the feed, and as such the research findings with model feed solutions may not be directly applicable to industrial applications (Swaisgood, 1996). Research on the interactions between aromas and the nonvolatile substances is mainly conducted from a sensory perspective: the aroma level is perceived to be reduced if less aroma is released due to binding to or associating with the nonvolatile components. Depending on the hydrophobicities of the aromas, a certain portion of them can be dissolved by dairy fats, which help prevent volatilization of volatile aromas (Leland, 1997; De Roos, 1997; Hatchwell, 1996). Milk proteins can bind numerous aroma compounds via reversible physical binding (e.g., hydrophobic interactions) or irreversible chemical binding (e.g., covalent coupling) (Fischer and Widder, 1997; Kuhn, Considine, and Singh, 2006; Kuhn et al., 2007; Guichard and Langourieux, 2000; Hansen and Booker, 1996). While carbohydrates can decrease the volatility of aroma compounds by intermolecular attractions, or increase their volatility by salting-out, lactose can generally bind aromas by hydrogen bonding due to its abundant hydroxyl groups, resulting in a reduction in the volatility of the aromas (Kellam, 1998; Godshall, 1997). Therefore, at a given aroma concentration in the feed solution, the presence of proteins, lactose, and fats will suppress the saturated vapor pressure of the aromas, thereby lowering the driving force for pervaporative transport of the aromas through the membrane. This will compromise the effectiveness of pervaporation to extract and concentrate the aromas present in the system. One should look into this aspect in extracting dairy aromas for practical applications.

Overington, Wong, and Harrison (2011) carried out a comprehensive study on the effects of nonvolatile dairy components on aroma recovery by pervaporation using a PDMS membrane. These nonvolatile components can influence both the driving force for permeation and the mass transfer coefficient. It is shown that a small amount of fat added to the feed solution will reduce the aroma permeation rate significantly. Even a small amount of fat is detrimental to the separation performance. For pervaporative extraction of esters and ketones, the milk proteins can reduce the driving forces for aroma permeation by 56–94%, and the impact seems to be more significant for long-chain and hydrophobic aroma compounds. Interestingly, they have little impact on the permeation flux of water. When both milk protein and fat are present in the feed, the separation performance for aroma extraction is affected by the protein more significantly. Similar to proteins, lactose does not have an evident negative effect on water permeation, but the aroma flux and thus the enrichment factor are lowered. These trends, however, cannot be generalized because the impacts of the nonvolatile substances on aroma extraction depend on the nature and magnitude of their interactions with the aromas as well as the type of membranes used. In contrast, lactose has been reported to have no effect on the permeation flux or selectivity for dairy aroma diacetyl using a 25-μm thick PDMS-polycarbonate copolymer membrane (Rajagopalan, Cheryan, and Matsuura, 1994) or methyl thiobutanoate using a 70-μm thick GKSS PEBA 40 membrane (Baudot, Marin, and Spinnler, 1996). In the latter case, even the addition of salts (NaCl, Na_2HPO_4, KH_2PO_4), sodium lactate, and amino acids to the feed solution close to a real culture medium does not change the flux and enrichment factor.

7.3.4 Influence of operating conditions on dairy aroma recovery pervaporation

At a given feed composition, other operating conditions (i.e., temperature, hydrodynamics of the feed liquid on the upstream side, and permeate pressure on the downstream side) can significantly affect the efficiency of pervaporation separation.

In general, a higher feed temperature favors both permeation fluxes of aroma and water. As discussed above, the increase in the permeation flux is attributed to an increased driving force for permeation arising from the higher vapor pressure of permeant in the feed, as well as enhanced permeability (though in certain cases the permeability may decrease with temperature if the increased diffusivity due to enhanced thermal motion of polymer chains and increasingly energized penetrant is insufficient to compensate for the decrease in the solubility due to exothermic sorption). Olsson and Tragardh (1999b), who studied pervaporation of several esters using a POMS membrane at different temperatures, found that the increase in aroma flux with an increase in temperature is mainly due to the increased membrane permeability, while the increase in water flux is attributed to its increased driving force. Similar results are observed for separation of acids, esters, and ketones from aqueous solutions with PDMS and POMS membranes (Baudot, Marin, and Spinnler, 1996; Overington et al., 2009).

The overall temperature dependence of permeation flux can be measured by the apparent activation energy (Overington et al., 2009; Beaumelle, Marin, and Gibert, 1992a), which is normally in the range of 20–60 kJ/mol (Feng and Huang, 1996). Obviously, the effect of temperature on the selectivity of pervaporation separation depends on the difference in the apparent activation energies of the permeating species (i.e., aroma and water) to be separated. If the activation energy of an aroma compound is higher than that of water, then the aroma selectivity will be increased with an increase in temperature (Sampranpiboon et al., 2000b; Mujiburohman and Feng, 2007; Lipnizki, Olsson, and Tragardh, 2002). For aroma compounds (e.g., ethyl buanoate (Sampranpiboon et al., 2000a; Mohammadi, Kikhavandi, and Moghbeli, 2008), ethyl hexanoate (Sampranpiboon et al., 2000a), and ethyl acetate (Feng and Huang, 1996)) with a lower activation energy than water, their selectivity will be lower at a higher operating temperature. In spite that a high temperature favors both permeation flux and selectivity, the operating temperature should still be limited so as to prevent thermal degradation of the aroma compounds.

The hydrodynamic conditions of the feed solution are another important aspect in pervaporation. Since aroma compounds permeate through the membrane preferentially, water molecules will be built up in the liquid boundary layer near the membrane surface (Feng and Huang, 1997; She and Hwang, 2004; Cussler, 1997), which is referred to as a "concentration polarization." This is shown schematically in Figure 7.10. Obviously, because of the concentration polarization, the concentration of aroma compounds that the membrane "sees" on the feed side is lower than the aroma concentration in the bulk liquid, whereas the opposite is true for water. The liquid boundary layer thus presents an additional resistance to the mass transfer of aroma compounds, while it does not affect water permeation significantly for dilute solutions. For pervaporation separation of minor components from water, the boundary-layer mass transfer may become dominant over the membrane itself if the hydrodynamic conditions of the feed solution are not properly controlled (Jiang, Vane, and Sikdar, 1997; Lipnizki and Tragardh, 2001).

The extent of concentration polarization is determined by both the permselectivity of the membrane and the hydrodynamic conditions of the feed on the upstream side. Fluid management techniques to reduce the boundary-layer thickness by promoting good mixing of the feed solution near the membrane surface are important in order to minimize concentration polarization. It is also helpful to reduce temperature polarization in the boundary layer, though it is normally not a significant issue because of the relatively low permeation rate in pervaporation. Thus, it is often recommended to use a relatively high turbulence on the feed side. A study on aroma extraction using a PDMS membrane packed in a plate and frame module showed that a Reynolds number of 500 will be needed in order to make concentration polarization negligible (Beaumelle, Marin, and Gilbert, 1992b).

Generally, water flux is not strongly affected by the hydrodynamic conditions of the feed solution at low aroma concentrations because water concentration on the membrane surface is not considerably different from the water concentration in the bulk solution (She and Hwang, 2004).

As the driving force for pervaporative transport is provided by applying vacuum on the downstream side, the permeation flux is directly affected by the permeate pressure. As expected, the lower the permeate pressure, the higher the permeation flux (Olsson and Tragardh, 1999a; Aroujalian and

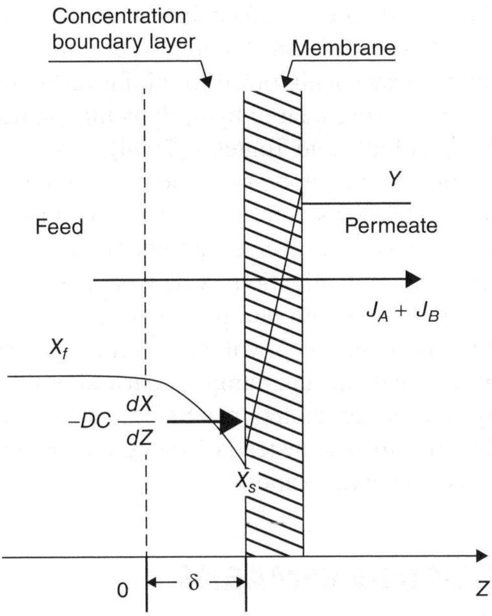

Figure 7.10 Schematic of boundary layer effect in pervaporation transport.

Raisi, 2007; Olsson, Tragardh, and Tragardh, 2001; Raisi, Aroujalian, and Kaghazchi, 2008; Fouda et al., 1993). For aroma extraction, an increase in the permeate pressure will lower both the partial fluxes of aroma and water. Thus, the pervaporation selectivity is not strongly influenced by the permeate pressure as compared to the permeation flux, unless the aroma is a high boiler with a low saturated vapor pressure on the feed side. It has been shown that the effect of permeate pressure on pervaporative extraction of a high-boiling aroma γ-decalactone (boiling point at 281 °C) is much more pronounced than extraction of low-boiling aromas (e.g., ethyl acetate and diacetyl, boiling points at 77 °C and 88 °C, respectively) (Baudot, Souchon, and Marin, 1999). Aromas with moderate boiling points (e.g., methyl thiobutanoate) displayed an intermediate behavior; the separation factor may increase or decrease with an increase in the permeate pressure, depending on the membranes used (Baudot, Souchon, and Marin, 1999).

In industrial operations, the permeate pressure may range from several hundred to a few thousand pascals, which is a much lower vacuum than that normally used in the laboratory (Baudot, Souchon, and Marin, 1999; Pereira et al., 2006). Therefore, caution should be exercised in extrapolating laboratory bench-scale pervaporation results in process design. In addition, when hollow fiber membranes are used, there will be a pressure build-up along the fiber length on the permeate side (especially for shell side feed configurations), which should be taken into account in module scale-up (Feng and Huang, 1995). Attention should also be paid to the pumping and piping aspects for permeate withdrawal to minimize the negative effect of pressure build-up (e.g., using a large hydrodynamic diameter and minimal length) (Willemsen, Dijkink, and Togtema, 2004).

In aroma extraction by pervaporation, the primary energy consumption element is the vacuum pump used for permeate evacuation. Conventionally, the permeate vapor needs to be condensed under vacuum before reaching the pump inlet to avoid contamination. A new development in the vacuum pump industry is the dry vacuum pump, which runs completely dry, without using any lubricant in the swept volume (Collins, 2012). They are more efficient than conventional vacuum pumps and use significantly less energy. More importantly, the permeate vapor can be condensed and collected at the pump outlet under atmospheric instead of vacuum conditions. It appears to be appropriate to use in pervaporation.

7.4 CONCLUDING REMARKS

The current state of pervaporation for extraction of dairy aroma compounds is reviewed, with an emphasis on three aspects: the nature of the aroma compounds, selection of membrane materials, and the effects of the process parameters. Among the large number of dairy aroma compounds identified, only about 30 aromas have been investigated extensively for pervaporative extraction from aqueous solutions using organophilic membranes. Ester aromas can be extracted effectively because of their hydrophobicity. Among the various membrane types, PDMS is a versatile material with good selectivity that can be used standalone or incorporated with hydrophobic zeolites to enhance the membrane performance. With the advances in materials science, research efforts have been made to develop novel membranes with improved permselectivity, and the membrane research has now been extended from the PDMS to other materials.

Most of the research to date on pervaporative recovery of aroma compounds has been carried out with binary aroma–water systems. While it is convenient for membrane testing and evaluation in the lab, the research findings so obtained cannot be simply extrapolated to real systems that involve multicomponent aromas as coupling effects are likely to occur. In addition, the presence of dairy proteins, fats, and lactose can hinder pervaporative extraction of aroma compounds as well.

For a given membrane and feed aroma composition, the separation performance is also affected by the operating temperature, permeate pressure, and hydrodynamic conditions of the feed solution. This is why there is quite a bit of variation in the membrane performance data reported in the literature, and these data should be used with care. Nonetheless, pervaporation separation offers the advantages of mild processing temperatures, low energy consumption, modular equipment design, and easy process operation. Its potential for recovery of dairy aromas as natural flavors is apparent in view of the increasing demand for natural aroma substances.

Nomenclature

D_i = diffusivity coefficient of species i
D_0 = preexponential factor of diffusivity
E = activation energy
E_D = activation energy of diffusion
E_P = activation energy of permeation
ΔH = enthalpy change
ΔH_V = heat of evaporation of the permeant
i = species i
j = species j
J = permeation flux
J_0 = preexponential factor of flux
l = effective thickness of membrane
p^p = permeate pressure
p^{sat} = saturated vapor pressure
P_i = permeability coefficient of species i
P_0 = preexponential factor for permeability
R = universal gas constant
S_i = solubility coefficient species i
S_0 = preexponential factor for solubility
T = temperature
X = concentration of permeant (in mol fraction) in feed
Y = concentration of permeant (in mol fraction) in permeate

Greek letters

α = separation factor
α_{ij}^0 = membrane permselectivity
β = enrichment factor
γ = activity coefficient

References

Aroujalian, A. and Raisi, A. (2007) Recovery of volatile aroma components from orange juice by pervaporation. *Journal of Membrane Science*, **303**, 154–161.

Bai, Y., Qian, J., An, Q., Zhu, Z., and Zhang, P. (2007) Pervaporation characteristics of ethylene–vinyl acetate copolymer membranes with different composition for recovery of ethyl acetate from aqueous solution. *Journal of Membrane Science*, **305**, 152–159.

Bai, Y., Qian, J., Zhang, C., Zhang, L., An, Q., and Chen, H. (2008) Cross-linked HTPB-based polyurethaneurea membranes for recovery of ethyl acetate from aqueous solution by pervaporation. *Journal of Membrane Science*, **325**, 932–939.

Baudot, A. and Marin, M. (1996) Dairy aroma compounds recovery by pervaporation. *Journal of Membrane Science*, **120**, 207–220.

Baudot, A. and Marin, M. (1997) Pervaporation of aroma compounds: comparison of membrane performances with vapour–liquid equilibria and engineering aspects of process improvement. *Food and Bioproduct Process*, **75**, 117–142.

Baudot, A. and Marin, M. (1999) Improved recovery of an ester flavor compound by pervaporation coupled with a flash condensation. *Industry Engineering and Chemistry Research*, **38**, 4458–4469.

Baudot, A., Marin, M., and Spinnler, H.E. (1996) The recovery of sulfur aroma compounds of biological origin by pervaporation, in *Flavour Science. Recent Developments* (eds. A.J. Taylor and D.S. Mottram), Royal Society of Chemistry, Cambridge, UK, pp. 301–304.

Baudot, A., Souchon, I., and Marin, M. (1999) Total permeate pressure influence on the selectivity of the pervaporation of aroma compounds. *Journal of Membrane Science*, **158**, 167–185.

Beaumelle, D., Marin, M., and Gibert, H. (1992a) Pervaporation of aroma compounds in water–ethanol mixtures: experimental analysis of mass transfer. *Journal of Food Engineering*, **16**, 293–307.

Beaumelle, D., Marin, M., and Gilbert, H. (1992b) Plate and frame modification: improvement of pervaporation efficiency regarding aroma compound transfer, in *Sixth International Conference on Pervaporation Process in the Chemical Industry* (ed. R. Bakish), Bakish Material Corporation, Englewood, New Jersey, pp. 223–232.

Binning, R.C., Lee, R.J., Jennings, J.F., and Martin, E.C. (1961) Separation of liquid mixtures. *Industry and Engineering Chemistry*, **53**, 45–50.

Boddeker, K.W. (1997) Pervaporation at the vapor pressure limit: vanillin. *Journal of Membrane Science*, **137**, 155–158.

Borjesson, J., Karlsson, H.O.E., and Tragardh, G. (1996) Pervaporation of a model apple juice aroma solution: comparison of membrane performance. *Journal of Membrane Science*, **119**, 229–239.

Bowen, T.C., Li, S., Noble, R.D., and Falconer, J.L. (2003) Driving force for pervaporation through zeolite membranes. *Journal of Membrane Science*, **225**, 165–176.

Cheng, H. (2010) Volatile flavor compounds in yogurt: a review. *Critical Reviews in Food Science and Nutrition*, **50**, 938–950.

Collins, D. (2012) Choosing process vacuum pumps, http://research.che.tamu.edu/groups/Seminario/numerical-topics/vacuum%20Pumps.pdf.

Cussler, E.L. (1997) *Diffusion Mass Transfer in Fluid Systems*, second edition, Cambridge University Press, Cambridge.

De Roos, K.B. (1997) How lipids influence food flavor. *Food Technology*, **51**, 60–62.

Djebbar, M.K., Nguyen, Q., Clement, R., and Germain, Y. (1998) Pervaporation of aqueous ester solutions through hydrophobic poly(ether-block-amide) copolymer membranes. *Journal of Membrane Science*, **146**, 125–133.

Dotremont, C., Brabants, B., Geeroms, K., Mewis, J., and Vandecasteele, C. (1995) Sorption and diffusion of chlorinated hydrocarbons in silicatlite-filled PDMS membranes. *Journal of Membrane Science*, **104**, 109–117.

Farber, L. (1935) Applications of pervaporation. *Science*, **82**, 158.

Feng, X. and Huang, R.Y.M. (1995) Permeate pressure build-up in shellside-fed hollow fiber pervaporation membranes. *The Canadian Journal of Chemical Engineering*, **73**, 833–843.

Feng, X. and Huang, R.Y.M. (1996) Estimation of activation energy for permeation in pervaporation processes. *Journal of Membrane Science*, **118**, 127–131.

Feng, X. and Huang, R.Y.M. (1997) Liquid separation by membrane pervaporation: a review. *Industry Engineering and Chemistry Research*, **36**, 1048–1066.

Fischer, N. and Widder, S. (1997) How proteins influence food flavor. *Food Technology*, **51**, 68–70.

Fleming, H.L. (1990) Membrane pervaporation: separation of organic/aqueous mixtures. *Separation Science and Technology*, **25**, 1239–1255.

Flesher Jr,, J.R. (1986) Pebax polyether block amide – A new family of engineering thermoplastic elastomers, in High Performance Polymers: Their Origin and Development (eds. R.B. Seymour and G.S. Kirshenbaum), Elsevier, pp. 401–408.

Fouda, A., Bai, J., Zhang, S.Q., Kutowy, O., and Matsuura, T. (1993) Membrane separation of low volatile organic compounds by pervaporation and vapor permeation. *Desalination*, **90**, 209–233.

Godshall, M.A. (1997) How carbohydrates influence food flavor. *Food Technology*, **51**, 63–67.

Graham, T. (1866) On the absorption and dialytic separation of gases by colloid septa. Part I. Action of a septum of caoutchouc. *Philosophical Transactions of the Royal Society*, **156**, 399–439.

Gu, J., Zhang, X., Bai, Y., Yang, L., Zhang, C., and Sun, Y. (2013) ZSM-5 filled polyether block amide membranes for separating EA from aqueous solution by pervaporation. *International Journal of Polymer Science*, **2013**, 1–10.

Guichard, E. and Langourieux, S. (2000) Interactions between β-lactoglobulin and flavour compounds. *Food Chemistry*, **71**, 301–308.

Hansen, A.P. and Booker, D.C. (1996) Flavor interaction with casein and whey protein, in Flavor-Food Interactions (eds. R.J. McGorrin and J.V. Leland), American Chemical Society, Washington, DC, pp. 75–89.

Hatchwell, L.C. (1996) Implications of fat on flavor, in Flavor–Food Interactions (eds. R/J/ McGorrin and L.V. Leland), American Chemical Society, Washington, DC, pp. 14–23.

Huang, R.Y.M., Moon, G.Y., and Pal, R. (2002) Ethylene propylene diene monomer (EPDM) membranes for the pervaporation separation of aroma compound from water. *Industry Engineering and Chemistry Research*, **41**, 531–537.

Huddleston, J.G., Visser, A.E., Reichert, W.M., Willauer, H.D., Broker, G.A., and Rogers, R.D. (2001) Characterization and comparison of hydrophilic and hydrophobic room temperature ionic liquids incorporating the imidazolium cation. *Green Chemistry*, **3**, 156–164.

Isci, A., Sahin, S., and Sumnu, G. (2006) Recovery of strawberry aroma compounds by pervaporation. *Journal of Food Engineering*, **75**, 36–42.

Ishii, N., Matsumura, M., Kataoka, H., Tanaka, H., and Araki, K. (1995) Diacetyl fermentation coupled with pervaporation using oleyl alcohol supported liquid membrane. *Bioprocess Engineering*, **13**, 119–123.

Ji, W., Sikdar, S.K., and Hwang, S.T. (1995) Sorption, diffusion and permeation of 1,1,1-trichloroethane through adsorbent-filled polymeric membranes. *Journal of Membrane Science*, **103**, 243–255.

Jiang, J.S., Vane, L.M., and Sikdar, S.K. (1997) Recovery of VOCs from surfactant solutions by pervaporation. *Journal of Membrane Science*, **136**, 233–247.

Jullok, N., Martinez, R., Wouters, C., Luis, P., Sanz, M.T., and Van der Bruggen, B. (2013) A biologically inspired hydrophobic membrane for application in pervaporation. *Langmuir*, **29**, 1510–1516.

Kanani, D.M., Nikhade, B.P., Balakrishnan, P., Singh, G., and Pangarkar, V.G. (2003) Recovery of valuable tea aroma components by pervaporation. *Industry Engineering and Chemistry Research*, **42**, 6924–6932.

Karlsson, H.E. and Tragardh, G. (1998) Aroma recovery during beverage processing. *Journal of Food Engineering*, **34**, 159–178.

Kellam, S. (1998) The manufacture of lactose, in *Chemical Processes in New Zealand* (eds. J. Packer, J. Robertson, and H. Wansbrough), second edition, New Zealand Institute of Chemistry, Auckland, New Zealand.

Kuhn, J., Considine, T., and Singh, H. (2006) Interactions of milk proteins and volatile flavor compounds: implications in the development of protein foods, *Journal of Food Science*, **71**, R72–82.

Kuhn, J., Zhu, X.Q., Considine, T., and Singh, H. (2007) Binding of 2-nonanone and milk proteins in aqueous model systems. *Journal of Agricultural and Food Chemistry*, **55**, 3599–3604.

Lamer, T., Rohart, M.S., Voilley, A., and Baussart, H. (1994) Influence of sorption and diffusion of aroma compounds in silicone rubber on their extraction by pervaporation. *Journal of Membrane Science*, **90**, 251–263.

Lamer, T. and Voilley, A. (1991) Influence of different parameters on the pervaporation of aroma compounds, in *Proceedings of the 5th International Conference on Pervaporation Processes in the Chemical Industry* (ed. R. Bakish), Bakish Materials Corporation, Englewood, NJ, pp. 110–122.

Leland, J.V. (1997) Flavor interactions: the greater whole. *Food Technology*, **51**, 75–80.

Lipnizki, F., Olsson, J., and Tragardh, G. (2002) Scale-up of pervaporation for the recovery of natural aroma compounds in the food industry. Part 1: Simulation and performance. *Journal of Food Engineering*, **54**, 183–195.

Lipnizki, F. and Tragardh, G. (2001) Modelling of pervaporation: models to analyze and predict the mass transport in pervaporation. *Separation and Purification Reviews*, **30**, 49–125.

Martinez, R., Sanz, M.T., and Beltran, S. (2011) Concentration by pervaporation of representative brown crab volatile compounds from dilute model solutions. *Journal of Food Engineering*, **105**, 98–104.

McGorrin, R.J. (2001) Advance in dairy flavor chemistry, in *Food Flavors and Chemistry: Advance of the New Millennium* (Eds A.M. Spanier,F. Shahidi, T.H. Parliment, et al.), The Royal Society of Chemistry, Cambridge, UK, pp. 67–84.

Mishima, S. and Nakagawa, T. (2000) Sorption and diffusion of volatile organic compounds in fluoroalkyl methacrylate-grafted PDMS membrane. *Journal of Applied Polymer Science*, **75**, 773–783.

Mishima, S. and Nakagawa, T. (2002) Pervaporation of volatile organic compounds/ water mixtures through poly(1H,1H,9H-hexadecafluorononyl methacrylate)-filled poly(1-trimethylsilyl-1-propyne) membranes. *Journal of Applied Polymer Science*, **83**, 1054–1060.

Mohammadi, T., Kikhavandi, T., and Moghbeli, M. (2008) Synthesis and characterization of poly (ether-block-amide) membranes for the pervaporation of organic/aqueous mixtures. *Journal of Applied Polymer Science*, **107**, 1917–1923.

Mujiburohman, M. and Feng, X. (2007) Permselectivity, solubility and diffusivity of propyl propionate/water mixtures in poly(ether block amide) membranes. *Journal of Membrane Science*, **300**, 95–103.

Nguyen, Q.T., Bendjama, Z., Clement, R., and Ping, Z. (2000) Poly(dimethylsiloxane) crosslinked in different conditions. Part II. Pervaporation of water-ethyl acetate mixtures. *Physical Chemistry Chemical Physics*, **2**, 395–400.

Nursten, H.E. (1997) The flavour of milk and dairy products: I. Milk of different kinds, milk powder, butter and cream. *International Journal of Dairy Technology*, **50**, 48–56.

Olsson, J. and Tragardh, G. (1999a) Influence of feed flow velocity on pervaporative aroma recovery from a model solution of apple juice aroma compounds. *Journal of Food Engineering*, **39**, 107–115.

Olsson, J. and Tragardh, G. (1999b) Influence of temperature on membrane permeability during pervaporative aroma recovery. *Separation Science and Technology*, **34**, 1643–1659.

Olsson, J., Tragardh, G., and Tragardh, C. (2001) Pervaporation of volatile organics from water II. Influence of permeate pressure on partial fluxes. *Journal of Membrane Science*, **186**, 239–247.

Ortiz, I. (2008) Separation and concentration of bilberry impact aroma compound from dilute model solution by pervaporation. *Journal of Chemical Technology and Biotechnology*, **982**, 973–982.

Osorio-Galindo, M., Iborra-Clar, A., Alcaina-Miranda, I., and Ribes-Greus, A. (2001) Characterization of poly(dimethyl siloxane)–poly(methyl hydrogen siloxane) composite membranes for organic water pervaporation separation. *Journal of Applied Polymer Science*, **81**, 546–556.

Overington, A.R., Wong, M., and Harrison, J.A. (2011) Effect of feed pH and non-volatile dairy components on flavour concentration by pervaporation. *Journal of Food Engineering*, **107**, 60–70.

Overington, A., Wong, M., Harrison, J. and Ferreira, L. (2008) Concentration of dairy flavour compounds using pervaporation. *International Dairy Journal*, **18**, 835–848.

Overington, A.R., Wong, M., Harrison, J.A., and Ferreira, L.B. (2009) Estimation of mass transfer rates through hydrophobic pervaporation membranes. *Separation Science and Technology*, **44**, 787–816.

Panek, D. and Konieczny, K. (2007) Preparation and applying the membranes with carbon black to pervaporation of toluene from the diluted aqueous solutions. *Separation and Purification Technology*, **57**, 507–512.

Parliament, T.H. and McGorrin, R.J. (2000) Critical flavor compounds in dairy products, in *Flavor Chemistry* (eds. S.J. Risch and C.T. Ho), American Chemical Society, Washington, DC, pp. 44–71.

Peng, M. and Liu, S. (2003) Recovery of aroma compounds from dilute model blueberry solution by pervaporation. *Journal of Food Science*, **68**, 2706–2710.

Pereira, C.C., Rufino, J.M., Habert, A.C., Nobrega, R., Cabral, L.M.C., and Borges, C.P. (2002) Membrane for processing tropical fruit juice. *Desalination*, **148**, 57–60.

Pereira, C.C., Rufino, J.R.M., Habert, A.C., Nobrega, R., Cabral, L.M.C., and Borges, C.P. (2005) Aroma compounds recovery of tropical fruit juice by pervaporation: membrane material selection and process evaluation. *Journal of Food Engineering*, **66**, 77–87.

Pereira, C.C., Ribeiro Jr,, C.P., Nobrega, R., and Borges, C.P. (2006) Pervaporative recovery of volatile aroma compounds from fruit juices. *Journal of Membrane Science*, **274**, 1–23.

Raisi, A. and Aroujalian, A. (2011) Aroma compound recovery by hydrophobic pervaporation: the effect of membrane thickness and coupling phenomena. *Separation and Purification Technology*, **82**, 53–62.

Raisi, A., Aroujalian, A., and Kaghazchi, T. (2008) Multicomponent pervaporation process for volatile aroma compounds recovery from pomegranate juice. *Journal of Membrane Science*, **322**, 339–348.

Rajagopalan, N., Cheryan, M. and Matsuura, T. (1994) Recovery of diacetyl by pervaporation. *Biotechnology Techniques*, **8**, 869-72.

Rautenbach, R. and Helmus, F.P. (1994) Some considerations on mass-transfer resistances in solution-diffusion-type membrane processes. *Journal of Membrane Science*, **87**, 171–181.

Routray, W. and Mishra, H.N. (2011) Scientific and technical aspects of yogurt aroma and taste: a review. *Comprehensive Reviews in Food Science and Food Safety*, **10**, 208–220.

Sampranpiboon, P., Jiraratananon R., Uttapap, D., Feng, X., and Huang, R.Y.M. (2000a) Separation of aroma compounds from aqueous solutions by pervaporation using polyoctylmethyl siloxane (POMS) and polydimethyl siloxane (PDMS) membranes. *Journal of Membrane Science*, **174**, 55–65.

Sampranpiboon, P., Jiraratananon R., Uttapap, D., Feng, X., and Huang, R.Y.M. (2000b) Pervaporation separation of ethyl butyrate and isopropanol with polyether block amide (PEBA) membranes. *Journal of Membrane Science*, **173**, 53–59.

Schafer, T. (2004) Coupled pervaporation/mass spectrometry for investigating membrane mass transport phenomena. *Journal of Membrane Science*, **241**, 197–205.

Shao, P. and Huang, R.Y.M. (2007) Polymeric membrane pervaporation. *Journal of Membrane Science*, **287**, 162–179.

She, M. and Hwang, S.T. (2004) Concentration of dilute flavor compounds by pervaporation: permeate pressure effect and boundary layer resistance modeling. *Journal of Membrane Science*, **236**, 193–202.

She, M. and Hwang, S.T. (2006a) Effects of concentration, temperature, and coupling on pervaporation of dilute flavor organics. *Journal of Membrane Science*, **271**, 16–28.

She, M. and Hwang, S.T. (2006b) Recovery of key components from real flavor concentrates by pervaporation. *Journal of Membrane Science*, **279**, 8–93.

Sheldon, R.A. and van Rantwijk, F. (2008) Ionic liquids as media for enzymatic transformations, in Organic Synthesis with Enzymes in Non-Aqueous Media (eds. G. Carrea and S. Riva), Wiley-VCH, Weinheim, pp. 227–254.

Shepherd, A., Habert, A.C., and Borges, C.P. (2002) Hollow fibre modules for orange juice aroma recovery using pervaporation. *Desalination*, **148**, 111–114.

Shirazi, Y., Ghadimi, A., and Mohammadi, T. (2012) Recovery of alcohols from water using polydimethylsiloxane-silica nanocomposite membranes: characterization and pervaporation performance. *Journal of Applied Polymer Science*, **124**, 2871–2882.

Slater, C.S. (1997) Recovery of ethyl acetate from process effluents using pervaporation technology. *Journal of Environmental Science and Health*, **A32**, 1339–1352.

Song, K.H. and Lee, K.R. (2005) Pervaporation of flavors with hydrophobic membrane. *Korean Journal of Chemical Engineering*, **22**, 735–739.

Song, K.H., Lee, K.R., and Rim, J.M. (2004) Pervaporation of esters with hydrophobic membrane. *Korean Journal of Chemical Engineering*, **21**, 693–698.

Swaisgood, H.E. (1996) Characteristics of milk, in Food Chemistry (ed. O.R. Fennema), Marcel Dekker, third edition, New York, pp. 841–878.

Tanaka, S., Chao, Y., Araki, S., and Miyake, Y. (2010) Pervaporation characteristics of pore-filling PDMS/PMHS membranes for recovery of ethylacetate from aqueous solution. *Journal of Membrane Science*, **348**, 383–388.

Te Hennepe, H.J.C., Smolders, C.A., Bargeman, D., and Mulder, M.H.V. (1991) Exclusion and tortuosity effects for alcohol/water separation by zeolite-filled PDMS membranes. *Separation Science and Technology*, **26**, 585–596.

Tian, X. and Jiang, X. (2008) Poly(vinylidene fluoride-co-hexafluoropropene) (PVDF-HFP) membranes for ethyl acetate removal from water. *Journal of Hazardous Materials*, **153**, 128–135.

Toso, B., Procida, G., and Stefanon, B. (2002) Determination of volatile compounds in cows' milk using headspace GC-MS. *Journal of Dairy Research*, **69**, 569–577.

Trifunovic, O. and Tragardh, G. (2005) The influence of support layer on mass transport of homologous series of alcohols and esters through composite pervaporation membranes. *Journal of Membrane Science*, **259**, 122–134.

Trifunovic, O. and Tragardh, G. (2006) Mass transport of aliphatic alcohols and esters through hydrophobic pervaporation membranes. *Separation and Purification Technology*, **50**, 51–61.

Urbach, G. (1997) The flavour of milk and dairy products: II. Cheese: contribution of volatile compounds. *International Journal of Dairy Technology*, **50**, 79–89.

Vankelecom, I.F.J., De Beukelaer, S., and Uytterhoeven, J.B. (1997) Sorption and pervaporation of aroma compounds using zeolite-filled PDMS membranes. *Journal of Physical Chemistry B*, **101**, 5186–5190.

Voilley, A., Charbit, G., and Gobert, F. (1990) Recovery and separation of 1-octene-3-ol from aqueous solutions by pervaporation through silicone membrane. *Journal of Food Science*, **55**, 1399–1402.

Voilley, A., Schmidt, B., Simatos, D., and Baudron, S. (1988) Extraction of aroma compounds by pervaporation technique, in *Proceedings of the 3rd International Conference on Pervaporation Processes in the Chemical Industry* (ed. R. Bakish), Bakish Materials Corporation, Englewood, NJ, pp. 429–438.

Voilley, A., Lamer, T., Nguyen, T., and Simatos, D. (1989) Extraction of aroma compounds by pervaporation, in *Proceedings of the 4th International Conference on Pervaporation Processes in the Chemical Industry* (ed. R. Bakish), Bakish Materials Corporation, Englewood, NJ, pp. 332–343.

Wang, X., Deng, X., Bai, Z., Zhang, X., Feng, X., and Huang, W. (2014) The synthesis of super-hydrophilic and acid-proof Ge–ZSM-5 membranes by simultaneous incorporation of Ge and Al into a Silicalite-1 framework. *Journal of Membrane Science*, **468**, 202–208.

Willemsen, J.H.A., Dijkink, B.H., and Togtema, A. (2004) Organophilic pervaporation for aroma isolation – Industrial and commercial prospects. *Membrane Technology*, **2004**, 5–10.

Wu, Y., Tan, H., Zhang, D., and Li, T. (2011) Pervaporation of acetaldehyde aqueous solution through zeolite 3A-polyurethane (PU) composite membrane. *Separation Science and Technology*, **46**, 1908–1914.

Wu, Y., Tan, H., Li, D., and Jin, Y. (2012) Pervaporation of aqueous solution of acetaldehyde through ZSM-5 filled PDMS composite membrane. *Chinese Journal of Chemical Engineering*, **20**, 625–632.

Yongquan, D., Ming, W., Lin, C., and Mingjun, L. (2012) Preparation, characterization of P(VDF-HFP)/[bmim]BF4 ionic liquids hybrid membranes and their pervaporation performance for ethyl acetate recovery from water. *Desalination*, **295**, 53–60.

Zhu, B., Tian, X., and Xu, Y. (2005) Recovering ethyl acetate from aqueous solution using P(VDF-co-HFP) membrane based pervaporation. *Desalination*, **184**, 71–78.

8 Membrane chromatography: current applications, future opportunities, and challenges

Raja Ghosh

Department of Chemical Engineering, McMaster University, Hamilton, Ontario, Canada

8.1 INTRODUCTION

Membrane chromatography, a bioseparation technique that involves the use of a stack of synthetic membrane as chromatographic media was first described in the late 1980s (Brandt et al., 1988). It was not until the 1990s that significant numbers of research publications on membrane chromatography started to appear in scientific journals (Tennikova, Bleha, and Švec, 1991; Tennikova and Švec, 1993; Champluvier and Kula, 1991; Suen and Etzel, 1994). It is therefore a relatively new separation technique compared to its competitors and, as with any new technique, it faces a whole gamut of challenges, both real technical ones and those based on perception. The working principles of membrane chromatography, including the binding mechanisms and transport phenomena, have been discussed in some research and review articles (Tennikova, Švec, and Belenkii, 1990; Charcosset, 1998; Thömmes and Kula, 1995; Zeng and Ruckenstein, 1999; Zhou and Tressel, 2006; Roper and Lightfoot, 1995; Ghosh, 2002). As will be discussed over the next few paragraphs, membrane chromatography has some very obvious advantages over its competitors.

The main advantage of membrane chromatography over column chromatography is the speed of separation (Tennikova, Švec, and Belenkii, 1990; Charcosset, 1998; Thömmes and Kula, 1995; Zeng and Ruckenstein, 1999; Zhou and Tressel, 2006; Roper and Lightfoot, 1995; Ghosh, 2002). Being largely based on convective mass transport, the time taken for solute molecules to travel to their binding sites during adsorption and away from them during elution is significantly lower than that in column chromatography, where such transport is generally diffusion-limited

Membrane Processing for Dairy Ingredient Separation, First Edition.
Edited by Kang Hu and James M. Dickson.

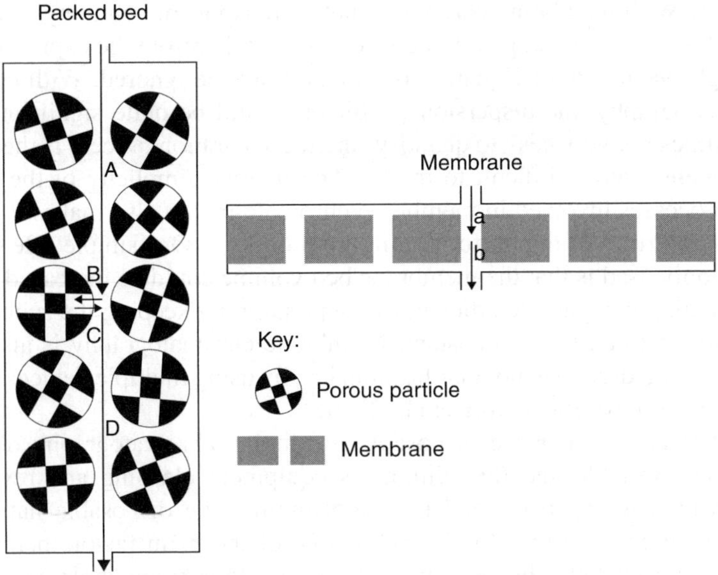

Figure 8.1 Comparison of mass transport in particle-based packed bed and membrane adsorbers (A: convection of solute from inlet to particle, B: pore diffusion of solute to binding site, C: pore diffusion of solute away from binding site, D: convection of solute from particle to outlet, a: convection of solute from inlet to binding site, b: convection of solute from binding site to outlet).

(see Figure 8.1). Depending on the scale of operation and the type of device, membrane chromatography could therefore be faster by more than one order of magnitude. The shorter process time contributes towards higher productivity due to the possibility of carrying out many more separation cycles within a given time. Also, the recovery of biological products such as a protein is further enhanced through a decrease in product degradation by proteolysis, denaturation, and aggregation, all of which increase with processing time.

The predominance of convection also means that membrane chromatography is relatively simpler to model than column chromatography. The general working equation for any fixed-bed-based adsorption–desorption process, which includes both column and membrane chromatography, is given below:

$$u\frac{\partial c}{\partial z} + \varepsilon\frac{\partial c}{\partial t} + (1-\varepsilon)\frac{\partial q}{\partial t} = E\frac{\partial^2 c}{\partial z^2} \tag{8.1}$$

where u is the fluid velocity through the bed, ε is the porosity of the bed, c is the free solute concentration, q is the bound solute concentration, t is the time, z is the distance along the axial direction, and E is a dispersion coefficient that primarily depends on the diffusivity of the solute within the bed. With membrane chromatography, transport of solute by diffusion is negligible when compared to convection and therefore the expression on the right-hand side of Equation (8.1) can easily be ignored. With column chromatography, the dispersion coefficient could be quite significant and sometimes not very easy to quantify, and the separation process is therefore significantly more difficult to model. The relative simplicity of the transport processes involved in membrane chromatography also makes it easier to develop rules of thumb for scaling-up processes. One simple rule that is commonly used is that the membrane bed volume could be increased while maintaining comparable efficiency of separation by keeping the superficial velocity and bed height constant. Membrane chromatography is also frequently scaled-up in a modular fashion, i.e., by using multiple devices either in series or in parallel (Mora et al., 2006).

Relatively inexpensive disposable membrane chromatography devices are now available and this eliminates equipment cleaning and revalidation, which is expensive and time-consuming. The disposable nature of membrane chromatography devices is one of the main factors behind its increased acceptance by users in recent years. Membrane chromatography is certainly gaining in popularity, as evident from some recent publications (Ghosh et al., 2013a, 2013b; Bhut et al., 2011a, 2011b; Saufi and Fee, 2013; Dimartino, Boi, and Sarti, 2011; von Lieres, Wang, and Ulbright, 2010). This chapter reviews some of the current applications and potential future opportunities of membrane chromatography, while highlighting some of the key challenges.

8.2 CURRENT APPLICATIONS

Adsorptive membranes for chromatographic separations are commonly available in the following ligand chemistries:

a. Anion exchange: e.g., quaternary ammonium (Q), primary amine (PA)
b. Cation exchange: e.g., sulfonic acid (S)
c. Affinity: e.g., protein-A
d. Hydrophobic interaction: e.g., environment responsive hydrogel based, phenyl.

Membrane chromatography devices are available in three different formats:

a. Stacked disc (see Figure 8.2)
b. Spiral wound radial flow (see Figure 8.3)
c. Pleated radial flow.

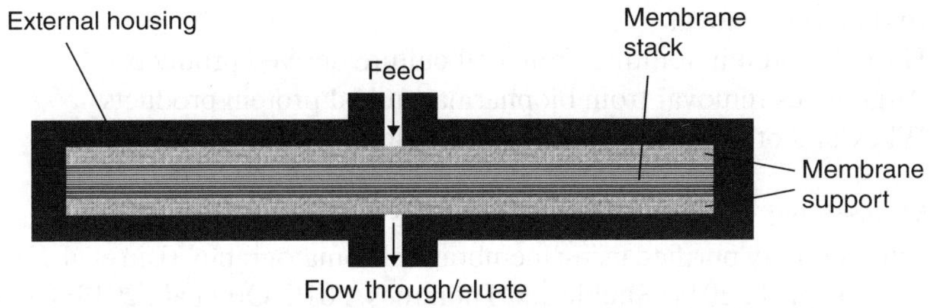

Figure 8.2 A stacked-disc type membrane chromatography device.

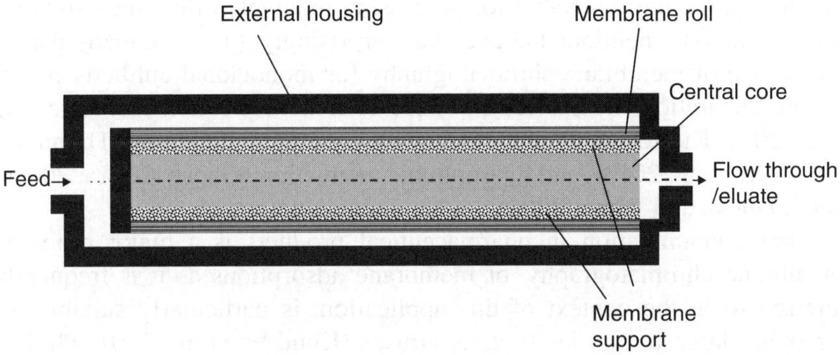

Figure 8.3 Arrangement of membrane roll within a spiral-wound radial flow membrane device.

The design of stacked disc devices is based on syringe-type microfilters. These are simple and relatively easy to fabricate and are good for preliminary process development work. These devices, however, suffer from poor scalability and the circular shape of the membrane leads to significant membrane wastage. Radial flow devices are more suitable for large-scale production and are relatively more scalable. However, they tend to have complicated designs, have larger dead volumes on both the feed and permeate side, and the effective membrane volume to device ratio is extremely low due to the presence of a large central core needed to support the membrane.

Most current applications of membrane chromatography are in the areas of biopharmaceutical and protein food production. The major applications include:

a. Monoclonal antibody purification
b. Virus removal from pharmaceutical products
c. Purification of viruses and virus-like particles
d. DNA removal from pharmaceutical products

e. Endotoxin removal
f. Host cell protein removal from cell culture derived products
g. Aggregates removal from biopharmaceutical protein products
h. Whey and other milk protein fractionation.

Monoclonal antibodies were one of the first categories of molecules to be commercially purified using membrane chromatography (Liu et al., 2011; Kuczewski et al., 2011; Shukla and Thömmes, 2010; Orr et al., 2013; Ghosh, 2001a, 2001b, 2005; Ghosh and Wang, 2006). In the 1990s, the typical monoclonal antibody titer in the cell culture supernatant was at the 1 g/L level. One of the popular perceptions about membrane chromatography is that it is particularly suitable for processing relatively dilute feed-streams, such as the one mentioned above. Not surprisingly, there are many papers on the use of membrane chromatography for monoclonal antibody purification. Separation chemistries that were used include cation exchange (Liu et al., 2011; Kuczewski et al., 2011), anion exchange (Shukla and Thömmes, 2010; Orr et al., 2013), and hydrophobic interaction (Ghosh, 2001a, 2001b, 2005; Ghosh and Wang, 2006).

Viral contamination in pharmaceutical products is a major problem. Membrane chromatography, or membrane adsorptions as it is frequently referred to in the context of this application, is particularly suitable for removing large impurities such as viruses (Knudsen et al., 2001; Phillips et al., 2005; El-Hadidy, Peldszus, and van Dyke, 2013). Viruses are generally negatively charged at neutral pH due to their high nucleic acid content. Anion exchange membrane chromatography is therefore commonly used for removing viruses from pharmaceutical products. Membrane chromatography is also used for purifying virus and virus-like particles, which are used as vaccines and transfecting agents (Segura, Kamen, and Garnier, 2011; Pato et al., 2014).

Cell culture derived products are frequently contaminated by media proteins, host cell DNA, and host cell proteins (HCPs), and other leachates. These are typically present in low concentrations and could easily be removed using membrane chromatography (Thömmes and Etzel, 2007; Low, O'Leary, and Pujar, 2007; Yoo and Ghosh, 2012). Most of these impurities are negatively charged at neutral pH and are removed using anion exchange chromatography. Positively charged HCP are removed using cation exchange membrane chromatography. In most applications, anion and cation exchange membrane chromatography are used sequentially to get rid of the above impurities.

In most of the above applications, the bind and elute mode is applied whereby impurities are bound on the membrane while the target products are allowed to flow through. With monoclonal antibody purification, both modes could be used.

8.3 FUTURE OPPORTUNITIES

As discussed in the previous section, most current applications of membrane chromatography is in the biopharmaceutical industry. It is anticipated that most future opportunities will continue to be in the same sector. Cost-effective production of biopharmaceuticals is already a major issue due to the advent of biosimilars, that is, generic biopharmaceuticals. The reason for this is that a large number of blockbuster biopharmaceuticals have become off-patent in the past few years. Also, the pace of development of new biopharmaceuticals has considerably slowed down and it is therefore anticipated that over the next few years the biopharmaceutical market will be dominated by biosimilars. In the context of biopharmaceutical production, downstream processing is frequently identified as the capacity bottleneck. Membrane chromatography has been shown to be suitable for producing pure and stable biopharmaceutical products in a consistent, scalable, and cost-effective manner and is therefore one of the main solutions for addressing this issue.

A relatively recent development in the biopharmaceutical industry is the entry into the product line by chemically modified biopharmaceuticals such as PEGylated proteins. These modified products have been shown to have some significant advantages over the native protein drugs. As with biosimilars, cost-effectiveness will be of critical importance in the manufacture for PEGylated proteins. It is therefore anticipated that membrane chromatography will play an important role in the production of PEGylated proteins. The feasibility of purifying PEGylated proteins using membrane chromatography has been shown in several recent papers (Yu and Ghosh, 2010a; Yu, Shang, and Ghosh, 2010).

Speed is one of the key attributes of membrane chromatography. This makes it particularly suitable for rapid biopharmaceutical analysis. Some of the recent developments in the area of rapid analysis of pharmaceutical products have centered on ultra-performance liquid chromatography (UPLC, which involves the use of ultra-fine nonporous particles as chromatographic media. While both speed and resolution are excellent with UPLC, these processes are typically carried out at very high pressures, necessitating the use of expensive and specialized equipment. Membrane chromatography provides a low-cost option for rapid biopharmaceutical analysis. Several recent papers have demonstrated the suitability of using membrane chromatography for a diverse range of analytical applications: analysis of PEGylated proteins (Yu, Shang, and Ghosh, 2010), detection and analysis of antibody aggregates (Wang, Hale, and Ghosh, 2006; Wang and Ghosh, 2008), antibody subclass determination (Wang and Ghosh, 2009), and screening of pharmaceutical compounds (Raiado-Pereira et al., 2014; Liu, Testa, and Fahr, 2011).

The scope of application of membrane chromatography goes far beyond biopharmaceutical purification and analysis. The predominance of convective mass transport that makes a stack of membranes attractive for carrying out chromatographic separations also makes it suitable for use as media for carrying out biochemical reactions. Membrane adsorber based reactors have been shown to be suitable for carrying out different types of enzymatic and biological reactions (Yu and Ghosh, 2010b, 2011). One of the attractive features of membrane adsorbers is that these are suitable for carrying out integrated reaction and separation processes (Shang, Yu, and Ghosh, 2011). It is also anticipated that some of the new developments and breakthroughs in the area of membrane chromatography will take place in the realm of device design and fabrication. Some of the pertinent issues related to membrane chromatography devices are elaborated in the next section.

8.4 CHALLENGES

There are quite a few real and perceived disadvantages of membrane chromatography. Devices currently available are still quite expensive and there is significant scepticism and resistance amongst potential users. Some of these factors could be described as teething troubles, that is, on account of the relative newness of the technology. Way back in 2002, some of the challenges facing membrane chromatography were summarized by the author (Ghosh, 2002). The list included low binding capacity of membranes, broad pore-size distribution, uneven membrane thickness, and poor flow distribution in modules. The solute binding capacity on a volume basis of most conventional membranes used in chromatographic separations continues to be low. This is due to the fact that the specific surface of membrane-like structures are lower than porous particles. Also, while both convex and concave binding surfaces are available on particulate chromatographic media, the binding surface within membranes is predominantly concave. This leads to higher steric hindrance during binding and consequently a lower binding capacity. This issue has been addressed through the development of hydrogel-type membranes (see Figure 8.4), within which significantly greater binding surface areas exist and consequently the binding capacities are comparable to if not better than that in particulate chromatographic media (Kanani et al., 2007). Some of the other issues such as pore-size distribution and uneven membrane thickness have been addressed. Flow distribution in devices continues to be a problem and the author feels that the efficiency of separation in membrane chromatography could be significantly improved by addressing this issue.

Poorly designed devices result is shallow breakthrough curves, which are indicative of poor binding capacity utilization. Breakthrough curves

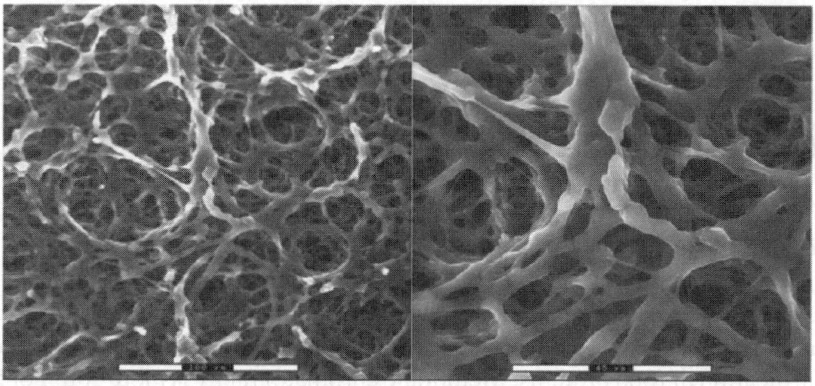

Figure 8.4 ESEM images of high-binding capacity hydrogel-based Natrix HD-S Membrane (A: 400X magnification, B: 1000X magnification) (Image courtesy of Natrix Separations Inc.)

typically obtained with membrane adsorbers tend to be shallow with effluent concentration change from breakthrough to exhaustion typically taking tens of bed volumes. While efficient device designing should have been a priority area, it has somehow remained in a blind spot. Up to this point, the membrane material has been the primary driver for innovation in membrane chromatography, with the focus being on new material development. Device and process engineering have generally not been given as much importance. However, a membrane will perform only as well as its device will allow it. There have been relatively few reports on innovative designs for membrane chromatography devices (Ghosh and Wong, 2006).

One of the factors that could be responsible for this undervaluing of device design is the disconnect between the end user and the product development process. An interesting comparison could be made between membrane chromatography and a more mature separation technology such as distillation. In distillation, custom design of equipment for specific processes is generally the norm. In membrane chromatography, on the other hand, custom design is virtually unheard of and the end user selects from available off-the-shelf devices, which are used more in a "plug and play" mode. More user involvement could potentially lead to better designed products. A good device would ideally give a sharper solute breakthrough, have a low void volume, and a simple design with a well-defined flow path. While disposable devices have greatly contributed toward the acceptance of membrane chromatography in the biopharmaceutical industry, the option of reusable devices should also be explored. More research is needed in the areas of transport phenomena and modelling and simulation of membrane chromatography, with closer cooperation between users, membrane companies, and academia.

8.5 CONCLUSION

Membrane chromatography is a fast separation technique that is gaining acceptance in the biopharmaceutical industry. It has been shown to be a viable alternative to column chromatography in certain niche application segments such as purification of monoclonal antibodies, removal of viruses, endotoxins, and host cell proteins from biopharmaceutical products, and purification of food proteins. Membrane chromatography is more commonly operated in the flow-through mode. Some of the major issues with membranes such as low binding capacity and nonuniformity have been addressed. However, there is significant room for improvement with regards to device design and process engineering. The application of membrane chromatography in newer areas such as integrated reaction–separation processes and analytical bioseparations need to be explored. More research in areas such as transport phenomena and modelling and simulation of membrane chromatography through closer cooperation between users, membrane companies, and academia is desirable.

References

Bhut, B.V., Weaver, J., Carter, A.R., et al. (2011a) The role of polymer nanolayer architecture on the separation performance of anion-exchange membrane adsorbers: I. Protein separations. *Biotechnology and Bioengineering*, **108** (11), 2645–2653.

Bhut, B.V., Weaver, J., Carter, A.R., et al. (2011b) The role of polymer nanolayer architecture on the separation performance of anion-exchange membrane adsorbers: Part II. DNA and virus separations. *Biotechnology and Bioengineering*, **108** (11), 2654–2660.

Brandt, S., Goffe. R.A., Kessler, SB., et al. (1988) Membrane-based affinity technology for commercial scale purifications. *Bio/Technology*, **6**, 779–782.

Champluvier, B. and Kula, M.R. (1991) Microfiltration membranes as pseudo-affinity adsorbents: modification and comparison with gel beads. *Journal of Chromatography A*, **539** (2), 315–325.

Charcosset, C. (1998) Purification of proteins by membrane chromatography. *Journal of Chemical Technology and Biotechnology*, **71**(2), 95–110.

Dimartino, S., Boi, C., and Sarti, G.C. (2011) A validated model for the simulation of protein purification through affinity membrane chromatography. *Journal of Chromatography A*, **1218** (13), 1677–1690.

El-Hadidy, A.M., Peldszus, S., and van Dyke, M.I. (2013) An evaluation of virus removal mechanisms by ultrafiltration membranes using MS2 and ϕX174 bacteriophage. *Separation and Purification Technology*, **120**, 215–223.

Ghosh, R. (2001a) Separation of proteins using hydrophobic interaction membrane chromatography. *Journal of Chromatography A*, **923** (1), 59–64.

Ghosh, R. (2001b) Bioseparation using supported liquid membrane chromatography. *Journal of Membrane Science*, **192** (1), 243–247.

Ghosh, R. (2002) Protein separation using membrane chromatography: opportunities and challenges. *Journal of Chromatography A*, **952** (1–2), 13–27.

Ghosh, R. (2005) Fractionation of human plasma proteins by hydrophobic interaction membrane chromatography. *Journal of Membrane Science*, **260** (1), 112–118.

Ghosh, R. and Wang, L. (2006) Purification of humanized monoclonal antibody by hydrophobic interaction membrane chromatography. *Journal of Chromatography A*, **1107** (1), 104–109.

Ghosh, R. and Wong, T. (2006) Effect of module design on the efficiency of membrane chromatographic separation processes. *Journal of Membrane Science*, **281** (1), 532–540.

Ghosh, P., Vahedipour, K., Lin, M., Vogel, J.H., et al. (2013a) Computational fluid dynamic simulation of axial and radial flow membrane chromatography: mechanisms of non-ideality and validation of the zonal rate model. *Journal of Chromatography A*, **1305**, 114–122.

Ghosh, P., Vahedipour, K., Lin, M., Vogel, J.H., et al. (2013b) Zonal rate model for axial and radial flow membrane chromatography. Part I: Knowledge transfer across operating conditions and scales. *Biotechnology and Bioengineering*, **110** (4), 1129–1141.

Kanani, D.M., Komkova, E., Wong, T., et al. (2007) Separation of human plasma proteins HSA and HIgG using high-capacity macroporous gel-filled membranes. *Biochemical Engineering Journal*, **35** (3), 295–300.

Knudsen, H.L., Fahrner, R.L., Xu, Y., et al. (2001) Membrane ion-exchange chromatography for process-scale antibody purification. *Journal of Chromatography A*, **907** (1), 145–154.

Kuczewski, M., Schirmer, E., Lain, B., and Zarbis-Papastoitsis, G. (2011) A single-use purification process for the production of a monoclonal antibody produced in a PER. C6 human cell line. *Biotechnology Journal*, **6** (1), 56–65.

Liu, X., Testa, B., and Fahr, A. (2011) Lipophilicity and its relationship with passive drug permeation. *Pharmaceutical Research*, **28** (5), 962–977.

Liu, H.F., McCooey, B., Duarte, T., et al. (2011) Exploration of overloaded cation exchange chromatography for monoclonal antibody purification. *Journal of Chromatography A*, **1218** (39), 6943–6952.

Low, D., O'Leary, R., and Pujar, N.S. (2007) Future of antibody purification. *Journal of Chromatography B*, **848** (1), 48–63.

Mora, J., Sinclair, A., Delmdahl, N., and Gottschalk, U. (2006) Disposable membrane chromatography. *BioProcess International*, **4**, 38–43.

Orr, V., Zhong, L., Moo-Young, M., and Chou, C.P. (2013) Recent advances in bioprocessing application of membrane chromatography. *Biotechnology Advances*, **31** (4), 450–465.

Pato, T.P., Souza, M.C.O., Silva, A., et al. (2014) Development of a membrane adsorber based capture step for the purification of yellow fever virus. *Vaccine*, **32** (24), 2789–2793.

Phillips, M., Cormier, J., Ferrence, J., et al. (2005) Performance of a membrane adsorber for trace impurity removal in biotechnology manufacturing. *Journal of Chromatography A*, **1078** (1), 74–82.

Raiado-Pereira, L., De La Vega, J., Prazeres, D.M.F., and Mateus, M. (2014) Development of a phenyl membrane chromatography-based process yielding pharmaceutical grade plasmid deoxyribonucleic acid for mammalian cells transfection. *Journal of Chromatography A*, **1337**, 67–74.

Roper, D.K. and Lightfoot, E.N. (1995) Separation of biomolecules using adsorptive membranes, *Journal of Chromatography A*, **702** (1–2), 3–26.

Saufi, S.M. and Fee, C.J. (2013) Mixed matrix membrane chromatography based on hydrophobic interaction for whey protein fractionation. *Journal of Membrane Science*, **444**, 157–163.

Segura, M.M., Kamen, A.A., and Garnier, A. (2011) Overview of current scalable methods for purification of viral vectors, in Viral Vectors for Gene Therapy, Humana Press, pp. 89–116.

Shang, X., Yu, D., and Ghosh, R. (2011) Integrated solid-phase synthesis and purification of PEGylated protein. *Biomacromolecules*, **12** (7), 2772–2779.

Shukla, A.A. and Thömmes, J. (2010) Recent advances in large-scale production of monoclonal antibodies and related proteins. *Trends in Biotechnology*, **28** (5), 253–261.

Suena, S. and Etzel, M.R. (1994) Sorption kinetics and breakthrough curves for pepsin and chymosin using pepstatin A affinity membranes. *Journal of Chromatography A*, **686** (2), 179–192.

Tennikova, T.B., Bleha, M., and Švec, F. (1991) High-performance membrane chromatography of proteins, a novel method of protein separation. *Journal of Chromatography A*, **555** (1–2), 97–107.

Tennikova, T.B. and Švec, F. (1993) High-performance membrane chromatography: highly efficient separation method for proteins in ion-exchange, hydrophobic interaction and reversed-phase modes. *Journal of Chromatography A*, **646** (2), 279–288.

Tennikova, T.B., Švec, F., and Belenkii, B.G. (1990) High-performance membrane chromatography. A novel method of protein separation. *Journal of Liquid Chromatography*, **13** (1), 63–70.

Thömmes, J. and Etzel, M. (2007) Alternatives to chromatographic separations. *Biotechnology Progress*, **23** (1), 42–45.

Thömmes, J. and Kula, M.R. (1995) Membrane chromatography – An integrative concept in the downstream processing of proteins. *Biotechnology Progress*, **11** (4), 357–367.

von Lieres, E., Wang, J., and Ulbricht, M. (2010) Model based quantification of internal flow distributions from breakthrough curves of flat sheet membrane chromatography modules. *Chemical Engineering and Technology*, **33** (6), 960–968.

Wang, L. and Ghosh, R. (2008) Fractionation of monoclonal antibody aggregates using membrane chromatography. *Journal of Membrane Science*, **318** (1), 311–316.

Wang, L. and Ghosh, R. (2009) Feasibility study for the fractionation of the major human immunoglobulin G subclasses using hydrophobic interaction membrane chromatography. *Analytical Chemistry*, **82** (1), 452–455.

Wang, L., Hale, G., and Ghosh, R. (2006) Non-size-based membrane chromatographic separation and analysis of monoclonal antibody aggregates. *Analytical Chemistry*, **78** (19), 6863–6867.

Yoo, S.M. and Ghosh, R. (2012) Simultaneous removal of leached protein-A and aggregates from monoclonal antibody using hydrophobic interaction membrane chromatography. *Journal of Membrane Science*, **390**, 263–269.

Yu, D. and Ghosh, R. (2010a) Purification of PEGylated protein using membrane chromatography. *Journal of Pharmaceutical Sciences*, **99** (8), 3326–3333.

Yu, D. and Ghosh, R. (2010b) Membrane bioreactor separator system for integrated IgG fragmentation and Fab purification. *Journal of Immunological Methods*, **359** (1), 37–41.

Yu, D. and Ghosh, R. (2011) Enzymatic fragmentation of cation exchange membrane bound immunoglobulin G. *Biotechnology Progress*, **27** (1), 61–66.

Yu, D., Shang, X., and Ghosh, R. (2010) Fractionation of different PEGylated forms of a protein by chromatography using environment-responsive membranes. *Journal of Chromatography A*, **1217** (35), 5595–5601.

Zeng, X. and Ruckenstein, E. (1999) Membrane chromatography: preparation and applications to protein separation. *Biotechnology Progress*, **15** (6), 1003–1019.

Zhou, J.X. and Tressel, T. (2006) Basic concepts in Q membrane chromatography for large-scale antibody production. *Biotechnology Progress*, **22** (2), 341–349.

9 Electrodialysis applications on dairy ingredients separation

Laurent Bazinet

Institute of Nutrition and Functional Foods (INAF), Dairy Research Center (STELA), Department of Food Sciences and Nutrition, Université Laval, Québec, Canada

9.1 INTRODUCTION

Milk is a source of ingredients that have been used in the food industry over the years. Amongst these dairy ingredients, milk proteins (casein and whey proteins) are extensively used. Caseins and caseinates are finding applications amongst others in cheese, baked products, coffee creamers, and ice creams while whey proteins are included in beverages, confectionary, dressings, and novel dairy products (Stathopoulos, 2008). Lactose is added in the formulation of various products, such as sweetened condensed milk, frozen milk products, confectionary products, baby foods, and also as flavor enhancers (Fox and McSweeney, 1998; Stathopoulos, 2008). Milk is also a source of added-value ingredients such as bioactive peptides (Pihlanto-Leppälä, 2001; Gauthier and Pouliot, 2003; Korhonen and Pihlanto, 2006; Bazinet and Firdaous, 2013) and organic acids (Fosset and Tomé, 2002). A variety of activities are associated with milk-derived peptides such as opioid agonistic and antagonistic activity, angiotensin-converting enzyme (ACE) inhibitory activity, antithrombotic activity, immunomodulating activity, and antimicrobial activities (Korhonen and Pihlanto, 2006; Pihlanto-Leppälä, 2001; Kitts and Weiler, 2003; Agyei and Danquah, 2011; Bazinet and Firdaous, 2013). Lactose derivatives such as lactulose, lactitol, and oligosaccharides stimulate the growth of bifidobacteria (Fosset and Tomé, 2002).

Conventional electrodialysis (ED) is a membrane-based technology that allows the separation of particles according to their charge without pressure application. The earliest studies on ED were carried out in 1890, when Maigrot and Sabates patented a three-compartment apparatus in which the membrane, used as a physical barrier, prevented the mixing of the ED

Membrane Processing for Dairy Ingredient Separation, First Edition.
Edited by Kang Hu and James M. Dickson.

products, without having a selective action on the migration of ions (Jain and Reed, 1985). The present largest area of application for ED in the dairy industry is in the de-ashing of milk whey to obtain valuable raw materials for baby foods (Batchelder, 1987; Fidaleo and Moresi, 2006). However, in the last ten years, the progress in terms of membrane material, membrane types, mass transfer understanding, and membrane fouling has contributed to the development of new electrodialytic applications, the increase in dairy process ecoefficiency, and the production of innovative added-value ingredients (Bazinet, 2005; Fidaleo and Moresi, 2006; Pourcelly and Bazinet, 2008).

In this chapter, the principle of conventional ED as well as mass transport phenomenon through an ion-exchange membrane will be first described. Then, hybrid ED processes using bipolar (BM) and filtration membranes (FM) will be highlighted. After that, recent results on conventional ED, EDBM, and EDFM processes for the separation, purification, and fractionation of dairy ingredients from milk, whey, or milk by-products will be presented. The basic applications of ED processes for milk or milk by-product de-ashing will not be presented in this chapter.

9.2 ELECTRODIALYSIS

9.2.1 Definition

ED is a membrane separation process based on the migration of electrically charged ionic species through permselective membranes, under the influence of an electric field (Bazinet, 2005; Bazinet and Firdaous, 2013). This electric field is created by the application of a potential difference between two electrodes immersed in an aqueous solution enriched in mineral or organic ionic species (Leitz, 1986). The electrical energy is mainly used for the transfer of charged species. During this process, the chemical nature of species in solution remains unchanged.

9.2.2 Electrodialysis system

The basic conventional ED system is composed of anion-exchange membranes (AEMs) and cation-exchange membranes (CEMs) arranged in an alternating pattern. A frame spacer equipped with a grill, serving both to support the membrane and to promote turbulence, is placed between the AEM and the CEM to allow circulation of a solution between them (Figure 9.1). Four solutions are circulating at the same time into the electrodialyser: the solution to be treated (diluate or desalted solution), the salt solution (concentrate), and the two solutions at the electrodes (electrolytes). Each solution

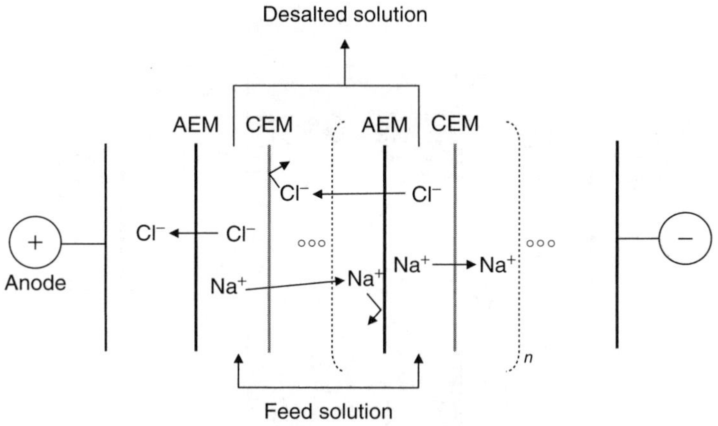

Figure 9.1 Schematic representation of a conventional electrodialysis cell. AEM: anion-exchange membrane, CEM: cation-exchange membrane.

compartment is connected by an external circuit to a separate reservoir to allow recirculation of the solutions during treatments. The electrodes, at the extremities of the system, are connected to a current generator and allow the transfer of the electricity.

9.2.3 Membranes

In ED systems, three kinds of membranes could be used according to the purpose of the process: the homopolar, the bipolar, and the filtration membranes.

9.2.3.1 Homopolar membranes

Ion-exchange membranes, which are composed of laminated ion-exchange resins, are the main parts of the ED process and are called homopolar because they are permeable to only one type of ion. They are made of a macromolecular material that carries charged groups (Figure 9.2). In aqueous solutions, ions fixed to the membranes are electronically neutralized by mobile charges of the opposite sign, called compensator ions or counter-ions (Bazinet, 2005; Bazinet and Firdaous, 2009).

Under the influence of an electrical field, cations migrate toward the cathode. They leave the dilution compartment by crossing the CEM and are then retained in the concentration compartment because they cannot pass the anionic membrane (AEM) (Figure 9.1). Simultaneously, the anions migrate toward the anode. They leave the dilution compartment, pass through the AEM, and are retained in the concentration compartment, as they cannot go through the CEM (Baker, 2004; Bazinet, 2005).

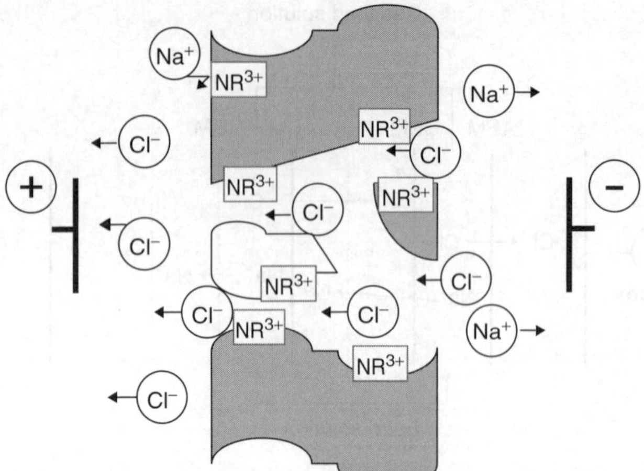

Figure 9.2 Principle of an anion-exchange membrane.

9.2.3.2 Bipolar membranes

Bipolar membranes appeared commercially at the end of the 1980s. They are composed of three parts: an anion-exchange layer, a cation-exchange layer, and an intermediary hydrophilic layer at the center. Bipolar membranes dissociate water molecules under an electric field (Mani, 1991) and consequently generate a flow of protons H^+ at its cationic interface and a flow of hydroxyl ions OH^- at its anion-exchange interface (Figure 9.3) (Pourcelly and Bazinet, 2008). The pH of a solution circulating inside an ED cell at the cationic side of the bipolar membrane, where the H^+ are generated, will decrease (or production of HCl; see Figure 9.3). In a similar way, the pH of a solution circulating at the anionic side of the bipolar membrane, where the OH^- are generated, will increase (or production of NaOH; see Figure 9.3) (Bazinet, 2005).

9.2.3.3 Filtration membranes

ED with a filtration membrane (EDFM), a technique developed and patented recently by Bazinet et al. (2005), has demonstrated very high selectivity for the separation of organic charged biomolecules. In this system, one or more ultrafiltration membrane is stacked as a molecular barrier in a conventional ED cell (Figure 9.4). EDFM couples size exclusion capabilities of UF or porous membranes with the charge selectivity of ED. EDFM acts by attracting the molecules, according to their charge, through filtration membranes with the proper molecular weight cut-off

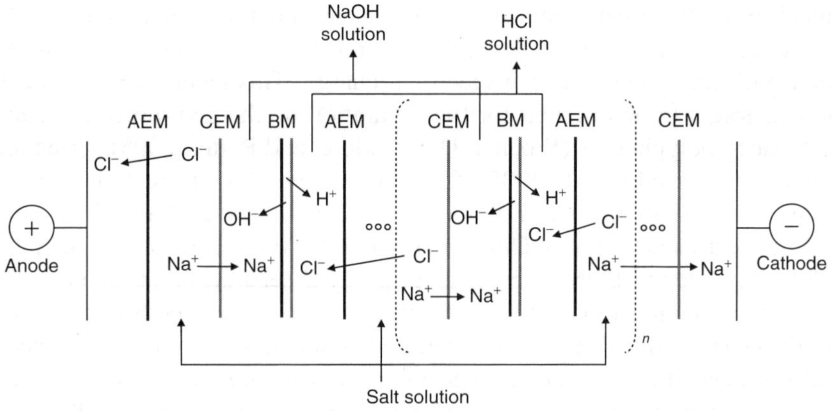

Figure 9.3 Electrodialysis with bipolar membrane cell configuration. BM: bipolar membrane, AEM: anion-exchange membrane, CEM: cation-exchange membrane.

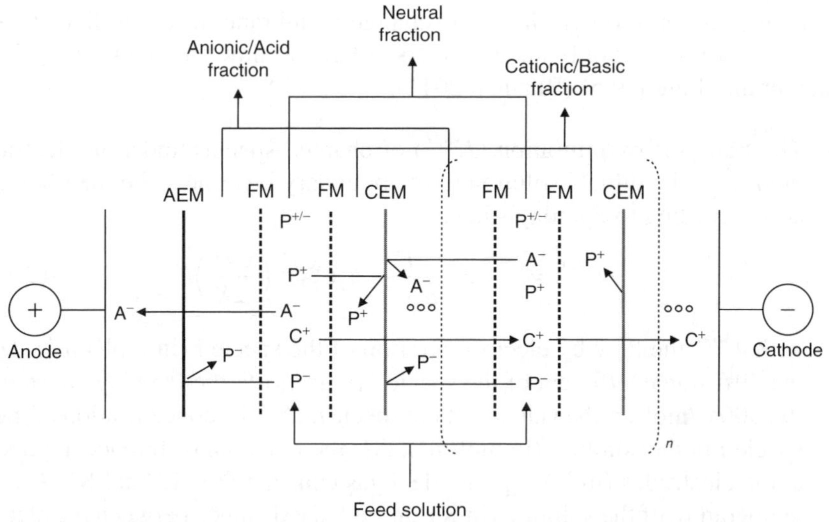

Figure 9.4 Electrodialysis with filtration membrane cell configuration. FM: filtration membrane, AEM: anion-exchange membrane, CEM: cation-exchange membrane.

(MWCO) range according to the fraction of interest. In this application, although a filtration membrane is used, no pressure is applied in the ED cell; only the charged molecules migrate under the effect of the electric field and the neutral molecules theoretically stay in the primary solution and do not reach or pass the filtration membrane (Bazinet and Firdaous, 2013). In addition, EDFM allows the accumulation of molecules at the interface of the membrane to diminish since no pressure is applied. In consequence, the process allows the separation of molecules according to their molecular size and their electrical charge. This technology has been demonstrated to be efficient for the separation of charged molecules such as tobacco polyphenols (Bazinet, DeGrandpré, and Porter, 2005), green tea polyphenols (Labbé et al., 2005), for bioactive peptides purification (Poulin, Amiot, and Bazinet, 2006; Firdaous et al., 2010; Doyen et al., 2012), and for the electroseparation of chitosan oligomers (Aider, Brunet, and Bazinet, 2008). The use of EDFM is an interesting way to separate bioactive peptides and other charged molecules of interest from complex feedstocks in the food, pharmaceutical, fine chemical, and fermentation industries, since it did not need the complete change of the ED configuration, and the ED module is commercially available (Poulin, Amiot, and Bazinet, 2006).

9.2.4 Mass transport

During conventional ED, the mass transfer from a diluate solution to the concentrate solution through the ion-exchange membrane occurs in five steps due to the formation of boundary layers of laminar flow on both sides of the membrane (Figure 9.5) (Bazinet, 2011):

1. The transport by migration (J_i^{Migr}) of charged species under an electric field from the diluate solution to the boundary layer near the membrane acts according to Faraday's law:

$$J_i^{Migr} = -\ D_i\ \frac{F}{R_U\ T}\ z_i C_i^{Sol}\ \left(\frac{\Delta U}{\Delta X}\right) \tag{9.1}$$

   with J_i^{Migr} the flow by electromigration of the specie i (in mol/s m^2), D_i the diffusion coefficient of the specie i (in m^2/s), F the Faraday constant (96 500 C/mol), z_i the valence of the specie i, C_i^{Sol} the concentration of the specie i in the solution (in mol/m^3), ΔU the potential difference applied at the electrodes (in V), R_U the ideal gas constant (8314 J/mol K), T the temperature of the solution (in K), and ΔX the distance between the electrodes (in m).
2. The transport by diffusion (J_i^{Diff}) through the partially desalinated boundary layer of the membrane given by Fick's first law is

$$J_i^{Diff} = \frac{-D_i\ (C_i^{Sol} - C_i^{Memb})}{\delta} \tag{9.2}$$

   with J_i^{Diff} the flow by diffusion of the specie i (in mol/s m^2), C_i^{Memb} the concentration of the specie i in the boundary layer at the interface of the membrane on the diluate side (in mol/m^3), and δ the thickness of the boundary layer (in m).

3. The transport through the membrane (J_i^{Tot}) given by the Nernst–Planck equation (Buck, 1984) is

$$J_i^{Tot} = -D_i \left(\frac{\Delta C_i}{\Delta X} - z_i \frac{F}{R_U T} C_i \frac{\Delta U}{\Delta X} \right) + C_i v \tag{9.3}$$

   with J_i^{Tot} the flux of the specie i through the membrane (in mol/m^2 s), C_i the concentration of the specie i (in mol/m^3), and v the convection velocity (in m/s).

   Although the Nernst–Planck equation did not take into account the coupling of the fluxes between species migrating through the membrane, this equation describes major factors having an effect on the transfer of ions. To describe correctly the transport of charged species through the membranes, it is necessary to couple the Nernst–Planck equation to the electroneutrality condition

$$\sum z_i \ C_i = 0 \tag{9.4}$$

   or, to consider the number of transport of the anions and the cations through the same membrane,

$$\sum (Nbt^- + Nbt^+) = 1 \tag{9.5}$$

   with Nbt^- the transport number of the anions and Nbt^+ the transport number of the cations.
4. The transport by diffusion (J_i^{Diff}) through the partially mineralized boundary layer on the other side of the membrane is given, as previously, by Fick's first law.
5. Finally, the transport by migration (J_i^{Migr}) of the charged species from the mineralized membrane boundary layer to the concentrate solution is given as previously by Faraday's law.

9.3 APPLICATIONS TO DAIRY INGREDIENTS

9.3.1 Proteins

9.3.1.1 Caseins

9.3.1.1.1 Coupling ED with chemical acidification

The use of coupling ED with chemical acidification was first proposed by Laiteries Triballat (1979). In this industrial procedure, the skim milk is acidified in an ED cell by batch processing to a pH ranging from 4.9 and 5.0. The pH decrease of the solution is due to the generation of protons by dissociation of water molecules at the interface of ion-exchange membranes.

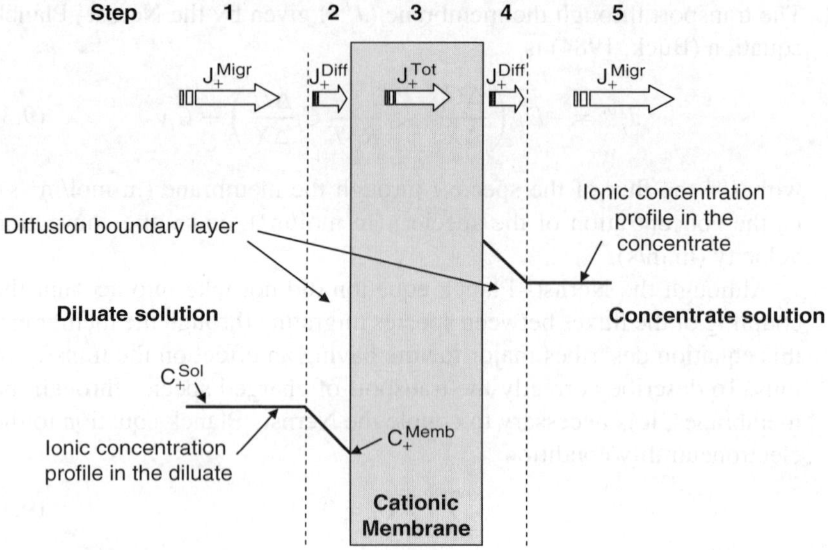

Figure 9.5 Ionic concentration profiles in boundary layers close to a cation-exchange membrane under an electric field and the different mass transfer mechanisms involved in cation migration (adapted from Bazinet, 2011).

In fact, if the pH of the solution is decreased under the pH value range of 4.9–5.0, the local pH of part of the skim milk in contact with the cationic membrane reaches the isoelectric point of the casein and coagulates immediately. Consequently, in an industrial application, it is necessary to follow the pH of the milk treated by ED and to recirculate the milk in a closed loop in the ED cell to obtain this controlled decrease of the pH. This first step completed, the skim milk is separated from the ED cell circuit. The final protein precipitation is obtained by addition of diluate acid until pH 4.6 is reached. Afterwards, the curd is classically prepared by separation from the whey, washing, pressing, and drying to produce casein powder (Bazinet, 2005; Bazinet, Lachkar, and Pouliot, 2012).

9.3.1.1.2 SAFIR process

The SAFIR process using ED was proposed by Bolzer (1985) for the preparation of acid caseinates from fresh milk. In this application, after storage and centrifugal skimming, the milk was circulated in a three-compartment ED cell, at 4 °C, to undergo a transfer and a stoichiometric exchange of protons against milk constitutive cations. A major portion of the cations were eliminated (Na^+, K^+, Ca^{2+}, and Mg^{2+}) and simultaneously the pH of the solution was decreased proportionately due to the generation of protons by water molecule dissociation.

As mentioned by the author, a main reservoir containing caseinate at pH 2.5 was used as pH control of the process. A known quantity of milk was added to the acid caseinate to reach a pH of 3.5. The mixture was then circulated in a ED cell, of which the outlet pH was about 1.8 to 2.4. Part of this very acid caseinate was used to maintain the pH of the control pH reservoir, while the remainder was supplied by pipe to the production process. Under these conditions, the milk added to the pH control rapidly diffuses into the latter and rapidly reaches a pH below 3.5. As a result, the clotting that occurs does not give rise to large aggregates. Micelles formed were sufficiently fine to pass again into a colloidal solution. Thus, a sufficiently homogeneous liquid was obtained to allow transfer to other treatment stages (Bazinet, 2005; Bazinet, Lachkar, and Pouliot, 2012).

9.3.1.1.3 Bipolar membrane electroacidification

A process using EDBM was first proposed by Bazinet et al. in 1999 to precipitate milk casein. In this process, the skim milk was circulated in a three-pair cell configuration and the bipolar membrane electroacidification was performed in a batch process using a current density of 20 mA/cm^2 (Bazinet et al., 1999). The pH of the solution was lowered from pH 6.6 to 4.4. Lowering the pH to 4.4 allowed a precipitation of 80–85% of the total protein. The remaining proteins were whey proteins soluble at pH 4.4. This technology allowed the production of bovine milk casein isolates with a similar or higher protein content than those of commercial isolates (97 versus 95% on a dry basis). The protein profiles of the isolates obtained by HPLC showed that caseins represented 98.5% of the total protein precipitated. Nevertheless, the chemical composition and the casein fraction composition of the isolates were the same for both chemical and electrochemical acidification (Bazinet et al., 2000). Bazinet et al. (2002) suggested adding a certain amount of salt to the skim milk in order to obtain a better electrical efficiency. The combination of KCl at an ionic strength of 0.5 M was the best combination with a 45% decrease in energy consumption in comparison with EDBM of skim milk without salt added.

Following these works, Mier, Ibañez, and Ortiz (2008) studied using the EDBM method the influence of the initial milk concentration (2.5–10 wt%) and current density (100–400 A/m^2) as well as the influence of the initial milk concentration in the same range in the conventional chemical acidification. As expected, the time required to decrease the pH of the milk solution rose with a lower current density as the driving force of the process lessened. The operation time also rose with increasing an initial milk concentration as the amount of protons required to decrease the pH was higher. It was also observed that the purity of the obtained casein rises when there is a lower initial milk concentration and increases slightly with current densities.

The EDBM appears to be a new alternative process for production of high-purity casein bovine milk isolate. The major disadvantages of EDBM are the high cost of equipment and the possibility of fouling of casein curd in the cell. The development of this technology, and its application in the dairy industry, and more generally in the food industry, would help to minimize the cost of the bipolar membranes and ED cells.

9.3.1.2 Whey proteins

9.3.1.2.1 Coupling ultrafiltration and ED with pH chemical adjustment

In this method, proposed by Amundson and Watanawanichakorn (1982) and Slack, Amundson, and Hill (1986), the whey proteins are first concentrated by UF, to partially remove water, salts, lactose, and other low molecular weight compounds. To extract nondesirable low molecular ions such as Na^{+}, K^{+}, Ca^{2+}, and Mg^{2+}, the pH concentrate is adjusted at pH 4.65 with concentrated HCl or NaOH before ED demineralization. The pH of the demineralized concentrate is then readjusted to pH 4.65, if necessary, with 0.1 N HCl or NaOH, and a precipitate containing mainly β-lactoglobulin (β-lg) is formed. This precipitate is separated from the α-lactalbumin (α-la) enriched fraction by simple centrifugation. Using this method, the protein solutions are desalted with a minimal loss of solute (Czytko, Ishii, and Kawai, 1989). However, with a maximum limiting 90% volume reduction of fresh acid whey (about 2.9% protein) by ultrafiltration and demineralization by ED at pH 4.65, only 33.51% of the acid whey proteins was recovered (Bazinet, 2005).

9.3.1.2.2 Coupling ED, pH chemical adjustment, and thermal treatment

Stack et al. (1995) patented a new process derived from the methods of Amundson and Watanawanichakorn (1982), Slack, Amundson, and Hill (1986), and Pearce (1983, 1987, 1988). The Pearce process is a worldwide well-known process based on a thermal separation of whey protein. In this process, the whey is heated to between 55 and 70 °C for at least 30 seconds to allow selective aggregation of the α-la fraction. The flocculated α-la is recovered by centrifugation while β-lg remains soluble under the conditions used, along with other whey constituents. In their derived process, Stack et al. (1995) sought to provide an efficient integrated process for treating whey for the recovery of its constituents, particularly a considerably pure β-lg fraction, an enriched α-la fraction, and lactose. In a first step the raw whey is treated by ED to reduce its mineral content up to 70% of the required demineralization. Then a cation exchanger is used to remove sodium, potassium, magnesium, and particularly calcium. Next, the whey is subjected to a heat treatment at a temperature ranging between 71 and 98 °C for 50 to 95 seconds. At this temperature and under the prevalent ionic conditions, the β-lg remains soluble to a large extent while the α-la undergoes a change that nevertheless results in no flocculation of the α-la protein molecules (Bazinet, 2005). Thus, after the heat treatment step, the whey proteins exist as substantially soluble protein. The whey is then concentrated by a two-stage process to between 55 and 63% in order to remove lactose by crystallization after cooling. In the next stage, the pH of the liquor is adjusted to a pH of between 4.3 and 4.7 at a temperature less than 10 °C and then heating to a temperature of between 35 and 54 °C for 1 to 3 hours. The α-la component of this whey liquor is flocculated at these temperatures (Bazinet, 2005). Stack et al. (1995) did not mention the yield of recovering for both whey protein fractions.

9.3.1.2.3 Bipolar membrane electroacidification
Based on previous results obtained by Amundson and Watanawanichakorn (1982) and Slack, Amundson, and Hill (1986), Bazinet, Ippersiel, and Mahdavi (2004a, 2004b) demonstrated the feasibility of EDBM to fractionate whey proteins. A whey protein isolate (WPI) solution at different concentrations was circulated in a one-pair cell configuration on the cationic layer of the bipolar membrane. Electroacidification was carried out in a batch process using a current density of 20 mA/cm^2 and, after reaching 60 V, the voltage was maintained constant at 60 V. The pH of the solution was lowered from pH 6.9 to 4.6. According to the initial concentration of the WPI solution treated by EDBM, the yield and purity of the protein fraction were different. With a 5% WPI initial concentration, this technology allowed the separation of a 98% pure β-lg fraction at a 44.0% yield of recovering. The 10% protein concentration was the best combination of electrodialytic parameters and protein recovery : separation of a 95.3% pure β-lg fraction at a 53.4% yield of recovering. The β-lg-enriched fraction contained 2.7% of α-la for a 98% total protein purity.

9.3.1.3 Lactoferrin

Bovine lactoferrin (LF), an 80 kDa iron-binding glycoprotein, present in whey at an average concentration of 0.1%, has been reported to have biological properties such as anti-inflammatory, antimicrobial, and immunostimulatory activities.

To overcome the disadvantages of low selectivity and solution contamination of pressure-driven and chromatographic processes, ED with a filtration membrane (EDFM) was tested by N'Diaye et al. (2010). To set the optimum conditions for electroseparation of this molecule, its electrophoretic mobility was measured on a large range of pH (pH 3 to 12), and the optimal electrophoretic mobility of LF was determined at pH 3.0 with a value of 1.5×10^{-8} m^2/V s in a 2 g/L KCl solution and of 3.0×10^{-8} m^2/V s in distilled water. Thereafter, a 0.1% LF solution at pH 3.0 was treated by EDFM with an ultrafiltration membrane of 500 kDa molecular weight cut-off. EDFM treatments were carried out using 200 mL of LF solutions in a batch process using a constant voltage difference of 20 V. A migration rate going up to 46% was obtained after 4 hours of treatment. Thus, it appears that the EDFM process could allow the separation of large proteins, such as LF, from the solution (N'Diaye et al., 2010). However, the deposit observed on the cation-exchange membrane and the decrease of the migration rate of the LF after consecutive EDFM treatments have suggested fouling on the cation exchange membrane since LF is a high molecular weight protein. The cleaning of membranes to remove the deposit should be considered for further industrial application.

9.3.2 Lactose

Cloidt and Lehmann (2007) and Evdokimov et al. (2009) proposed ED processes for the production of lactose from whey. In their patent, Evdomikov et al. (2009) proposed to desalinate the skimmed whey first by nanofiltration at a pressure of 2.0–2.5 MPa until a dry substance content of 18–20% was reached. In a second step, the demineralization was achieved by ED. The demineralized whey was then acidified at pH 4.4–4.6 and kept at a temperature of 90–95 °C to precipitate proteins. The solution separated from the protein was then cooled and condensed at temperatures between 55–60 °C to reach a final dry substance content of 60–65%. Lactose was then crystallized in syrup during 20–22 hours while cooled down to 8–10 °C. Crystals were separated by centrifugation at this temperature and dried at 120–140 °C. Milk sugar was obtained at a 95–97% purity. The authors claimed that their process increases the speed of obtaining milk sugar in comparison with known ways. The main differences in the process proposed by Cloidt and Lehmann (2007) is that (1) the whey protein separation is achieved by ultrafiltration instead of acid precipitation, (2) a part of the partially desugarized permeate that is isolated from the lactose concentrate is fed back to the lactose concentrate, prior to the evaporation step, and (3) according to the authors the ED step is used for water removal from the permeate instead of demineralization (Bazinet and Firdaous, 2011). This application by Cloidt and Lehman (2007) and water removal by ED is quite surprising but no other information was presented in their European patent application concerning the effectiveness of this phenomenon. For the removal of water they also proposed the use of membrane separation, such as reverse osmosis, which appears to be more efficient and more realistic for such an application (Bazinet and Firdaous, 2011).

9.3.3 Organic acids

9.3.3.1 Lactic acid

In order to recover lactic acid from fermentation broth, Norddahl (2001) suggested a two-stage ED lactic acid recovery process. The fermentation liquid was first subjected to a clarification by ultrafiltration. The ultrafiltration permeate was then treated in an ion-exchange unit to replace magnesium and calcium ions with sodium ions, thus preventing a slow irreversible scaling of the membranes in a subsequent ED treatment. The resulting eluate from the ion exchange operation was then concentrated by a conventional ED process to improve the global yield of the process (purification in terms of organic constituents such as residual sugar and amino acids). The resulting concentrate was then subjected to ED with bipolar membranes. The salts formed were converted into lactic acid, inorganic acids, and an ammonium hydroxide solution. Ammonium lactate was converted into ammonium hydroxide and lactic acid in two separate streams, a base stream and an acid stream. The ammonium hydroxide solution was led back to the fermentation reactor to regulate pH. The acid solution was led to a strong cationic ion exchanger where cations were exchanged with hydrogen ions. The eluate from the cation exchange operation was then treated in a weak anionic ion exchanger in the hydroxide form, adsorbing the lactic acid. Finally, the lactic acid was purified and concentrated to the desired concentration using a falling film multistage vacuum evaporator or in a compression evaporator. According to the author, this process based on the combination of the above-described steps allowed a quite high overall recovery rate of lactic acid, about 85–90%, based on the amount of sugar used in the fermentation. A similar two-stage ED process to purify salts and organic acids was patented by Mani (2001). However, the succession of unit operations is quite different, with especially nanofiltration suggested as a solution to retain divalent ions prior to ED processing (Bazinet and Firdaous, 2011).

A more simplified lactic acid recovery electrodialytic process was recently proposed by Baets and Groot (2007). Broth containing multivalent ions and lactate ions negatively charged were directly introduced into a first compartment ED apparatus, which was limited by an anion exchange membrane and a cathode. Multivalent ions (magnesium, calcium, iron, etc.) were converted to obtain a residual stream comprising the hydroxides of the multivalent ions and lactate ions which were transported through the anion exchange membrane into a second compartment limited by an anion-exchange membrane and an anode, after which the lactate ions were neutralized to lactic acid. According to the invention, the ED process uses in the first compartment a second membrane, being an anion selective membrane or a bipolar membrane having its cation-selective side directed to the cathode (Figure 9.6). The anion-exchange membrane through which

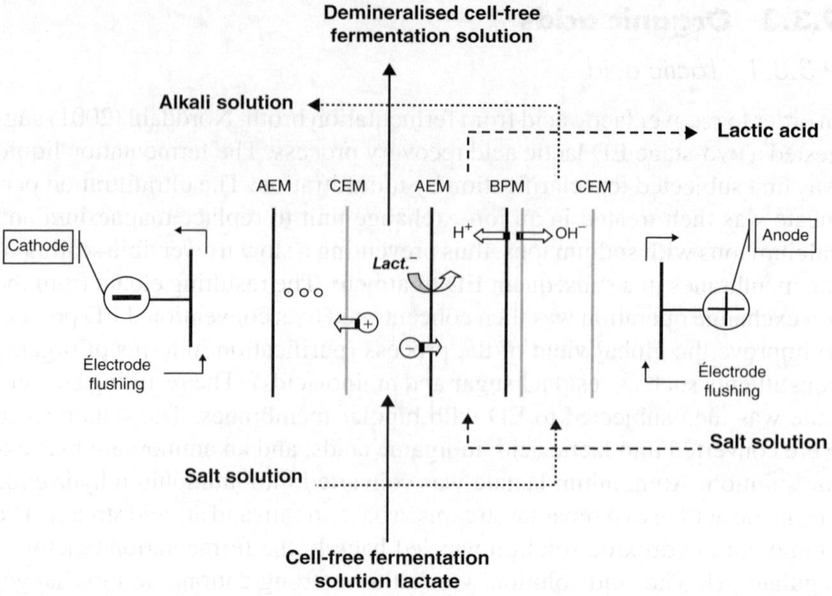

Figure 9.6 Schematic diagram of a bipolar membrane electrodialysis unit using a three-compartment configuration for lactate migration. BM: bipolar membrane, AEM: anion-exchange membrane, CEM: cation-exchange membrane.

the lactate ions were transported is preferably monovalent permselective. After isolation, the lactic acid may be further purified by distillation, extraction, filtration, or evaporation. The procedure described in this patent leads to a lactic acid end product that is less expensive and without the introduction of additional chemicals (Bazinet and Firdaous, 2011).

For lactic acid production, Boergardts, Krischke, and Troesch (1998) proposed an integrated three-stage bioprocess for dairy sewage treatment, using a bipolar membrane ED. In the first stage, the wastewater was pretreated with base to precipitate out the majority of Ca^{2+} and Mg^{2+} ions and PO_4^{3-} ions present. In the second stage of the process, the pretreated wastewater was fed to a fermentation reactor and the fermentation broth formed was given a secondary purification. In the fermenter, the lactose present and any other sugars were converted homofermentatively to lactic acid using lactic acid bacteria. To lead to significantly higher productivity, the authors used a cell retention system operated by a microfiltration membrane unit fed by the circulating stream from the fermenter. However, the cells could be removed by membrane processes or by centrifugation. In the third stage, the permeate was fed to the ED system via a nanofiltration unit or via a selective ion exchanger to remove any calcium residues still present. In this third stage, the concentration of the lactic acid in the wastewater was reduced. In their invention, Boergardts, Krischke, and Troesch (1998) suggested that the ED could be carried out in two stages. The fermentation broth was subjected in a first stage to a bipolar membrane ED in which the broth was continuously reduced in concentration to about 10 to 15 g/L diluate concentration. In a second stage, a conventional ED was carried-out in which the concentration wass further decreased to the desired wastewater concentration, preferably in a batch procedure. The sodium lactate produced in the second stage can be returned to the feed stream of the first ED stage in order to increase the starting concentration of lactate there. According to Boergardts, Krischke, and Troesch (1998), the free lactic acid may be produced at a concentration of approximately 200 g/L.

The use of ED with bipolar membranes also allowed the isolation of free acid in high concentration and purity directly from the fermentation solution. As a third product in this case, alkali solution was formed, which could serve to elevate the pH during pretreatment of the wastewater. Alternately, the alkali solution formed could be used to regulate the pH of the fermenter. The EDBM configuration was the same as the one used by Van Nispen and Jonker (1991), wherein the fermentation solution filtrate circulates in the diluate circuit.

9.3.3.2 Lactobionic acid

Lactobionic acid (LBA, 4-*O*-β-D-galactopyranosyl-D-gluconic acid) is an aldonic acid obtained from the oxidation of lactose, with high potential applications as an ingredient in foods, pharmaceutical, and chemical products. The main commercial application of LBA is as an ingredient of the solutions employed for stabilizing organs prior to transplantation (Nordkvist, Nielsen, and Villadsen, 2007) and represents a new ingredient in skin care products featuring potent antioxidant, wound healing, humectant anti-aging, and keratinizing properties (Berardesca et al., 1997; Yu, Du, and Chen, 2011). LBA is an excellent carrier for calcium supplementation and has recently been employed in the design of functional nanomaterials for carrying drugs (Luo et al., 2011; Guo et al., 2012 ; Gutiérrez, Hamoudi, and Belkacemi, 2012).

Very recently, Gutiérrez et al. (2013) investigated the viability of producing LBA from a sodium lactobionate (LBNa) solution obtained by the aerobic oxidation of lactose over gold nanoparticles supported on mesoporous silica, using a three-compartment ED with a bipolar membrane (EDBM) stack with an effective area of 0.01 m^2. The results indicated that LBNa was continuously split in LB^- and Na^+ ions, which formed LBA and NaOH in the acid and base compartments, respectively. A demineralization rate of about 50% was achieved after applying a voltage difference of 5.0 V during 100 min followed by 5.5 V during 80 min, while the LBA concentration increased about 2.5 times in the acid compartment. The specific energy consumption was 0.57 kW h/kg over the whole stack. Under the low voltage difference applied to the EDBM stack (5.0–5.5 V), the current efficiency after 180 min was about 42% and the conversion of LBNa into LBA was circa 50%. In spite of the low yield and current efficiency obtained in this preliminary study compared to the data given in the literature for the recovery of other acids by means of the EDBM, at the investigated operating conditions, the LBA recovery was more than twice that reported previously by Peretti, Silveira, and Zeni (2009). Although sodium was found in minor amounts (17.5 ppm) in the acid compartment, these preliminary results show for the first time the viability of producing LBA by means of an integral process comprising the catalytic oxidation of lactose and the EDBM, with the advantage that the NaOH could be efficiently recovered for its reutilization in further oxidation reactions. Further improvement of the process in terms of yield and current efficiency may be possible by adding more cell units to the ED system, which increases the effective area by increasing the concentration of the solutions circulating through the stack and by using a higher voltage difference during the operation of the system.

9.3.4 Bioactive peptides

β-Lactoglobulin (β-lg), the major whey protein, can release different bioactive peptidic sequences by enzymatic hydrolysis (Pihlanto-Leppälä, Rokka, and Korhonen, 1998 ; Bazinet and Firdaous, 2009, 2013). However, higher functionality would require the purification of these bioactive peptides from the protein hydrolysate. Amongst others, β-lg 142–148 has been reported to be the most potent antihypertensive sequence.

9.3.4.1 Separation of β-lg bioactive peptides by EDFM

Poulin, Amiot, and Bazinet (2006) were the first to evaluate the feasibility of separating basic/cationic and acid/anionic peptides from a β-lg hydrolysate using one or several 20 kDa ultrafiltration (UF) membranes stacked in an ED cell. The authors demonstrated that the EDFM was a selective method of bioactive peptide separation since amongst a total of 40 peptides in the raw hydrolysate, only 13 were recovered in the separated adjacent solutions (anionic recovery compartment, KCl 1, and cationic recovery compartment, KCl 2). Amongst these 13 migrating peptides, three acid/anionic peptides migrated only in one compartment (KCl 1), while three basic/cationic peptides migrated only in the second compartment (KCl 2), and that was regardless of the pH conditions of the hydrolysate solution. Furthermore, the highest migration was obtained for an antihypertensive peptide (β-lg 142–148) with a value of 10.75%. According to the first measurement to evaluate UF membrane fouling, it appeared that the integrity of the UF membrane was kept and that the EDFM would minimize its fouling.

In further works, Poulin, Amiot, and Bazinet (2007, 2008) studied the possible EDFM fractionation and concentration of the ACE-inhibitory sequence β-lg 142–148 by studying the influence of the effective membrane surface, the strength of the electrical field, the feed solution flow rate and the membrane conditioning before treatment. Two different membrane surfaces (10 and 40 cm^2), three values of electrical field strength (2.75, 5.5, and 11 V/cm), and four flow rates (100, 150, 200 or 250 mL/min) were studied to determine the global influence of those process parameters on the peptide migration rate and permeate composition. The increase of the membrane area resulted in a fourfold increase of the peptide concentration. Moreover, raising the electrical field strength also resulted in increasing considerably the peptide concentration in the permeate solution. In the best conditions, a total migration of 29% was obtained for peptide β-lg 142–148 after only 90 minutes of processing with a relative concentration of 30%. However, at a high electrical field and membrane area, the final migration rate was affected due to a retro-migration phenomenon linked with pH changes. It appeared that the feed solution flow rate had no impact on the yield of the process but induced changes in the selectivity. In fact, increases in the flow rate decreased the migration of the peptides with limited electrophoretic mobility. Increasing the field strength and membrane surface was also associated with faster and more important changes in solution conductivity and pH, which had a strong influence on peptide migration (Poulin, Amiot, and Bazinet, 2007).

9.3.4.2 Production of β-lg bioactive peptides by simultaneous hydrolysis and separation by EDFM

Very recently, Doyen, Husson, and Bazinet (2013) investigated the feasibility of combining enzymatic hydrolysis of β-lg with the separation of released peptides in an EDFM process (Figure 9.7). The hydrolysis of β-lactoglobulin protein and the fractionation of generated peptides were performed in one step in an ED cell with ultrafiltration membranes stacked. After 240 min of treatment, amongst the 31 peptides detected in the β-lactoglobulin hydrolysate, 16 and 4 peptides were detected, respectively, in the anionic and cationic peptide recovery compartments. Amongst these 16 anionic peptides, two hypocholesterolemic, three antihypertensive, and one antibacterial peptides were recovered and concentrated with migration rates ranging between 5.5 and 66.0%. Amongst the four cationic peptides, the peptide sequence ALPMHIR (β-lg 142–148), identified as lactokinin and known to exert an important antihypertensive effect, was recovered with an estimated 66% migration rate. In the authors' knowledge, it was the first time that hydrolysis was conducted under an electric field to simultaneously produce separate anionic and cationic peptides.

9.3.4.3 ACE inhibiting peptide recovery from lactic acid fermentation liquid

Kitamura and Ueyama (2001) proposed a purification method by ED for whey separated from lactic acid fermentation liquid. After being cultured by culturing lactic acid bacteria, or lactic acid bacteria and yeast, the supernatant part of the whey was obtained by solid–liquid separation or decantation. Whey thus obtained contains large amounts of angiotensin-converting enzyme (ACE) inhibitor peptides, mainly from β-lg as well as acids, ions, and proteins. In order to use the whey as an anti-high blood pressure agent or foods, a further ED purification step is proposed. The ED purification process is based on the basic/cationic character of the target ACE inhibitor peptides contained in the whey and loose-type anion-exchange membranes having a permeability of diffusion coefficient in the range of 3.0 to 9.0×10^{-6} cm/s. This type of membrane has rather large holes, through which charged substances having large molecular weights easily penetrate. Hence, acids including lactic acid, ions, and water were smoothly removed, and the ACE inhibitor peptides remained in the concentrate at high yields for a very short period of treatment. According to the authors, this purification method allows a very high yield of ACE inhibitor peptide recovery and a significant removal of impurities (Bazinet and Firdaous, 2011).

9.3.5 Phospholipids

Cheese whey contains 7–10 g/L of proteins and about 2–8 g/L of lipids (Lehmann and Wasen, 1990). These lipids are composed of almost 66% of nonpolar lipids and 33% of polar lipids. Polar lipids are mainly phospholipids with 34% phosphatidylethanolamine, 31% phosphatidylcholine, 15% sphingomyelins, 12% phosphatidylinositol, and 8% phosphatidylserine (Théodet and Gandemer, 1994).

Lin Teng Shee, Angers, and Bazinet (2005) used EDBM technology for acidification and decreasing the ionic strength of a fresh cheddar cheese whey. In this study, the EDBM process was carried out with or without a preliminary decrease of whey mineral salts content by conventional ED to obtain precipitates with a high level of lipids. After centrifugation of the treated whey, the composition of flocs and precipitation yields were determined. Whey centrifugation at 1000 g during 5 min (Process 1) allowed a 20.8% recovery of cheddar cheese whey initial lipids. Of the other components, proteins and lactose, very few precipitated (1.1% and 0.6%,

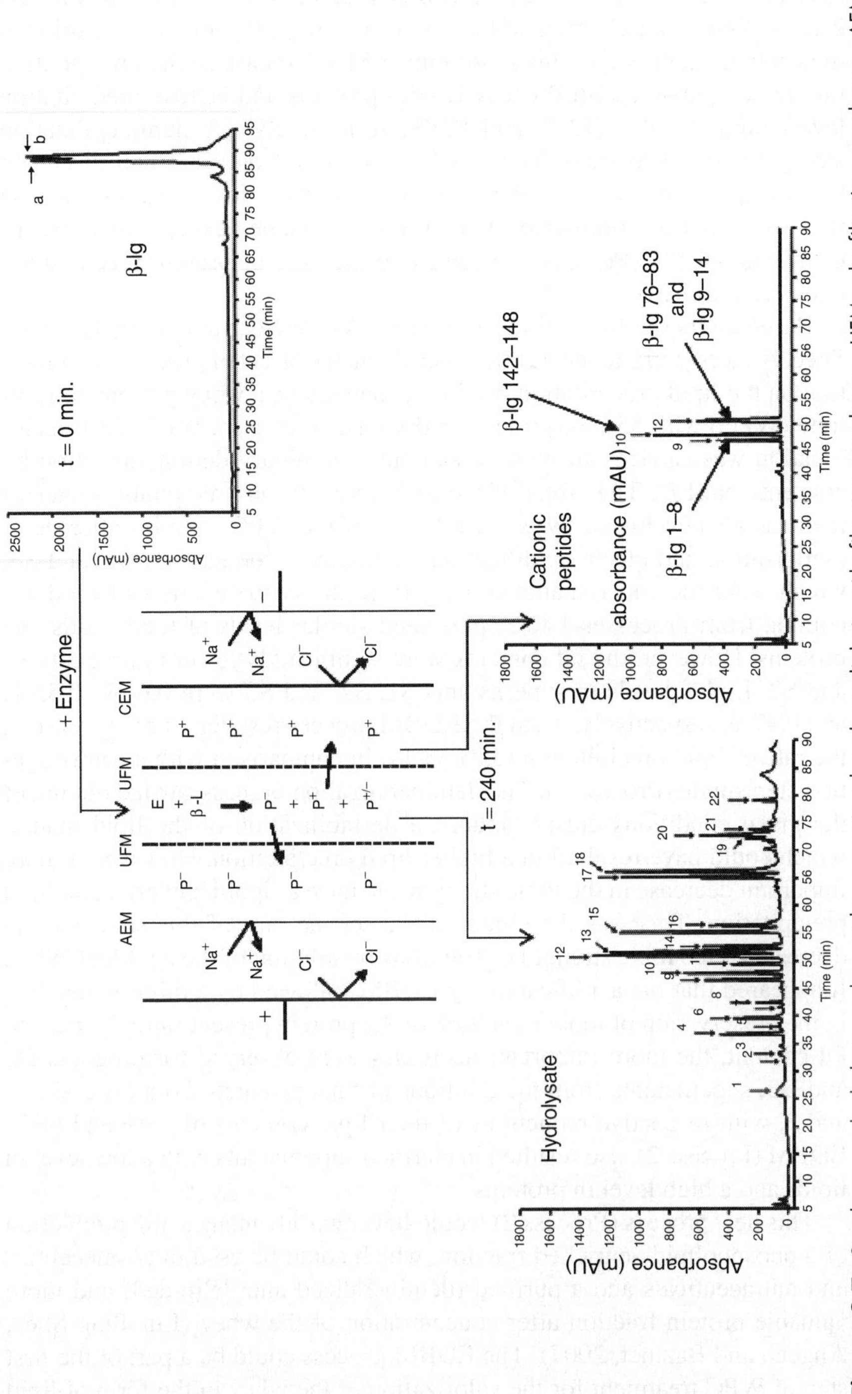

Figure 9.7 Simultaneous hydrolysis and peptide separation by electrodialysis with filtration membrane. UFM: ultrafiltration membrane, AEM: anion-exchange membrane, CEM: cation-exchange membrane.

respectively). A 32.1% whey lipid precipitation was obtained in Process 2 consisting of an electroacidification to reach a pH value of 3.7 prior to the centrifugation step. This represents a 54% increase in the precipitation rate in comparison with Process 1, with proteins and lactose precipitation levels quite similar (1.9% and 0.9%, respectively). A demineralization step prior to electroacidification (Process 3) had only a small effect on the precipitation level. Similar precipitation levels for lipids and lactose were obtained in comparison with Process 2 values except for proteins. Conventional ED (Process 3) allowed an increase of protein precipitation from 1.9% to 3.3%.

Based on these observations, further works were carried out by Lin Teng Shee, Angers, and Bazinet (2007) to evaluate the effect of protein concentration on the lipid precipitation. In this second study, a whey protein concentrate (WPC) with 55% of protein on dry basis was used. WPC electroacidification was carried out with or without preliminary demineralization by conventional ED. The effect of the ionic strength on precipitation rates of fats was also evaluated by water dilution of the WPC samples after demineralization and electroacidification treatments. Products S1, S2, and S3, which were the supernatants coming from the centrifugation of products coming from Processes 1 to 3, presented similar levels of total solids and proteins. However, the fat contents were significantly lower in products S1 and S2. Lipids levels in supernatants S1, S2, and S3 were 0.63%, 0.55%, and 0.47%, respectively. Then the EDBM process resulted in an increase in the rate of lipid precipitation (35% yield) in comparison with a centrifugation step single (Process 1). The demineralization process carried out under the given conditions did not induce a destabilization of the lipid matter, which could have resulted in a higher lipid precipitation. However, a more important decrease in the ionic strength can have a significant effect on lipid precipitation. This was the object of the second part of this study on the decrease of the ionic strength of solutions by dilution of the acidified WPC. It appeared that the acidification by EDBM followed by a dilution resulted in the preservation of more than 90% of the protein present initially. For the fat content, the more important decreases were observed for products D2 and D3, supernatants from the dilutions of final products from Processes 2 and 3, with respective reductions of their lipid contents of 73% and 66%. EDBM (Process 2) also resulted in clarified supernatants with a low level in lipids and a high level in proteins.

This new process (Process 2) would have two advantages, the production of a phospholipids-enriched fraction, which could be used in cosmeceutics and nutraceuticals and a purified (demineralized and delipided) and more valuable protein fraction after concentration of the whey (Lin Teng Shee, Angers, and Bazinet, 2007). The EDBM process could be a part of the first step of WPC treatment for the valorization of the whey in the form of lipid fractions of different nature. In comparison with the chemical acidification, the EDBM process has the advantage of offering continuous acidification without the addition of salt. This could allow, in an industrial process, the ability (1) to maintain a low ionic strength of the solution, for example, prior or following WPC dilution, and (2) to produce a purified WPC (with a low ash level).

9.4 CONCLUSION AND PERSPECTIVES

The use of electrodialytic systems in dairy production lines can increase the ecoefficiency of the global processes. Indeed, the in situ generation of base and acid by EDBM would decrease the needs for chemicals in caseins production and the risks associated with their handling and storage. In addition, the new electrodialytic processes developed recently have demonstrated valuable applications in the dairy industry for the production of high added-value compounds such as bioactive peptides and organic acids. These new processes allow a more selective separation or fractionation of milk or milk by-products in valuable compounds. Due to the increasing market for functional foods and nutraceuticals, ED hybrid processes could be used for the separation of some minor compounds of milk and/or to valorize milk by-products.

The perspectives in the field of ingredient separation by electrodialytic processes will be the development of new membrane materials and to use the effect of ionic strength decrease created by demineralization coupled with pH change by a bipolar membrane for developing new applications. The use of a filtration membrane could be an interesting way to separate molecules of interest and to decrease protein fouling. The combination of different molecular weight cut-off filtration membranes in an ED stack, as proposed by Doyen et al. (2014), could also be a promising technology for the simultaneous separation of molecules with different molecular weights. Finally, the use of binding membranes during ED could be a potential way to improve the separation of large molecules that could not migrate through filtration membranes under an electric field due to their low charge or electrophoretic mobilities.

Nomenclature

C_i = concentration of the specie i (in mol/m^3)

C_i^{Memb} = concentration of the specie i in the boundary layer at the interface of the membrane on the diluate side (in mol/m^3)

C_i^{Sol} = concentration of the specie i in the solution (in mol/m^3)

D_i = diffusion coefficient of the specie i (in m^2/s)
F = Faraday constant (96 500 C/mol)
J_i^{Diff} = flow by diffusion of the specie i (in mol/s m^2)
J_i^{Migr} = flow by electromigration of the specie i (in mol/s m^2)
J_i^{Tot} = flux of the specie i through the membrane (in mol/m^2 s)
Nbt^- = transport number of the anions
Nbt^+ = transport number of the cations
R_U = ideal gas constant (8314 J/mol K)
T = temperature of the solution (in K)
z_i = valence of the specie i
δ = thickness of the boundary layer (in m)
ΔU = potential difference applied at the electrodes (in V)
ΔX = distance between the electrodes (in m)
v = convection velocity (in m/s)

References

Agyei, D. and Danquah, M.K. (2011) Industrial-scale manufacturing of pharmaceutical-grade bioactive peptides. *Biotechnology Advances*, **29**, 272–277.

Aider, M., Brunet, S., and Bazinet, L. (2008) Electro-separation of chitosan oligomers by electrodialysis with ultrafiltration membrane (EDUF) and impact on electrodialytic parameters. *Journal of Membrane Science*, **309**, 222–232.

Amundson, C.H. and Watanawanichakorn, S. (1982) Production of enriched protein fractions of β-lactoglobulin and α-lactalbumin from cheese whey. *Journal of Food Processing and Preservation*, **6**, 55–71.

Baets, P. and Groot, W.J. (2007) Method of separating multivalent ions and lactate ions from a fermentation broth. US patent 7,217,349.

Baker, R.W. (2004) Ion exchange membrane processes – Electrodialysis, in *Membrane Technology and Applications*, second edition, John Wiley & Sons, Ltd, Chichester, England, pp. 393–422.

Batchelder, B.T. (1987) Electrodialysis applications in whey processing. *Bulletin of the International Dairy Federation*, **212**, 84–90.

Bazinet, L. (2005). Electrodialytic phenomena and their applications in the dairy industry: a review. *CRC Critical Review in Food Science and Nutrition*, **45**, 307–326.

Bazinet, L. (2011) Déminéralisation, enrichissement et stabilisation par procédés électromembranaires, in *Concepts de Génie Alimentaire: Procédés Associés et Applications à la Conservation des Aliments* (eds. L. Bazinet and F. Castaigne), Éditions Tec & Doc Lavoisier, Paris, pp. 527–568.

Bazinet, L., DeGrandpré, Y., and Porter, A. (2005) Electromigration of tobacco polyphenols. *Separation and Purification Technology*, **41**, 101–107.

Bazinet, L., and Firdaous, L. (2009) Membrane processes and devices for separation of bioactive peptides. *Recent Patents on Biotechnology*, **3**, 61–72.

Bazinet, L. and Firdaous, L. (2011) Recent patented applications of ion-exchange membranes in the agri-food sector. *Recent Patents on Chemical Engineering*, **4**, 207–216

Bazinet, L. and Firdaous, L. (2013) Separation of bioactive peptides by membrane processes: technologies and devices. *Recent Patents on Biotechnology*, **7**, 9–27.

Bazinet, L., Ippersiel, D., and Mahdavi, B. (2004a) Fractionation of whey protein by bipolar membrane electroacidification. *Innovative Food Sciences and Emerging Technology*, **5**, 17–25.

Bazinet, L., Ippersiel, D., and Mahdavi, B. (2004b) Effect of conductivity adjustment on the separation of whey protein by bipolar membrane electroacidification. *Journal of Agricultural and Food Chemistry*, **52**, 1980–1984.

Bazinet, L., Lachkar, D.P., and Pouliot, Y. (2012) Emerging membrane technologies and applications for added-value dairy ingredients, in *Membrane Technologies and Applications* (eds. K. Mohanty and M.K. Purkait), CRC Press, Taylor and Francis Group, Amsterdam, pp. 201–212.

Bazinet, L., Lamarche, F., Ippersiel, D., and Amiot, J. (1999) Bipolar membrane electro-acidification to produce bovine milk casein isolate. *Journal of Agricultural and Food Chemistry*, **47**, 5291–5296.

Bazinet, L., Lamarche, F., Ippersiel, D., et al. (2000) Comparison of electrochemical and chemical acidification of skim milk. *Journal of Food Science*, **65**, 1303–1307.

Bazinet, L., Gendron, C., Ippersiel, D., et al. (2002) Effect of type of added salt and ionic strength on physicochemical and functional properties of casein isolates produced by electroacidification. *Journal of Agricultural and Food Chemistry*, **50**, 6875–6881.

Bazinet, L., Amiot, J.. Poulin, J.-F., Tremblay, A., and Labbé, D. (2005) Process and system for separation of organic charged compounds. World patent application PCT/CA2005/000337 and CA patent 2,558,164.

Berardesca, E., Distante, F., Vignoli, G.P., Oresajo, C., and Green, B. (1997) Alpha hydroxyacids modulate stratum corneum barrier function. *Britain Journal of Dermatology*, **137**, 934–938.

Boergardts, P., Krischke, W., and Troesch, W. (1998) Process for purifying dairy wastewater. US patent 5,746,920.

Bolzer, R. (1985) Installation and process for the preparation of acid caseinates. US patent 4,559,119.

Buck, P. (1984) Microscopic base and limits for Nernst–Planck equation applied to membrane systems. *Journal of Membrane Science*, **17**, 1–62.

Cloidt, R. and Lehmann, H. (2007) Process for the preparation of lactose from whey. *EP patent* **1**,869,984-A1.

Czytko, M., Ishii, K., and Kawai, K. (1989) Continuous process for preparing organic acids by fermentation. US patent 4,882,277.

Doyen, A., Husson, E., and Bazinet, L. (2013) Use of an electrodialytic reactor for the simultaneous β-lactoglobulin enzymatic hydrolysis and fractionation of generated bioactive peptides. *Food Chemistry*, **136**, 1193–1202.

Doyen, A., Saucier, L., Beaulieu, L., Pouliot, Y., and Bazinet. L. (2012) Electroseparation of an antibacterial peptide fraction from snow crab by-products hydrolysate by electrodialysis with ultrafiltration membranes. *Food Chemistry*, **132**, 1177–1184.

Doyen, A., Udenigwe, C.C., Mitchell, P., Marette, A., Aluko, R.E., and Bazinet, L. (2014) Respective anti-diabetic and antihypertensive activities of two flaxseed protein hydrolysate fractions revealed following their simultaneous separation by electrodialysis with ultrafiltration membranes. *Food Chemistry*, **145**, 66–76.

Evdokimov, I.A., Dykalo, N.J., Permjakov, A.V., and Smirnov, E.R. (2009) Production method of milk sugar out of salty wheyfield. RU patent **2**,360,006.

Fidaleo, M. and Moresi, M. (2006) Electrodialysis applications in the food industry. *Advances in Food and Nutrition Research*, **51**, 265–360.

Firdaous, L., Dhulster, P., Amiot, J., Doyen, A., Lutin, F., Vézina, L.-P., and Bazinet, L. (2010) Investigation of the large scale bioseparation of an antihypertensive peptide from alfalfa white protein hydrolysate by an electromembrane process. *Journal of Membrane Science*, **355**, 175–181.

Fosset, S. and Tomé, D. (2002) Nutrition and health: nutraceuticals from milk, in *Encyclopedia of Dairy Sciences*, second edition, Elsevier, pp. 1062–1066.

Fox, P.F. and McSweeney, P.L.H. (1998) Dairy Chemistry and Biochemistry, Blackie Academic, London.

Gauthier, S.F., and Pouliot, Y. (2003) Functional and biological properties of peptides obtained by enzymatic hydrolysis of whey proteins. *Journal of Dairy Science*, **86**, 78–87.

Guo, R., Yao, Y., Cheng, G., et al. (2012) Synthesis of glycoconjugated poly(amindoamine) dendrimers for targeting human liver cancer cells. *RSC Advances*, **2**, 99–102.

Gutiérrez, L.F., Hamoudi, S., and Belkacemi, K. (2012) Lactobionic acid: a high value-added lactose derivative for food and pharmaceutical applications. *International Dairy Journal*, **26**, 103–111.

Gutiérrez, L.-P., Bazinet, L., Hamoudi, S. and Belkacemi, K. (2013) Production of lactobionic acid by means of a process comprising the catalytic oxidation of lactose and bipolar membrane electrodialysis: a concept-proof study. *Separation and Purification Technology*, **109**, 23–32.

Jain, S.M. and Reed, P.B. (1985) Electrodialysis, in *Comprehensive Biotechnology*, Vol. **2**: *The Principles of Biotechnology: Engineering Considerations* (eds. C.L. Cooney and A.E. Humphrey), Pergamon, New York, pp. 575–590.

Kitamura, S. and Ueyama, T. (2001) Method for purifying whey of lactic acid fermentation by electrodialysis. US patent 6,204,362.

Kitts, D.D. and Weiler, K. (2003) Bioactive proteins and peptides from food sources. Applications of bioprocesses used in isolation and recovery. *Current Pharmaceutical Design*, **9**, 1309–1323.

Korhonen, H. and Pihlanto, A. (2006) Bioactive peptides: production and functionality. *International Dairy Journal*, **16**, 945–960.

Labbé, D., Araya-Farias, M., Tremblay, A., and Bazinet, L. (2005) Electromigration feasability of green tea catechins. *Journal of Membrane Science*, **254**, 101–109.

Laiteries Triballat (1979) Procédé et installation pour la préparation de la caséine à partir du lait et produits ainsi obtenus. FR patent 2,428,626.

Lehmann, H. and Wasen, D. (1990) Method of dephospholipidating whey. US patent 4,897,279.

Leitz, F.B. (1986) Measurements and control in electrodialysis. *Desalination*, **59**, 381–401.

Lin Teng Shee, F., Angers, P., and Bazinet, L. (2005) Precipitation of cheddar cheese whey lipids by electrochemical acidification. *Journal of Agricultural and Food Chemistry*, **53**, 5635–5639.

Lin Teng Shee, F., Angers, P., and Bazinet, L. (2007) Delipidation of a whey protein concentrate by electroacidification with bipolar membranes (BMEA). *Journal of Agricultural and Food Chemistry*, **55**, 3985–3989.

Luo, Z., Cai, K.Y., Hu, Y., et al. (2011) Mesoporous silica nanoparticles end-capped with collagen: redox-responsive nanoreservoirs for targeted drug delivery. *Angewandte Chemie International Edition*, **50**, 640–643.

Mani, K.N. (1991) Electrodialysis water splitting technology. *Journal of Membrane Science*, **58**, 117–138.

Mani, K.N. (2001) Apparatus and process for electrodialysis of salts. US patent 6,221,225.

Mier, M.P., Ibañez, R., and Ortiz, I. (2008) Influence of process variables on the production of bovine milk casein by electrodialysis with bipolar membranes. *Biochemical Engineering Journal*, **40**, 304–311.

N'diaye, N., Pouliot, Y., Saucier, L., Beaulieu, L., and Bazinet, L. (2010) Electroseparation of bovine lactoferrin from model and whey solutions. *Separation and Purification Technology*, **74**, 93–99.

Norddahl, B. (2001) Fermentation production and isolation of lactic acid. US patent 6,319,382.

Norddahl, B., Eriksen, S., and Pedersen, F.M. (2001) Method for purifying lactic acid. *WO patent* **01**,92555-A1.

Nordkvist, M., Nielsen, P.M., and Villadsen, J. (2007) Oxidation of lactose to lactobionic acid by a *Microdochium nivale* carbohydrate oxidase: kinetics and operational stability. *Biotechnology and Bioengineering*, **97**, 694–707.

Pearce, R.J. (1983) Thermal separation of β-lactoglobulin and α-lactalbumin in bovine cheddar cheese whey. *Australian Journal of Dairy Technology*, **38**, 144–149.

Pearce, R.J. (1987) Fractionation of whey proteins. *Australian Journal of Dairy Technology*, **42**, 75–78.

Pearce, R.J. (1988) Fractionation of whey proteins. *Bulletin of the International Dairy Federation*, **212**, 150–153.

Peretti, F.A., Silveira, M.M., and Zeni, M. (2009) Use of electrodialysis technique for the separation of lactobionic acid produced by *Zymomonas mobilis*. *Desalination*, **245**, 626–630.

Pihlanto-Leppälä, A. (2001) Bioactive peptides derived from whey proteins: opioid and ace-inhibitory peptides. *Trends in Food Science and Technology*, **11**, 347–356.

Pihlanto-Leppälä, A., Rokka, T., and Korhonen, H. (1998) Angiotensin I converting enzyme inhibitory peptides derived from bovine milk proteins. *International Dairy Journal*, **8**, 325–331.

Poulin, J.-F., Amiot, J., and Bazinet, L. (2006) Simultaneous separation of acid and basic bioactive peptides by electrodialysis with ultrafiltration membrane. *Journal of Biotechnology*, **123**, 314–328.

Poulin, J.-F., Amiot, J., and Bazinet, L. (2007) Improved peptide fractionation by electrodialysis with ultrafiltration membrane: influence of ultrafiltration membrane stacking and electrical field strength. *Journal of Membrane Science*, **299**, 83–90.

Poulin, J.-F., Amiot, J., and Bazinet, L. (2008) Impact of feed solution flow rate on peptide fractionation by electrodialysis with ultrafiltration membrane. *Journal of Agricultural and Food Chemistry*, **56**, 2007–2011.

Pourcelly, G. and Bazinet, L. (2008). Developments of bipolar technology in food and bio-industries, in *Handbook of Membrane Separations: Chemical, Pharmaceutical and Biotechnological Applications* (eds. A.K. Pabby, S.S.H. Rizvi, and A.M. Sastre), CRC Press, Taylor & Francis Group, Boca Raton, FL, pp. 581–633.

Slack, A.W., Amundson, C.H., and Hill, C.G. (1986) Production of enriched β-lactoglobulin and α-lactalbumin whey protein fractions. *Journal of Food Processing and Preservation*, **10**, 19–30.

Stack, F.M., Hennessy, M., Mulvihill, D., and O'Kennedy, B.T. (1995) Procédé de fractionnement du petit-lait en ses éléments constitutifs. WO patent 9,534,216-C1.

Stathopoulos, C.E. (2008) Dairy-based ingredients, in Gluten-Free Cereal Products and Beverages, Elsevier, Amsterdam, pp. 217–236.

Théodet, C. and Gandemer, G. (1994) Devenir des lipides au cours de la clarification du lactosérum. *Lait*, **74**, 281–295.

Van Nispen, J.G.M. and Jonker, R. (1991) Process for the fermentative preparation of organic acids. US patent 5,002,881.

Yu, T., Du, Y.X., and Chen, B. (2011) Evaluation of clarithromycin lactobionate as a novel chiral selector for enantiomeric separation of basic drugs in capillary electrophoresis. *Electrophoresis*, **32**, 1898–1905.

Index

Membrane Processing for Dairy Ingredient Separation, First Edition.
Edited by Kang Hu and James M. Dickson.